Current Progress in Geosciences

Current Progress in Geosciences

Editor: Joe Carry

Callisto Reference
www.callistoreference.com

Callisto Reference,
118-35 Queens Blvd., Suite 400,
Forest Hills, NY 11375, USA

Visit us on the World Wide Web at:
www.callistoreference.com

ISBN: 978-1-64116-248-7 (Hardback)

The publisher's policy is to use permanent paper from mills that operate a sustainable forestry policy. Furthermore, the publisher ensures that the text paper and cover boards used have met acceptable environmental accreditation standards.

Printed in China.

Cataloging-in-Publication Data

Current progress in geosciences / edited by Joe Carry.
p. cm.
Includes bibliographical references and index.
ISBN 978-1-64116-248-7
1. Earth sciences. 2. Geology. 3. Geography. 4. Soil science. I. Carry, Joe.
QE26.3 .C87 2020
550--dc23

Table of Contents

Preface

The field of geoscience delves into the study of the physical constitution of the Earth and its atmosphere. It also encompasses all aspects of the planet, from floods to fossils, and raindrops to earthquakes. In geoscience, the hydrosphere, lithosphere, atmosphere and biosphere are the chief areas of study. The hydrosphere comprises of the total mass of water found under, on and above the surface of the Earth. The lithosphere is the rigid, outermost shell of the Earth that is characterized by its rigid mechanical properties. The atmosphere is the set of layers of gases that surround the Earth and is held in place by the force of gravity. The biosphere is the ecological system comprising of all life on Earth and their interactions with each other and the environment. The study of these are approached through specialized fields, such as atmospheric sciences, ecology, geophysics and geodesy, soil science, physical geography, geology and hydrology, oceanography and limnology. This book attempts to understand the multiple branches that fall under geoscience and how such concepts have practical applications. It aims to shed light on some of the unexplored aspects of geoscience and the recent researches in this field. Students, researchers, experts and all associated with this field will be benefitted by this book.

After months of intensive research and writing, this book is the end result of all who devoted their time and efforts in the initiation and progress of this book. It will surely be a source of reference in enhancing the required knowledge of the new developments in the area. During the course of developing this book, certain measures such as accuracy, authenticity and research focused analytical studies were given preference in order to produce a comprehensive book in the area of study.

This book would not have been possible without the efforts of the authors and the publisher. I extend my sincere thanks to them. Secondly, I express my gratitude to my family and well-wishers. And most importantly, I thank my students for constantly expressing their willingness and curiosity in enhancing their knowledge in the field, which encourages me to take up further research projects for the advancement of the area.

Editor

1

Volcanic Plume CO_2 Flux Measurements at Mount Etna by Mobile Differential Absorption Lidar

Simone Santoro [1,2], Stefano Parracino [2,3], Luca Fiorani [2], Roberto D'Aleo [1], Enzo Di Ferdinando [2], Gaetano Giudice [4], Giovanni Maio [5,†], Marcello Nuvoli [2] and Alessandro Aiuppa [1,4,*]

1 Dipartimento di Scienze della Terra e del Mare, Università di Palermo, 90123 Palermo, Italy; simone.santoro@unipa.it (S.S.); roberto.daleo01@unipa.it (R.D.)
2 Nuclear Fusion and Safety Technologies Department, ENEA (Italian National Agency for New Technologies, Energy and Sustainable Economic Development), 00044 Frascati, Italy; stefano.parracino@uniroma2.it (S.P.); luca.fiorani@enea.it (L.F.); enzo.diferdinando@enea.it (E.D.F.); marcello.nuvoli@enea.it (M.N.)
3 Department of Industrial Engineering, University of Rome "Tor Vergata", 00173 Rome, Italy
4 Istituto Nazionale di Geofisica e Vulcanologia, 90146 Palermo, Italy; gaetano.giudice@ingv.it
5 ARES Consortium, 00100 Rome, Italy; prof.maio@gmail.com
* Correspondence: alessandro.aiuppa@unipa.it
† Current address: Vitrociset SpA, 00156 Rome, Italy.

Abstract: Volcanic eruptions are often preceded by precursory increases in the volcanic carbon dioxide (CO_2) flux. Unfortunately, the traditional techniques used to measure volcanic CO_2 require near-vent, in situ plume measurements that are potentially hazardous for operators and expose instruments to extreme conditions. To overcome these limitations, the project BRIDGE (BRIDging the gap between Gas Emissions and geophysical observations at active volcanoes) received funding from the European Research Council, with the objective to develop a new generation of volcanic gas sensing instruments, including a novel DIAL-Lidar (Differential Absorption Light Detection and Ranging) for remote (e.g., distal) CO_2 observations. Here we report on the results of a field campaign carried out at Mt. Etna from 28 July 2016 to 1 August 2016, during which we used this novel DIAL-Lidar to retrieve spatially and temporally resolved profiles of excess CO_2 concentrations inside the volcanic plume. By vertically scanning the volcanic plume at different elevation angles and distances, an excess CO_2 concentration of tens of ppm (up to 30% above the atmospheric background of 400 ppm) was resolved from up to a 4 km distance from the plume itself. From this, the first remotely sensed volcanic CO_2 flux estimation from Etna's northeast crater was derived at ≈2850–3900 tons/day. This Lidar-based CO_2 flux is in fair agreement with that (≈2750 tons/day) obtained using conventional techniques requiring the in situ measurement of volcanic gas composition.

Keywords: volcanic plumes; volcanic CO_2 flux; remote sensing; Differential Absorption Lidar (DIAL)

---

1. Introduction

In the last two decades, there have been major advances in the instrumental monitoring of volcanic gas plume composition and fluxes [1]. These have included the first instrumental networks of scanning Differential Optical Absorption Spectrometers (DOAS) for volcanic SO_2 flux monitoring, the implementation of satellite-based volcanic gas observations, and the advent of sensor units for in situ gas monitoring [1–6]. Owing to this technical progress, volcanic gas plume composition and fluxes have increasingly been used to extract information on degassing mechanisms/processes [4], and to derive constraints on shallow volcano plumbing systems [5]. However, work still needs to be done to increase the number of volcanic gas species that can be detected in plumes, which remain few if compared to the countless number of chemicals quantified from fumarole direct sampling [1].

Studying volcanic gas plumes has additionally contributed to monitoring, and eventually allowed the prediction of volcano behavior [6]. In particular, it has been shown that, at open-vent persistently degassing volcanoes, volcanic eruptions are often preceded by anomalous increases of the volcanic CO_2 flux [7]. These initial observations have motivated attempts to systematically monitor the volcanic CO_2 flux, and to identify novel measurement strategies [8]. Until recently, however, attempts to remotely sense the volcanic CO_2 flux from distal locations have been limited in number [9,10], while the majority of the observations have involved in situ measurements in the proximity of hazardous active volcanic vents [3]. On Mt. Etna, for example, one of the largest volcanic CO_2 point sources on Earth [11], the volcanic CO_2 flux has systematically been measured since the mid-2000s by combining in situ measurement of the volcanic CO_2/SO_2 ratio (with portable or permanent Multi-Component Gas Analyzer Systems, Multi-GAS; [12–15]) with remotely sensed SO_2 fluxes [16–18]. No successful report exists, at least to the best of our knowledge, of spectroscopy-based detection of Etna's volcanic CO_2 flux from a remote (distal) location.

Within the context of the ERC (European Research Council) starting the grant project BRIDGE (BRIDging the gap between Gas Emissions and geophysical observations at active volcanoes), we designed a new DIAL (Differential Absorption Lidar) [19], with the specific objective to remotely sense the volcanic CO_2 flux. Lidars have only recently been introduced in volcanic gas studies. A CO_2 laser–based lidar was used at Mt. Etna in 2008 [20] and at Stromboli Volcano in 2009 [21] to measure the volcanic plume water vapor flux. More recently, lidars were first been used to target volcanic CO_2 [9,10,22–24]. Our lidar BILLI (BrIdge voLcanic LIdar) [22], for example, has recently been used to successfully retrieve three-dimensional tomographies of volcanic CO_2 in the plumes of Pozzuoli, Solfatara in 2014 [9] and Stromboli volcano in 2015 [23,24]. As such, although gas-sensing lidars remain far less exploited in volcanology than those targeting volcanic ash/particles [25,26], this novel application field may expand rapidly in the near future.

Here, we report on the first successful use of BILLI at Mt. Etna. We show that, in our July–August 2016 Etna experiment, the lidar successfully resolved a volcanic CO_2 signal of a few tens of ppm (in excess to the background air) from more than 4 km of distance, and with good spatial (5 m) and temporal (10 s) resolution. These results are used to derive the first "remote" assessment of Etna's volcanic CO_2 flux. Our observations open new perspectives for routine volcanic CO_2 flux monitoring via lidars.

2. Materials and Methods

2.1. Field Set-Up on Mt. Etna

Observations on Mt. Etna were conducted from 28 July to 1 August 2016, including an initial phase of instrumental setup (28–29 July). Successful CO_2 flux detections were obtained on 31 July, when optimal viewing conditions persisted over the day. The DIAL was mounted in a trailer loaded on a truck, parked at the INGV (Istituto Nazionale di Geofisica e Vulcanologia) observatory "Pizzi Deneri" (Figure 1). The observatory is located at 2823 m a.s.l., northeast of the summit crater of Mt. Etna (3329 m a.s.l.), and at about 3 km from the main degassing vents (Figure 1).

The lidar was used to scan the volcanic plume vertically, keeping a constant azimuth angle (230°) and varying the elevation angle from 7° to 14° (Figure 1). A full 7° to 14° vertical scan was completed in ~15 min, and one atmospheric profile every 10 s was recorded throughout. With this instrumental set-up, the volcanic gas plume of the Etna's northeast crater was investigated (Figure 1), plumes from other craters being either too dilute (southeast crater) or only partially visible (central craters).

At our measurement conditions, two rock surfaces, corresponding to the eastern, outer flanks of the central crater, were intercepted by the laser beam at distances of 1.6 and 2.1 km, and at elevation angles from 7° to 9°. These rock surfaces retro-reflected the laser beam, yielding strong return signals (see below, Figure 2). The volcanic plume, e.g., high in-plume excess CO_2 concentrations, was detected in between the two above rock surfaces, and in the 2.2–4.2 km distance range.

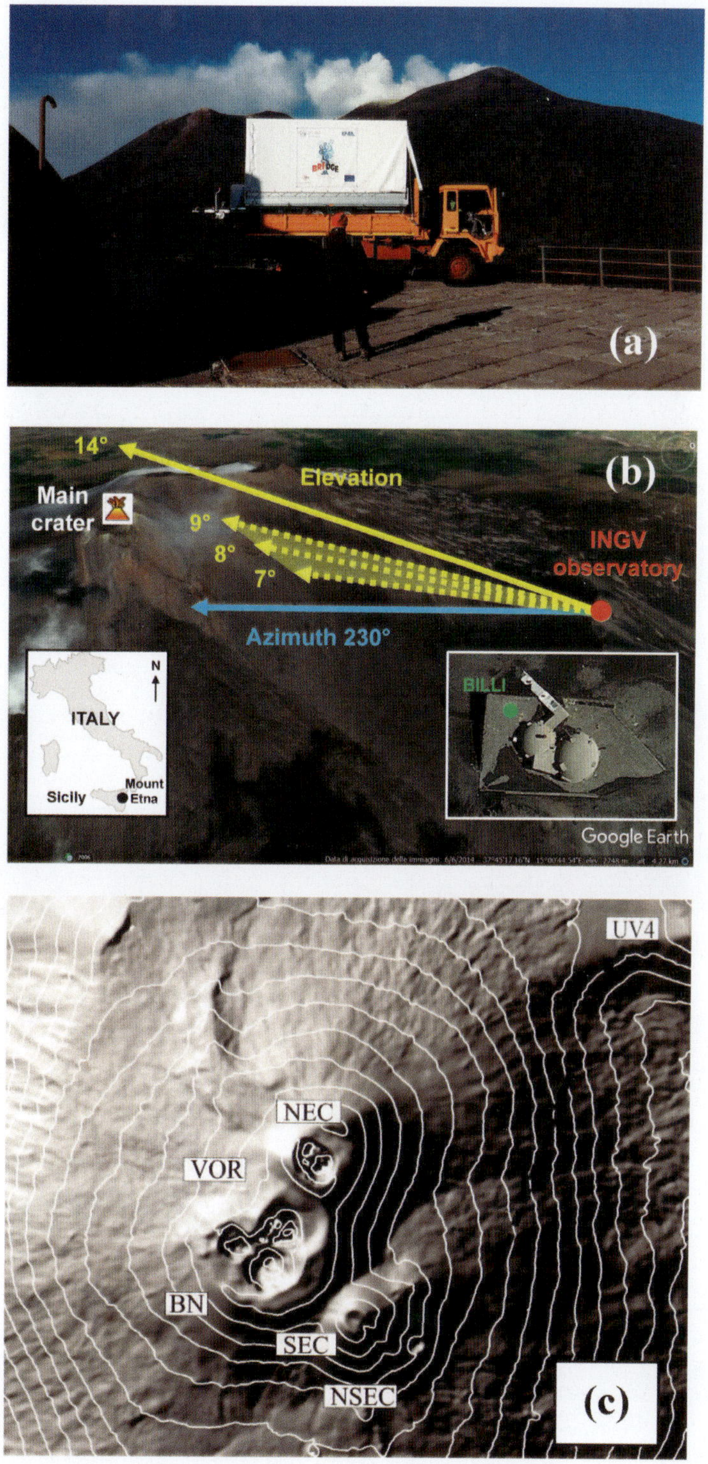

Figure 1. (**a**) The BILLI DIAL (BrIdge voLcanic Lidar, Differential Absorption Lidar) system mounted in a trailer (white) loaded on a truck (orange) at the INGV (Istituto Nazionale di Geofisica e Vulcanologia) observatory "Pizzi Deneri" (the volcanic plume of Mt. Etna is clearly visible); (**b**) Location of Mt. Etna in Sicily, southern Italy (left inset); the truck was parked at the INGV observatory "Pizzi Deneri" (right inset); the laser was fired at constant azimuth and different elevations. The volcanic plume of the northeast crater has been crossed by the laser beam. From 7° to 9° of elevation, rock faces were encountered; (**c**) A map of the summit area showing the active craters and the UV camera site (UV4). NEC: northeast crater; VOR: Voragine; BN: Bocca Nuova (VOR and BN are part of the Etna's central craters); SEC: southeast crater; NSEC: new southeast crater.

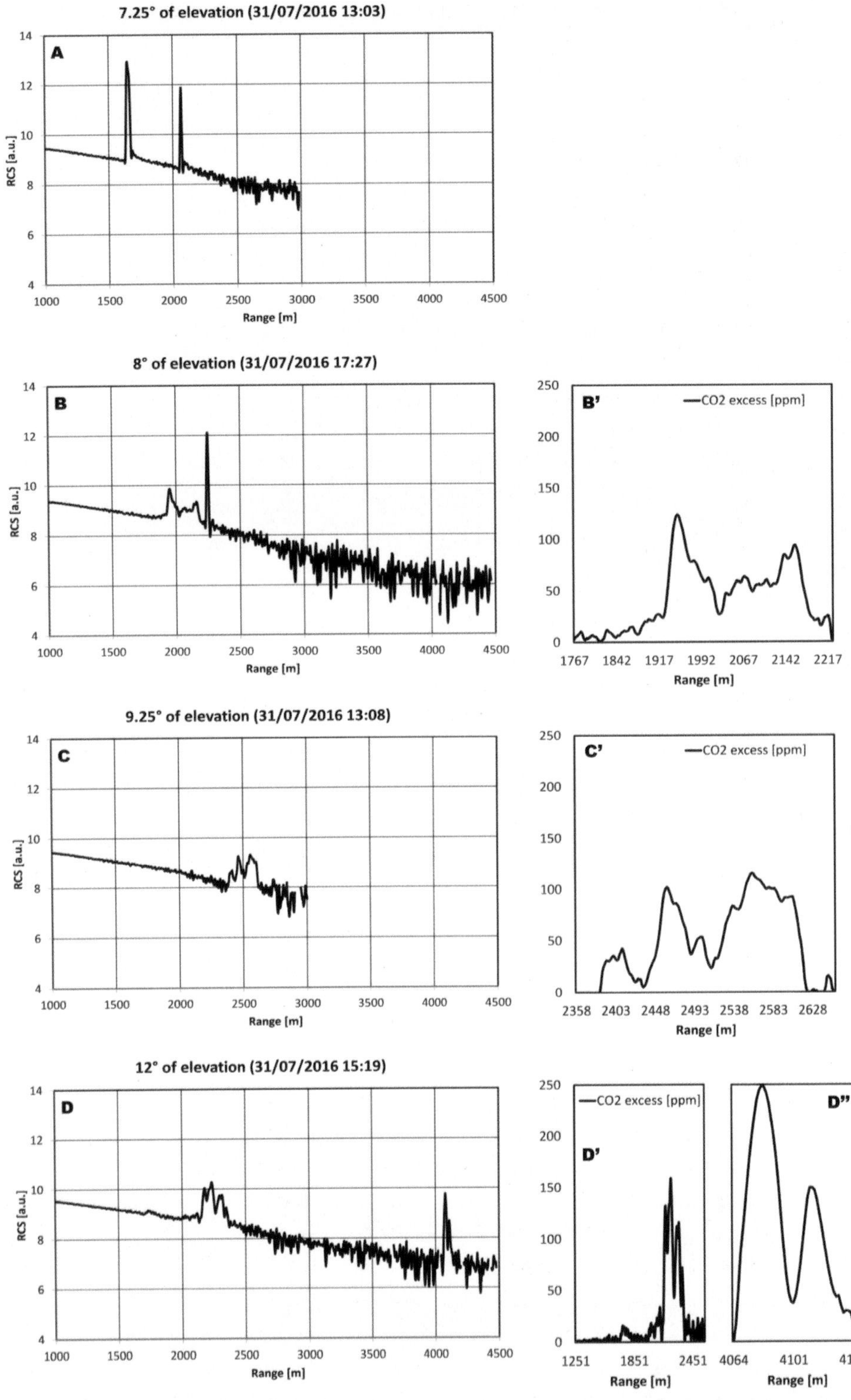

Figure 2. (**A**) Lidar return at 7.25° of elevation: two narrow and defined peaks due to beam backscattering from rock faces are clearly visible (beyond 3000 m only noise was recorded and the corresponding signal is not shown); (**B**) lidar return at 8° of elevation: a wide and jagged peak from the volcanic plume and a narrow and defined peak from a rock face are clearly visible; the CO_2 profile inside the volcanic plume is shown in (**B′**); (**C**) lidar return at 9.25° of elevation: a wide and jagged peak from the volcanic plume is clearly visible; the CO_2 profile inside the volcanic plume is shown in (**C′**) (beyond 3000 m only noise was recorded and the corresponding signal is not shown); (**D**) lidar return at 12° of elevation: two wide and jagged peaks from the volcanic plume are clearly visible; the CO_2 profiles inside the volcanic plume are shown in (**D′**,**D″**).

2.2. DIAL

The main components of a lidar are the transmitter (laser) and the receiver (telescope). A lidar is merely an optical radar [19]: a laser pulse is transmitted to the atmosphere, and some of its photons are backscattered to the telescope by air molecules and aerosols (droplets, particles etc.). The optical power corresponding to this photon flux is transformed into an electronic signal by photodetector and preamplifier, and converted in digital signal by an ADC (analog-to-digital converter).

The chemico-physical properties of the atmosphere along the laser beam, at distance R (range) from the lidar, can be inferred from analysis of the detected signal as a function of t, the time interval between emission and detection. R and t are linked by the relation $R = ct/2$, where c is the speed of light. The returned signal to the lidars' telescope, as a function of R (or t), then yields an atmospheric profile (Figure 2). In other words, an atmospheric profile is a range-resolved characterization of the lidar returned signal, which allows studying the air/plume optical properties along the light trajectory.

Air attenuates the laser pulse due to molecules and aerosol scattering and to the specific absorption of gases: if the laser wavelength coincides with absorption lines of a target gas, the attenuation will be stronger. A DIAL takes advantage of this effect: unlike a usual lidar, two wavelengths, ON and OFF, are transmitted, with only the former being absorbed by the target gas (Figure 3).

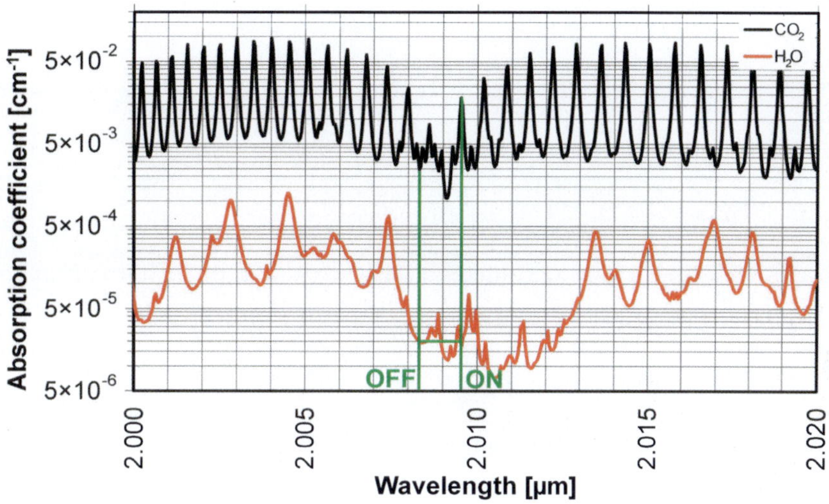

Figure 3. Carbon dioxide and water vapor absorption coefficients (from [27]) around the ON and OFF wavelengths (indicated in green).

If the absorption line is narrow, and ON and OFF wavelengths are close enough, the target gas concentration along the lidar optical path can be derived from the ratio between the OFF and ON signals. In this application, we selected the following wavelengths (Figure 3): ON, 2009.537 nm; OFF, 2008.484 nm. This selection was motivated by: (i) the CO_2 absorption is relatively low, thus allowing the system to probe far ranges (beyond 4 km). If a stronger line had been used, the ON laser pulse would have extinguished before; (ii) the beam energy (depending mainly on the dye efficiency curve) and the detector responsivity are near their maximum; and (iii) the H_2O absorption is very low (Figure 3). Moreover, ON and OFF have been chosen so that the differential absorption of H_2O is approximately zero (within the uncertainty of the spectroscopic data), thus minimizing the interference of water vapor to the carbon dioxide measurement.

In our case, the transmitter and the receiver are coaxial, and the lidar field-of-view can be aimed in the whole atmosphere thanks to a system made of two large elliptical mirrors [22]. This configuration allows the experimenters to scan the plume in both horizontal and vertical planes, thus measuring CO_2 concentrations both outside and inside the volcanic plume [9]. From this, by scanning the volcanic gas plume from different angles and viewing directions, the CO_2 distribution in a cross-section of the volcanic plume can be retrieved. This, combined with independent knowledge of plume speed and altitude, allows the CO_2 flux to be retrieved.

The reader is referred to previous work [9,22–24] for details on instrumental setup and data processing. The systematic error associated with the derived CO_2 concentrations is dominated by imprecision in wavelength setting [22]. This leads to inaccuracy in the differential absorption cross section, and thus in gas concentration. Thanks to a photo-acoustic cell filled with pure CO_2 at atmospheric pressure and temperature, the ON and OFF wavelengths were set before each atmospheric scan. The residual imprecision [23] of ± 0.02 cm^{-1} (half laser linewidth: half width at half maximum of the energy transmitted by the laser system (J) vs the wavenumber (cm^{-1})) implies a systematic error on CO_2 concentrations of 5.5% [24]. The statistical error of CO_2 measurement has been calculated by usual error propagation techniques from the standard deviation of the lidar signal. At 2.5 km, a mean range, it is about 2%, while it can exceed 5% at 4.2 km. Table 1 compares the instrumental set-up during the Mt. Etna field campaign, with those used at Solfatara [9,22], and Stromboli [23,24].

Table 1. Summary of field operational conditions at Pozzuoli Solfatara, Stromboli, and Mt. Etna (this study).

Campaign	Pozzuoli Solfatara	Stromboli Volcano	Mt. Etna
Latitude	40°49′46.28″N	38°48′06.69″N	37°45′57.28″N
Longitude	14°08′50.51″E	15°14′25.69″E	15°00′59.65″E
Period	13–17 October 2014	24–29 June 2015	28 July–1 August 2016
Azimuth scan	196°–234°	235.3°–253.6°	230°
Elevation scan	0°–18°	15.2°–27.4°	7°–14°

3. Results

Figure 2 shows examples of lidar returns obtained during our Etna campaign. Results are illustrated for four atmospheric profiles taken on 31 July (the best measurement day) at four distinct elevations, and are shown in the form of range vs. RCS (range corrected signal) scatter plots.

During its atmospheric propagation, the laser beam intensity approximately decreased:

- exponentially, due to atmospheric extinction, according to the Lambert-Beer law, and;
- as $1/R^2$, because the solid angle subtended by the receiver is A/R^2, where A is the telescope effective area.

For these reasons, it is a common practice in lidar science to express results using a RCS, this being the logarithm of the product of the signal times the square of the range. To improve the SNR (signal-to-noise ratio), the RCS was obtained by averaging 50 laser shots for each lidar return, and a 13-point Savitzky-Golay filter was applied [28].

During a vertical scan, the measured range-resolved RCS profiles varied as the laser elevation was sequentially increased. Below 7.25° elevation, the laser beam hit a first rock surface at about a 1.6 km distance. Laser beam retro-reflection at this rock surface produced, in the lidar return signal, a strong, narrow RCS peak at $R = 1.6$ km. At 7.25° elevation (Figure 2a), only part of the beam was intercepted by the $R = 1.6$ km rock surface, while the remaining part impinged on the rock surface at $R = 2.1$ km, producing a second narrow RCS peak. For geometrical reasons, an elevation increase corresponded to an increase in the range at which the rock surfaces were encountered, e.g., the second rock surface was encountered at $R = 2.1$ at 7.25° elevation, shown in Figure 2a, and at $R = 2.3$ km at 8° elevation, shown in Figure 2b. No rock surface was hit by the laser beam at elevations >9°, e.g., note the absence of narrow RCS peaks in Figure 2c,d.

Back-scattering of the laser beam by the volcanic plume produced wide and jagged RCS peaks, therefore very distinct from the narrow and defined peaks produced by beam retro-reflection at rock surfaces (compare the two peak shapes in Figure 2b).

The volcanic plume was detected at range distances in between the two rock surfaces up to a 9° elevation (e.g., Figure 2b), or beyond them at a 9° to 14° elevation (Figure 2c,d). A broad, irregular RCS peak in the lidar returns, corresponding to the volcanic plume, was resolved up to a maximum measurement range of 4.2 km (Figure 2d).

We used the procedure detailed in References [9,24] to convert the RCS profiles into range-resolved profiles of in-plume excess CO_2 concentrations (see Figure 2b′,c′,d′,d″). This procedure involves calculating the excess CO_2 concentration corresponding to each i-th ADC channel of the lidar profile from:

$$C_{CO2,i} = k\,RCS_i \tag{1}$$

$$k = \frac{\Delta C\,(R_1 - R_2)}{\Delta R \sum_i RCS_i} \tag{2}$$

where ΔR and RCS_i are, respectively, the range interval and range corrected signal corresponding to each ADC channel; R_1 and R_2 are the range distances of the two above rock surfaces; and ΔC is the average excess CO_2 concentration in the air/plume parcel between them (this is obtained from the intensity contrast of lidar returns produced by the two rock surfaces). The term "excess" implies that the reported CO_2 concentrations are after subtraction of the ambient atmospheric background, and therefore correspond to the "volcanic" CO_2 levels in the plume. The ambient atmospheric CO_2 background was obtained from the processing of lidar returns in the 0–1.6 km range distance, where no plume signal was detected (see Reference [9] for details of calculations).

At an 8° elevation, shown in Figure 2b′, the volcanic plume was evidenced by a band of excess CO_2 concentrations of $\leq$125 ppm. These excess CO_2 concentrations agree well with those derived by in situ in-plume measurements with conventional techniques (e.g., the Multi-GAS), from which in-plume CO_2 concentrations of tens to hundreds of ppm above ambient air are typically obtained [12]. The plume appears to be about 300 m thick; this relatively narrow plume's cross-section was probably justified by the fact that, at such an 8° elevation, the laser beam intercepted the volcanic plume at below the summit crater's rim altitude. Due to its close proximity to the crater slopes, the volcanic plume was, at least partly, protected from the local wind field, a fact that reduced its atmospheric dispersion. The volcanic plume was still relatively narrow at a ~9° elevation (Figure 2c′), where the laser beam pointed just above the summit crater's rim. At even higher elevations, the volcanic plume was wider and scattered by the wind, and the returned RCS often presented multiple peaks (Figure 2d,d′).

As explained before (Section 2), a sequence of atmospheric profiles was acquired as the lidar vertically scanned the horizon, from a 7° to 14° (max) elevation. All CO_2 profiles (e.g., Figure 2), taken at different elevations during a single lidar rotation sequence, were combined and integrated to obtain a CO_2 scan, examples of which are illustrated in Figure 4. On 31 July 2016, the most fruitful day, 19 scans were obtained. Each scan consisted of 24 profiles, all at a 230° azimuth. These profiles covered the elevation angle interval (between 7° and 13°) with an angular resolution of about 0.25°.

The results are illustrated in the form of contour maps of excess CO_2 concentrations, plotted as a function of the range and elevation. The colored spots correspond to areas of high excess CO_2 (the natural background is dark blue), and therefore illustrate the spatial distribution and temporal evolution of the volcanic plume (the yellow lines delimit the positions of the laser beam reflections off the rock surfaces). In all the maps we obtained (see examples in Figure 4), the structure of the plume was well resolved. The plume was tracked as a cluster of high CO_2 concentration spots, trending from about a 9° elevation and R = 2.4 km (the vent rim) to a 13° elevation and R = 2.5–2.9 km. As such, our CO_2 concentration maps were consistent with a gently lofting volcanic plume (Figure 4), with vertical and horizontal movements driven by thermal buoyancy and by the local wind field pattern. The maps indicate the plume was being dispersed away from the lidar during our observations, since the range of volcanic plume detection increased with the elevation in all the maps.

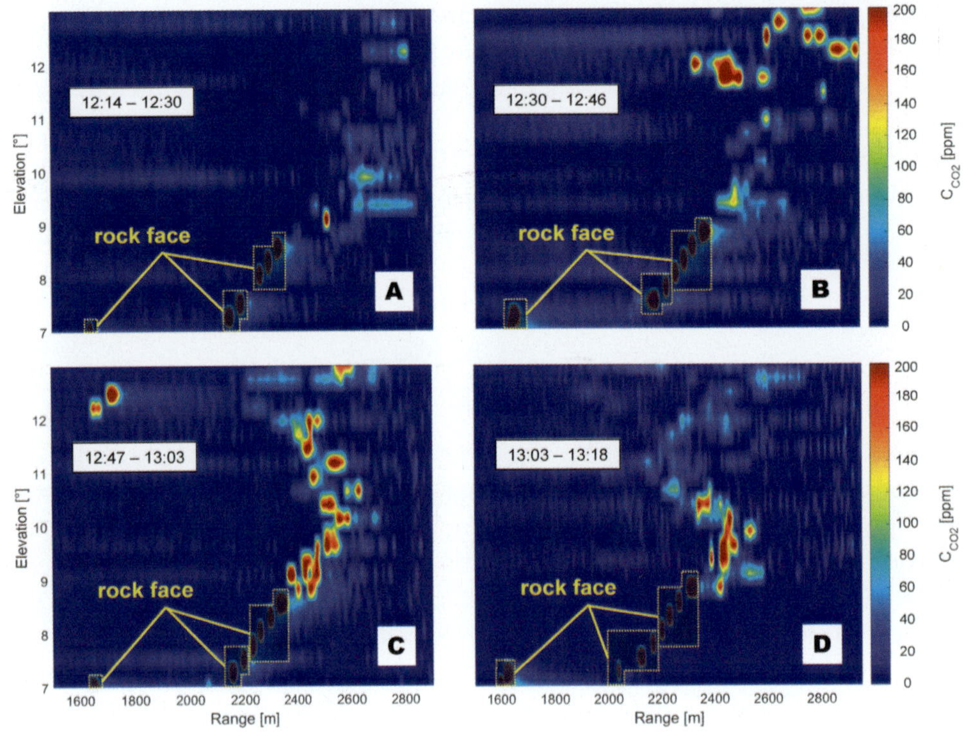

Figure 4. Vertical scans (fixed azimuth: 230°) of the volcanic plume (CO_2 excess) acquired on 31 July 2016; (**A**) from 12:14 p.m. to 12:30 p.m.; (**B**) from 12:30 p.m. to 12:46 p.m.; (**C**) from 12:47 p.m. to 1:03 p.m. and (**D**) from 1:03 p.m. to 1:18 p.m. (local civil time). At this azimuth, in the elevation interval between 7° and 9°, the laser beam is back-scattered by rock faces, thus causing signal peaks not due to the volcanic plume (rock faces are sorted out from real CO_2 by the correspondence of narrow peaks with certain range values).

The maps of Figure 4 set the basis for the calculation of the volcanic CO_2 flux. In analogy with previous work [9], we obtained the volcanic CO_2 flux by integrating the background-corrected (excess) CO_2 concentrations over a plume cross-section (from the maps of Figure 4), which allowed us to derive the plume CO_2 molecular density. This was then multiplied by the plume transport speed to obtain the CO_2 flux (ΦCO_2, in $Kg{\cdot}s^{-1}$), as:

$$\Phi_{CO_2} = v_p \cdot \frac{PM_{CO_2}}{10^3 N_A} \cdot N_{molCO2-total} \tag{3}$$

where v_p is the plume transport speed (in m/s); $NmolCO_{2-total}$ is the total-plume CO_2 molecular density (expressed in molecules$\cdot m^{-1}$); and PM_{CO2} and N_A are, respectively, the CO_2 molecular weight and Avogadro's constant. The term $N_{molCO2-total}$ was obtained by integrating the effective average excess CO_2 concentrations ($\overline{C_{exc,i}}$ [ppm]) over the entire plume cross-section, according to:

$$N_{molCO2-total} = N_h \cdot 10^{-6} \cdot \sum_i \overline{C_{exc,i}} \cdot A_i \tag{4}$$

where N_h is the atmospheric number density (molecules$\cdot m^{-3}$) at the crater's summit height, and A_i represents the i-th effective plume area.

The plume transport speed was inferred at 9.7 ± 0.8 m/s from the processing of plume images taken on the same day by the permanent UV camera system (UV4) in use at the Pizzi Deneri observatory since 2014; see Reference [29] for details on the instrument. The UV camera images were processed using an optical flow sub-routine using the Lukas/Kanade algorithm [30,31], integrated in the Vulcamera software [32] (same methodology as described in [29]).

Our derived CO_2 fluxes are illustrated in Figure 5. The CO_2 flux varied from 1235 to 8050 tons/day during the measurement interval, and averaged at 2850 $\pm$ 1800 tons/day.

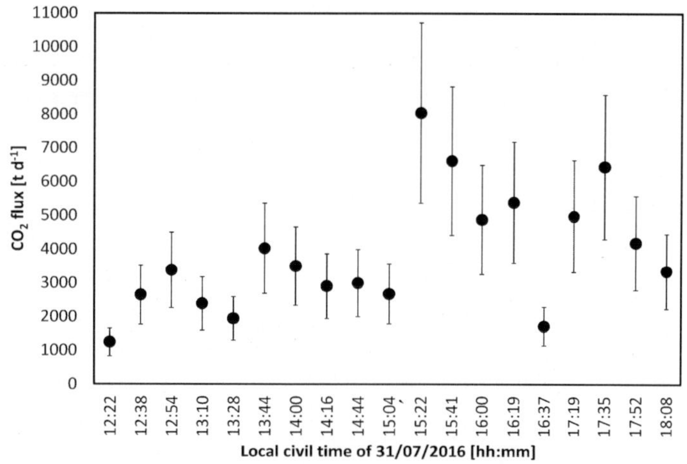

Figure 5. CO_2 flux from the northeast crater retrieved on 31 July 2016 from 12:22 p.m. to 6:08 p.m. (local civil time). The error bars indicate the inferred CO_2 flux error ($\pm$33%), as based upon the error propagation of the plume speed an in-plume integrated CO_2 amounts (procedure detailed in [24]).

4. Discussion

As long-term volcanic gas records have increased in number and quality over the last few decades [33], full empirical evidence has emerged for precursory increases of the volcanic CO_2 flux emissions prior to eruption of mafic to intermediate volcanoes [7]. However, remote direct measurements of the volcanic CO_2 flux, which are intrinsically safer for operators and more prone to provide continuous, long-term observations, have remained impossible until recently [9,10].

Our results here support the ability of the DIAL-Lidar BILLI to profile atmospheric CO_2 concentrations over large optical paths (Figures 2 and 4), and to remotely sense the CO_2 flux from distal (up to 4 km distant) sources (Figure 5). This instrument thus promises a real step ahead in the remote observation of volcanic gas emissions. Improved CO_2 flux measurements are not only vital for better gas-based volcano monitoring, but are also needed to better constrain the global volcanic CO_2 budget, which is still inaccurately known [8].

The volcanic CO_2 flux from Mt. Etna has been assessed in the past by either in-plume airborne CO_2 profiling [11], or by indirect methods involving in situ measurements of plume CO_2/SO_2 ratios, via either the Multi-GAS [3,12–15] or Fourier Transform InfraRed Spectrometers (FTIR; [16]). To the best of our knowledge, our results are the first to report a direct, remote quantification of Etna's CO_2 flux.

Our lidar results show that, in the circa 5-h-long observational widow, the CO_2 flux from Etna's northeast crater varied from 1235 to 8050 tons/day (Figure 5). The CO_2 flux was somewhat higher, typically >4000 tons/day and up to 8050 tons/day, after 3 p.m. local time, relative to the 12–3 p.m. period (<4000 tons/day) (Figure 5). No change in activity was yet observed at the northeast crater, which continued to exhibit quiescent degassing over the entire measurement interval. We therefore consider the observed variation as part of the normal fluctuations in degassing activity that occur at Etna, likely in response to temporal variations in the magma/gas transport rate in the volcano's feeding conduits [15–18]. By taking the arithmetic mean of the individual CO_2 flux measurements in Figure 5, we would obtain a time-averaged CO_2 flux of 2850 $\pm$ 1800 tons/day for 31 July 2016. In view of the non-stationary CO_2 emission behavior captured by our high-temporal resolution measurement (Figure 5), we also perform an independent exercise in which we calculated the total CO_2 output from the northeast crater by integrating (in the time domain) the available CO_2 flux measurements, each representative of 13–18 min of observation (the mean duration of scans was

15 min). From this, we obtained that ≈796 tons of CO_2 were cumulatively released during 5 h of observations, implying a time-averaged CO_2 flux of 3900 tons/day. This is about 30% higher than, but within one standard deviation of, the CO_2 flux obtained above from a simple arithmetic mean approach (2850 ± 1800 tons/day).

In the attempt to add confidence to our results, we compared our lidar-based CO_2 flux with independent estimates based upon a more conventional technique that involves a combination of SO_2 fluxes and plume CO_2/SO_2 ratios (Figure 6). Our permanent UV camera system (UV4) at Pizzi Deneri indicated, for the morning of the same 31 July, a time-averaged SO_2 flux of 645 ± 125 tons/day. This is the mean (±1 standard deviation) of 4 h of observations at a 0.5 Hz rate (Figure 6; same methodology as in [29]). Our inferred northeast crater's SO_2 flux (645 ± 125 tons/day) corresponded to about 30% of the total volcano's SO_2 emissions (≈2200 tons/day). These latter emissions were inferred using the same UV camera system, and were thus primarily determined by the central craters (not targeted by our DIAL-Lidar). The northeast crater's volcanic plume was in situ measured by a portable Multi-GAS instrument (the same as in [12,13]) two days later. These in situ observations yielded a (molar) CO_2/SO_2 ratio of ≈6, demonstrating the usual [12,16] CO_2-poor composition of the northeast crater (the simultaneously observed CO_2/SO_2 ratio of the central crater's plume was ≈16). We consider our Multi-GAS–derived composition on 2 August as still representative of the northeast crater's emissions on 31 July, since volcanic activity at that crater did not exhibit any substantial change in between the two days. By combining the two sets of data together, we converted the SO_2 flux time-series into a 4-h-long CO_2 flux time-series (Figure 6), from which an averaged (arithmetic mean) UV-Camera + MultiGAS CO_2 flux of ≈2750 tons/day was obtained. This is close to our lidar-based estimates above (≈2850–3900 tons/day). We caution that the two independent CO_2 flux time-series (from lidar and UV-Camera + MultiGAS) are not temporally overlapping, since the UV camera system ran only in the morning, when sunlight conditions were optimal [29], while our successful CO_2 flux measurement with the lidar started a few hours later in the afternoon. In addition, the UV-Camera + MultiGAS used a constant CO_2/SO_2 ratio (of six) throughout the entire UV camera temporal window, while it is valid only as a first approximation. However, the close CO_2 flux values inferred from lidar and UV-Camera + MultiGAS provide mutual validation for the two independent techniques.

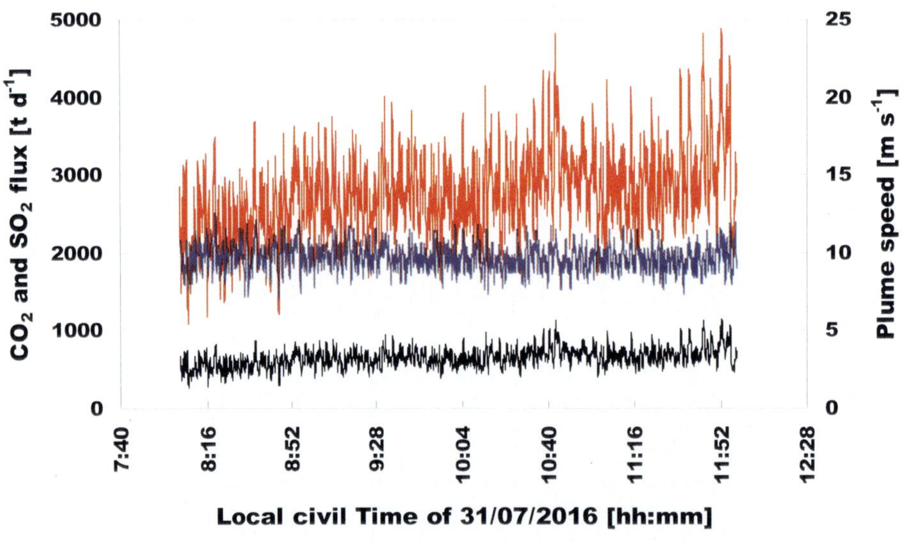

Figure 6. Time-series of CO_2 flux emissions from the northeast crater (in red) obtained from the UV-Camera + Multi-Gas technique. These were calculated by converting the SO_2 flux time-series (in black) obtained by the UV4 permanent UV camera system on 31 July 2016 (from 8 a.m. to 12 p.m., local civil time) using a CO_2/SO_2 ratio (molar) of six. The plume speed time-series calculated from the UV camera on the same 31 July is also shown (blue trend).

5. Conclusions

We have shown for the first time that the volcanic CO_2 flux can be detected with lidar from up to a 4 km distance. During our Mt. Etna field campaign, our DIAL-Lidar BILLI vertically scanned the volcanic plume while profiling CO_2 concentrations every 10 s, with a spatial resolution of 5 m. With this configuration, we successfully detected an excess volcanic CO_2 signal of a few tens of ppm, with relatively low systematic and statistical errors (5.5% and 2%, respectively). By integrating the results of the atmospheric profile taken at different heading angles, and covering a full scan of the plume, the volcanic CO_2 flux was derived (after integration, and in combination with the plume transport speed) at ≈2850–3900 tons/day. This lidar-based flux is close to that independently obtained by in situ observations of the volcanic plume (≈2750 tons/day), which combined Multi-GAS in situ sensing of the plume composition and remotely sensed (UV camera) SO_2 fluxes.

Clearly, additional field tests are required to validate our novel technique even further. Still, our results suggest BILLI is a major advance in ground-based observations of volcanic plumes. The instrument allows the remote measurements of volcanic CO_2 (and particles, if desired) from distal (safe) areas, and with unprecedented temporal resolution and high spatial coverage. Further development is now required to make this technology an operational tool for routine volcanic gas observations. Efforts are currently being undertaken to reduce the weight and power requirements (the current prototype is ~1100 kg and requires 6.5 kW), and to implement more user-friendly operational routines and software. These implementations are required to widen the application range of the lidar, and to allow its use in remote/harsh volcanic environments.

Acknowledgments: The authors are grateful to ENEA, in general, and Aldo Pizzuto, Roberta Fantoni and Antonio Palucci, in particular, for constant encouragement. They thank the staff of INGV-OE, and especially the Director Eugenio Privitera and Salvatore Consoli, for logistical support and for granting access to the INGV observatory "Pizzi Deneri". The support from the ERC project BRIDGE, n. 305377, is gratefully acknowledged. This work benefitted from the insightful comments of two anonymous reviewers.

Author Contributions: S.S., L.F and A.A. conceived and designed the experiments; S.S., L.F, R.D., E.D.F., M.N., G.G., and A.A performed the experiments; S.P., G.M., and R.D. analyzed the data; A.A. L.F. and S.S. wrote the paper with contributions from all co-authors.

Conflicts of Interest: The authors declare no conflict of interest.

References

1. Oppenheimer, C.; Fischer, T.P.; Scaillet, B. Volcanic Degassing: Process and Impact. In *Treatise on Geochemistry, The Crust*, 2nd ed.; Holland, H.D., Turekian, K.K., Eds.; Elsevier: Amsterdam, The Netherlands, 2014; Volume 4, pp. 111–179.
2. Carn, S.A.; Clarisse, L.; Prata, A.J. Multi-decadal satellite measurements of global volcanic degassing. *J. Volcanol. Geothermal Res.* **2016**, *311*, 99–134. [CrossRef]
3. Aiuppa, A. Volcanic gas monitoring. In *Volcanism and Global Environmental Change*; Schmidt, A., Fristad, K.E., Elkins-Tanton, L.T., Eds.; Cambridge University Press: Cambridge, UK, 2015; pp. 81–96.
4. Edmonds, M. New geochemical insights into volcanic degassing. *Philos. Trans. R. Soc. A Math. Phys. Eng. Sci.* **2008**, *366*, 4559–4579. [CrossRef] [PubMed]
5. Allard, P.; Carbonnelle, J.; Métrich, N.; Loyer, H.; Zettwoog, P. Sulphur output and magma degassing budget of Stromboli volcano. *Nature* **1994**, *368*, 326–330. [CrossRef]
6. Saccorotti, G.; Iguchi, M.; Aiuppa, A. In situ Volcano Monitoring: Present and Future. In *Volcanic Hazards, Risks and Disasters*; Elsevier: Amsterdam, The Netherlands, 2014; pp. 169–202.
7. Aiuppa, A.; Burton, M.; Caltabiano, T.; Giudice, G.; Guerrieri, S.; Liuzzo, M.; Murè, F.; Salerno, G. Unusually large magmatic CO_2 gas emissions prior to a basaltic paroxysm. *Geophys. Res. Lett.* **2010**, *37*, L17303. [CrossRef]
8. Burton, M.R.; Sawyer, G.M.; Granieri, D. Deep carbon emissions from volcanoes. *Rev. Mineral. Geochem.* **2013**, *75*, 323–354. [CrossRef]
9. Aiuppa, A.; Fiorani, L.; Santoro, S.; Parracino, S.; Nuvoli, M.; Chiodini, G.; Minopoli, C.; Tamburello, G. New ground-based lidar enables volcanic CO_2 flux measurements. *Sci. Rep.* **2015**, *5*, 13614. [CrossRef] [PubMed]

10. Queißer, M.; Granieri, D.; Burton, M. A new frontier in CO_2 flux measurements using a highly portable DIAL laser system. *Sci. Rep.* **2016**, *6*, 33834.
11. Allard, P.; Carbonnelle, J.; Dajlevic, D.; le Bronec, J.; Morel, P.; Robe, M.C.; Maurenas, J.M.; Faivre-Pierret, R.; Martin, D.; Sabroux, J.C.; et al. Eruptive and diffuse emissions of CO_2 from Mount Etna. *Nature* **1991**, *351*, 387–391. [CrossRef]
12. Aiuppa, A.; Federico, C.; Giudice, G.; Gurrieri, S.; Liuzzo, M.; Shinohara, H.; Favara, R.; Valenza, M. Rates of carbon dioxide plume degassing from Mount Etna volcano. *J. Geophys. Res.* **2006**, *111*, B09207. [CrossRef]
13. Aiuppa, A.; Giudice, G.; Gurrieri, S.; Liuzzo, M.; Burton, M.; Caltabiano, T.; McGonigle, A.J.S.; Salerno, G.; Shinohara, H.; Valenza, M. Total volatile flux from Mount Etna. *Geophys. Res. Lett.* **2008**, *35*, L24302. [CrossRef]
14. Patanè, D.; Aiuppa, A.; Aloisi, M.; Behncke, B.; Cannata, A.; Coltelli, M.; di Grazia, G.; Gambino, S.; Gurrieri, S.; Mattia, M.; et al. Insights into magma and fluid transfer at Mount Etna by a multiparametric approach: A model of the events leading to the 2011 eruptive cycle. *J. Geophys. Res. Solid Earth* **2013**, *118*, 3519–3539. [CrossRef]
15. Pering, T.D.; Tamburello, G.; McGonigle, A.J.S.; Aiuppa, A.; Cannata, A.; Giudice, G.; Patanè, D. High time resolution fluctuations in volcanic carbon dioxide degassing from Mount Etna. *J. Volcanol. Geotherm. Res.* **2014**, *270*, 115–121. [CrossRef]
16. La Spina, A.; Burton, M.; Salerno, G.G. Unravelling the processes controlling gas emissions from the central and northeast craters of Mt. Etna. *J. Volcanol. Geotherm. Res.* **2010**, *198*, 368–376. [CrossRef]
17. Caltabiano, T.; Burton, M.; Giammanco, S.; Allard, P.; Bruno, N.; Murè, F.; Romano, R. Volcanic Gas Emissions from the Summit Craters and Flanks of Mt. Etna, 1987–2000. In *Mt. Etna: Volcano Laboratory*; Bonaccorso, A., Calvari, S., Coltelli, M., del Negro, C., Falsaperla, S., Eds.; Geophysical Monograph Series; American Geophysical Union: Washington, DC, USA, 2004; Volume 143, pp. 111–128.
18. Allard, P.; Behncke, B.; D'Amico, S.; Neri, M.; Gambino, S. Mount Etna 1993–2005: Anatomy of an evolving eruptive cycle. *Earth-Sci. Rev.* **2006**, *78*, 85–114. [CrossRef]
19. Fiorani, L. Lidar application to litosphere, hydrosphere and atmosphere. In *Progress in Laser and Electro-Optics Research*; Koslovskiy, V.V., Ed.; Nova: New York, NY, USA, 2010; pp. 21–75.
20. Fiorani, L.; Colao, F.; Palucci, A. Measurement of Mount Etna plume by CO_2-laser-based lidar. *Opt. Lett.* **2009**, *34*, 800–802. [CrossRef] [PubMed]
21. Fiorani, L.; Colao, F.; Palucci, A.; Poreh, D.; Aiuppa, A.; Giudice, G. First-time lidar measurement of water vapor flux in a volcanic plume. *Opt. Commun.* **2011**, *284*, 1295–1298. [CrossRef]
22. Fiorani, L.; Santoro, S.; Parracino, S.; Nuvoli, M.; Minopoli, C.; Aiuppa, A. Volcanic CO_2 detection with a DFM/OPA-based lidar. *Opt. Lett.* **2015**, *40*, 1034–1036. [CrossRef] [PubMed]
23. Fiorani, L.; Santoro, S.; Parracino, S.; Maio, G.; Nuvoli, M.; Aiuppa, A. Early detection of volcanic hazard by lidar measurement of carbon dioxide. *Nat. Hazards* **2016**, *83*, S21–S29. [CrossRef]
24. Aiuppa, A.; Fiorani, L.; Santoro, S.; Parracino, S.; D'Aleo, R.; Liuzzo, M.; Maio, G.; Nuvoli, M. Advances in Dial-Lidar-based remote sensing of the volcanic CO_2 flux. *Front. Earth Sci.* **2017**, *5*, 15. [CrossRef]
25. Scollo, S.; Boselli, A.; Coltelli, M.; Leto, G.; Pisani, G.; Spinelli, N.; Wang, X. Monitoring Etna volcanic plumes using a scanning LiDAR. *Bull. Volcanol.* **2012**, *74*, 2383–2395. [CrossRef]
26. Wanga, X.; Boselli, A.; D'Avino, L.; Pisani, G.; Spinelli, N.; Amodeo, A.; Chaikovsky, A.; Wiegner, M.; Nickovic, S.; Papayannis, A.; et al. Volcanic dust characterization by EARLINET during Etna's eruptions in 2001–2002. *Atmos. Environ.* **2008**, *42*, 893–905. [CrossRef]
27. Rothman, L.S.; Gordon, I.E.; Babikov, Y.; Barbe, A.; Benner, D.C.; Bernath, P.F.; Birk, M.; Bizzocchi, L.; Boudon, V.; Brown, L.R.; et al. The HITRAN2012 molecular spectroscopic database. *J. Quant. Spectrosc. Radiat. Transf.* **2013**, *130*, 4–50. [CrossRef]
28. Schafer, R.W. What is a Savitzky-Golay Filter? *IEEE Signal Process Mag.* **2011**, *28*, 111–117. [CrossRef]
29. D'Aleo, R.; Bitetto, M.; Delle Donne, D.; Tamburello, G.; Battaglia, A.; Coltelli, M.; Patanè, D.; Prestifilippo, M.; Sciotto, M.; Aiuppa, A. Spatially resolved SO_2 flux emissions from Mt Etna. *Geophys. Res. Lett.* **2016**, *43*. [CrossRef] [PubMed]
30. Bruhn, A.; Weickert, J.; Schnörr, C. Lucas/Kanade Meets Horn/Schunck: Combining Local and Global Optic Flow Methods. *Int. J. Comput. Vis.* **2005**, *61*, 211–231. [CrossRef]
31. Peters, N.; Hoffmann, A.; Barnie, T.; Herzog, M.; Oppenheimer, C. Use of motion estimation algorithms for improved flux measurements using SO_2 cameras. *J. Volcanol. Geotherm. Res.* **2015**, *300*, 58–69. [CrossRef]

32. Tamburello, G.; Kantzas, E.P.; McGonigle, A.J.S.; Aiuppa, A. Vulcamera: A program for measuring volcanic SO_2 using UV cameras. *Ann. Geophys.* **2011**, *54*, 2. [CrossRef]
33. Fischer, T.P.; Chiodini, G. Volcanic, Magmatic and Hydrothermal Gas Discharges. In *Encyclopaedia of Volcanoes*, 2nd ed.; Elsevier: Amsterdam, The Netherlands, 2015; pp. 779–797.

2

Spatio-Temporal Mapping of Plate Boundary Faults in California Using Geodetic Imaging

Andrea Donnellan [1,*], Ramón Arrowsmith [2] and Stephen DeLong [3]

1 Jet Propulsion Laboratory, California Institute of Technology, Pasadena, CA 91109-8099, USA
2 School of Earth and Space Exploration, Arizona State University, Tempe, AZ 85287-6004, USA; ramon.arrowsmith@asu.edu
3 United States Geological Survey, Menlo Park, CA 94025, USA; sdelong@usgs.gov
* Correspondence: andrea.donnellan@jpl.nasa.gov

Abstract: The Pacific–North American plate boundary in California is composed of a 400-km-wide network of faults and zones of distributed deformation. Earthquakes, even large ones, can occur along individual or combinations of faults within the larger plate boundary system. While research often focuses on the primary and secondary faults, holistic study of the plate boundary is required to answer several fundamental questions. How do plate boundary motions partition across California faults? How do faults within the plate boundary interact during earthquakes? What fraction of strain accumulation is relieved aseismically and does this provide limits on fault rupture propagation? Geodetic imaging, broadly defined as measurement of crustal deformation and topography of the Earth's surface, enables assessment of topographic characteristics and the spatio-temporal behavior of the Earth's crust. We focus here on crustal deformation observed with continuous Global Positioning System (GPS) data and Interferometric Synthetic Aperture Radar (InSAR) from NASA's airborne UAVSAR platform, and on high-resolution topography acquired from lidar and Structure from Motion (SfM) methods. Combined, these measurements are used to identify active structures, past ruptures, transient motions, and distribution of deformation. The observations inform estimates of the mechanical and geometric properties of faults. We discuss five areas in California as examples of different fault behavior, fault maturity and times within the earthquake cycle: the M6.0 2014 South Napa earthquake rupture, the San Jacinto fault, the creeping and locked Carrizo sections of the San Andreas fault, the Landers rupture in the Eastern California Shear Zone, and the convergence of the Eastern California Shear Zone and San Andreas fault in southern California. These examples indicate that distribution of crustal deformation can be measured using interferometric synthetic aperture radar (InSAR), Global Navigation Satellite System (GNSS), and high-resolution topography and can improve our understanding of tectonic deformation and rupture characteristics within the broad plate boundary zone.

Keywords: deformation; topography; earthquake

1. Introduction

For effective seismic hazard assessment, it is necessary to understand how plate motions are partitioned across the plate boundary zone and to characterize potential earthquake rupture sources within it. Plate tectonics drive the accumulation and release of strain either via earthquakes or via aseismic slip such as creep or anelastic bulk deformation, but how faults interact within the plate boundary can be complex. Earthquakes occur within systems of faults [1,2] driven by plate tectonic motions. In this paper, we focus on the Pacific–North American plate boundary across California (Figure 1). We discuss deformation associated with repeated earthquakes and aseismic slip, concentrating on timescales of days to millennia. Initial deformation occurs quickly during and immediately following an earthquake. Tectonic motions can significantly reorganize the landscape as earthquakes occur over seismic cycles, leaving patterns in the land surface that reflect fault and geologic processes. Fundamental research questions about the landscape include: How do geopatterns on the Earth's surface arise and what do they tell us about processes? What are the transport laws that govern the evolution of the Earth's surface? [3]. Including understanding of landscape processes into interpretation of crustal deformation measurements enables a better assessment of the mechanisms contributing to deformation of the Earth's crust.

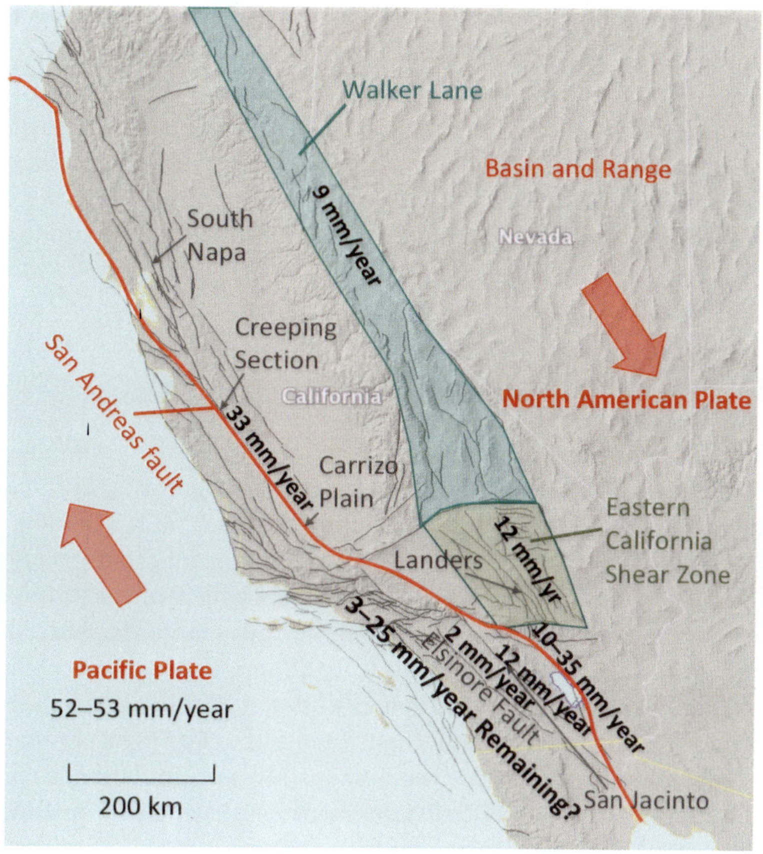

Figure 1. The Pacific and North America plate boundary accommodates a total relative plate motion of 52–53 mm/year [3,4]. The San Andreas fault (red line onshore) takes up most of the plate boundary motion of up to 66%. About 25% of the plate motion is accommodated by the Eastern California Shear Zone in the Mojave Desert [4] and the Walker Lane to the north [5], leaving 9% to be accommodated elsewhere. Estimated slip rate along the southern San Andreas fault has a wide range, representing substantial uncertainty. The estimated slip rate along the San Jacinto fault varies, but in this example we use 12 mm/year [6–8]. Slip rate along the Elsinore fault is approximately 2 mm/year [9] leaving a remaining 3–25 mm/year of deformation south of the Central and Western Transverse Ranges to accommodate the total plate motion. Numerous faults take up the entire plate boundary motion and the rates are more uncertain than the simplified model shown here.

Our motivation is to review emerging methods that allow for improved spatio-temporal understanding of fault zones and to call for new, more holistic, approaches to high-resolution topographic mapping of the entire greater San Andreas fault (SAF) system and beyond. We argue that it is necessary to infer slip distribution and rheologic properties at local and plate boundary scales in order to assess earthquake hazard, and that these properties are reflected in topography and crustal deformation. Faults connecting from the upper crust to deeper plate boundary motions and measurement at multiple scales best addresses the above questions. Measurement of crustal deformation and surface morphology can be accomplished by broader deployment of existing methods including synthetic aperture radar, lidar and Structure-from-Motion (SfM) from small unmanned aerial systems (sUAS) and by development of new airborne and spaceborne platforms to deploy next generation SAR, laser, and multispectral mapping technology. We focus on five distinct areas within the California Pacific–North American plate boundary to illustrate these processes.

The characteristics of plate boundaries include tectonic features on spatial scales ranging from local sub-meter rock damage zones and surface faulting, to kilometer-wide fault zones, and networks of faults hundreds of kilometers wide. Understanding the system of faults within a plate boundary is necessary to better forecast earthquakes. No predictable pattern of where or when the next earthquake will occur in a sequence of events has emerged. Earthquakes can occur sequentially along sections of a fault such as the 1939–1999 North Anatolian earthquake sequence [10,11] or between different faults such as occurred for the 1992 Big Bear/Landers earthquake sequence [12]. Seismic hazard analysis combines statistical analysis of seismicity, geodetically measured strain accumulation, historical earthquakes, and records of prehistoric earthquakes preserved in the geologic record. In an ideal case, the moment release from earthquakes exactly matches the strain accumulation from plate tectonics. However, geodetic imaging, broadly defined as measurement of crustal deformation and topography of the Earth's surface, indicates two complicating factors: (1) rates of deformation change over multi-year timescales and can switch modes, particularly after large earthquakes [13], such as from horizontal shortening to lateral shear; and (2) off-fault and aseismic deformation may account for a significant amount of total plate boundary motion.

Earthquake rupture characteristics are largely controlled by the geometry of fault networks. Earthquake magnitude correlates with earthquake rupture length [14]. Earthquake rupture propagation can be inhibited by barriers [15] or steps between fault segments [16]. Faults may be rooted to a single fault at depth but split into strands at the near surface in a flower structure [17]. The geometry of fault zones, the partitioning of deformation between seismic, aseismic and postseismic slip, and the nature of fault damage may be reflected in the topographic features of fault zones [18,19]. Here, we present three unresolved questions to argue that multiple scale measurements using different types of geodetic and geologic measurements are necessary to better understand earthquake processes. These are: (1) How do plate boundary motions distribute across California faults? (2) How do faults within the plate boundary interact? (3) What fraction of strain accumulation is relieved aseismically? We propose that observations at scales of fault zones (m), fault systems (km), and plate boundaries (10s–100s of km) can reveal the spatio-temporal factors controlling earthquake processes, and we discuss approaches to multi-scale measurement and observation that may characterize the importance of geometry and kinematics to infer of mechanical processes.

2. Materials and Methods

Several methods exist to measure velocities and strain rates at or near the surface of the Earth. We focus here on characterization of crustal deformation using geodetic measurements from GPS, airborne radar interferometry, and high-resolution topography from multiple collection platforms. Other methods used to characterize crustal deformation include spaceborne InSAR, leveling, triangulation and trilateration, borehole or laser strainmeters, and paleoseismology. The measurements are used to infer processes at depth by inverting the data or constructing forward models. For topographic measurements, we focus on ground and airborne lidar and Structure from Motion (SfM) photogrammetric measurements from small sUAS [20]. These measurements are complementary, capturing a wide range of spatial and temporal scales of crustal deformation and landscape evolution.

2.1. GPS

Global Positioning System (GPS) measurements provide precise position time series from which velocities from plate tectonic or other motions or displacements from earthquake can be determined [21]. GPS data for measuring crustal deformation were originally collected infrequently at a sparse set of stations in southern California [22,23]. As an example of early demonstration of the value of GPS measurements, modeling of deformation determined from GPS combined with triangulation measurements suggested that buried thrust faults bounded and dipped away from the Ventura basin [24]. The fault that ruptured in the subsequent 1994 M6.7 Northridge earthquake [25] was consistent with that identified through modeling the GPS measurements. In the following decade, a continuous network of about 250 GPS stations was implemented throughout Southern California [21]. This was followed by the Plate Boundary Observatory [26], which is part of the NSF-funded EarthScope program to explore the four-dimensional structure of the North American continent, and consists of about 1100 continuous GPS stations in the western United States [27].

Station spacing of the Plate Boundary Observatory continuous GPS network across the active faults of the plate boundary is on average about 10 km. In California, the Pacific–North American plate boundary is about 400 km wide, which means that at any given latitude about 40 GPS span the width of the zone. The GPS stations are generally concentrated closer to fault zones and sparser in less tectonically active areas such as the Central Valley of California. The accuracy of GPS measurements is about 1 mm/year for horizontal velocities [6] and about a factor of two lower for vertical velocities [28]. The daily position time series provide a high data rate for detecting slip from earthquakes, postseismic motions, and any other transient motions, such as subsidence from fluid withdrawal [29] which occur on timescales of months to years. Surface deformation measurements generally indicate fault motion at a depth on the order of the distance of the point of measurement from the fault [30]. That means that for 10 km station spacing for a fault that is creeping at depth, it is not possible to accurately determine the locking depth of the fault. In order to understand whether or by how much faults creep in the near surface, crustal deformation must be sampled at a much denser spacing.

2.2. *InSAR*

Interferometric Synthetic Aperture Radar (InSAR) uses differences between two images to produce a map of line of sight changes between points on the ground and the airborne or spaceborne instrument [31,32]. Measurement of displacements from an earthquake requires before and after images. The images are compared and if the ground has moved and the instrument repeats the image from the same place, the waves interfere producing a map of fringes that represent line of sight phase and thus range changes. The first SAR image of an earthquake was produced by the European Radar Satellite (ERS-1) for the 1992 M7.3 Landers earthquake [33]. An advantage of InSAR over GPS measurements is the spatial continuity of the measurements. The time between images can range from weeks to years, which means that InSAR typically does not capture rapid deformation processes. Furthermore, InSAR produces line of site measurements, so unless images from multiple vantage points are collected, decomposing the results into component horizontal and vertical motions requires constraints from field observations or GPS vector displacements.

Spaceborne InSAR data have been available from a variety of orbiting spacecraft and more are planned, including the NASA-ISRO Synthetic Aperture Radar (NISAR) mission [34] scheduled to launch in late 2021. Since 2009, NASA has operated UAVSAR, an airborne InSAR instrument that is mounted underneath a Gulfstream-III aircraft [35–37]. UAVSAR is an L-band instrument that images a 15 km $\times$ 90 km swath, and it has been operational since 2009 (Figure 2). It has been deployed globally, but the main focus has been in California where it supports a number of investigations. UAVSAR has been used to image deformation after three earthquakes ranging in magnitude from 5.2 to 7.2 [38–40]. UAVSAR produces images with 1 m spatial resolution for phase change images and 7 m for images that have been unwrapped. Unwrapped images convert the phase changes to line of sight range changes between the ground surface and the instrument. If the surface is too disrupted, the phases decorrelate and it is not possible to construct an image of phase changes in that disrupted area.

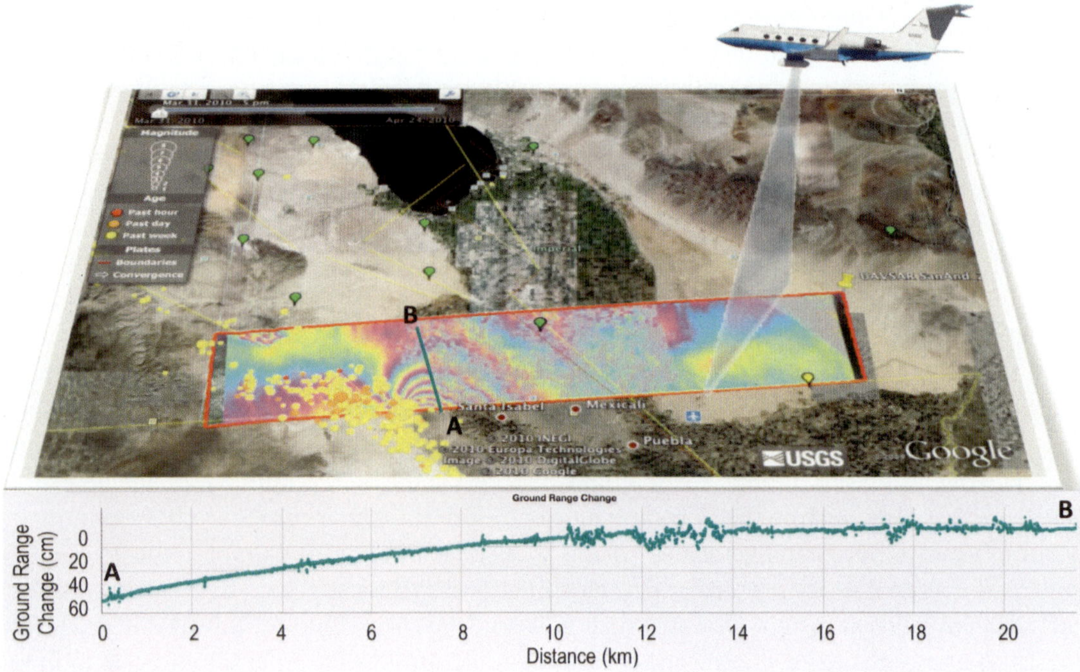

Figure 2. UAVSAR concept showing the airborne instrument. The oblique image is of the Salton Trough in southern California. The south (bottom) edge of the UAVSAR image lies at the US–Mexico border. The yellow dots represent aftershocks from the 2010 M7.2 El Mayor–Cucapah earthquake. The Salton Sea is shown at the north end of the image bounds the Imperial Valley. The width of the swat is about 15 km. The lobes of fringes represent about 60 cm of differential motion in the crust north of the main rupture as shown by the extracted ground range change in the profile plot with the location noted on the interferogram [38].

2.3. *Topography*

The Earth's surface is the interface between geomorphic and tectonic processes. Geomorphic processes act to change elevation through erosion and deposition while tectonic processes translate, depress and/or elevate the surface directly. At larger spatial scales, topography indicates cumulative interactions across mountain ranges, such as the balance of accretionary and erosional fluxes [41], for example, as manifest by bedrock channel forms [42]. Differential motions over hundreds of kilometer wavelengths are driven by variations in mantle buoyancy to produce "dynamic topography" (e.g., [43]). Investigations of active faulting, on the other hand, may require large ranges in inner and outer length scales. For example, an M8 earthquake may be 400 km in length, but the bulk of the displacements occur within a 1 km wide zone and fine displacements and geomorphic responses are at decimeter or finer scales. Topographic-based investigations of faulting include fault trace mapping, reconstructing slip histories, understanding geomorphic response to uplift, and 3D topographic differencing (e.g., [44–46]).

A major challenge is to acquire, handle, and process the right topographic data in time, space, and sampling density for the question. 3D heterogeneous sampling of a point cloud is distinct from 2D or "2.5D" regular or gridded samples from digital elevation models (DEM). The surface is generally defined as a digital surface model (DSM) for the first case and the inferred ground surface digital terrain model (DTM) for the second. Many space-based platforms provide global to regional DSM coverage with 10–100 m per pixel such as the Shuttle Radar Topography Mission [47], ASTER GDEM [48,49], Global Multi-Resolution Topography (GMRT), which includes bathymetry [50], TanDEM-X [51], etc. At local to site scales, topography is sampled with cm to meter/pixel, tends towards 3D representations, and is acquired from airborne, mobile, and ground-based platforms. This centimeter to decimeter sampling is essential for representing faulting and surface processes. Topographic data are thus collected from a range of platforms with increasingly finer resolutions, greater accuracy, and shorter repeat times. Understanding faulting processes requires the ability to discover, manage, share, and process topographic data. Community cyberinfrastructure increases the impact of investments in topography data and catalyzes scientific discovery (see for example, OpenTopography [52]).

2.3.1. Lidar

Laser scanning from airborne, mobile, and terrestrial platforms is the current benchmark method for acquiring topographic data of sufficient resolution to resolve fine scale fault zone features and to track crustal motions in three dimensions through time [53]. Airborne data are generally collected from fixed-wing or rotary blade aircraft and produce data with decimeter scale accuracy and sub-meter scale spatial resolution and are only partially hindered by vegetation [54]. Mobile data are gathered from vehicles, backpack systems and low airborne platforms such as balloons or sUAS and provide data with centimeter scale accuracy and precision. Terrestrial data are collected from stationary tripods, generating data with sub-centimeter scale accuracy and resolution.

While there are examples of widespread lidar coverage from airborne platforms (e.g., [55]), it remains expensive to map at the plate boundary scale and significant data gaps remain. The promise of repeat coverage over large spatial scales to quantify Earth surface changes through time is not possible given current priorities and resources. This challenge is exemplified by the limited progress towards nationwide coverage in the United States via the U.S. Geological Survey 3D Elevation Program [56].

Despite limitations in the use of lidar for understanding plate boundary process, several examples are indicative of the promise of remotely sensed topographic measurements to understand specific tectonic settings and processes. Coseismic deformation patterns of several earthquakes have been constrained using multitemporal airborne lidar data. These include the 2010 M7 El Mayor–Cucapah earthquake [57,58], the 2008 M6.9 Iwate-Miyagi Nairiku (Japan) and 2011 M7.1 Fukushima Hamadori (Japan) earthquakes [59], and the 2010 M7.1 Darfield (New Zealand) earthquake [60]. These earthquakes all had meter-scale near-field deformation, so could be quantified using airborne data that tend to have decimeter scale accuracy and resolution, but finer scale and more distributed deformation remains difficult to quantify with airborne lidar. We do however anticipate scientific response to recent and imminent large continental earthquakes will generate new studies reliant on repeat airborne lidar at progressively higher levels of detail.

Finer-scale deformation can be quantified using terrestrial and mobile lidar, but over somewhat limited spatial scales. Mobile laser scanning (e.g., [61]) has advantages over airborne lidar in speed of deployment and level of resolution, and over terrestrial methods in its ability to survey significantly larger areas fairly quickly but ground-based systems are more limited by vegetation cover than airborne systems.

Several studies have employed terrestrial laser scanning to quantify plate boundary process and to map fault zones. These include the 2004 M6.0 Parkfield, California earthquake; the 2004 M6.6 Niigita Ken Chuetsu earthquake in Japan [62]; the 2010 M7 El Mayor–Cucapah earthquake in Baja California, Mexico [63]; the 2009 M6.3 L'Aquila earthquake in Italy [64]; and the 2014 M6.0 South Napa earthquake [40,65].

Here we review a few published and previously unpublished studies that report on the use of remotely-sensed topographic data from laser scanning to study plate boundary processes. For more complete descriptions of the processing required to acquire and analyze these data, we direct the reader to a recent review of lidar and other high-resolution topographic data [53], and, for an example of terrestrial laser scanner collection and processing, a recent study of surface rupture associated with the South Napa earthquake [65].

2.3.2. Structure from Motion (SfM)

The advent of inexpensive digital cameras, sUAS, and fast computer processing power has enabled rapid growth in modeling three-dimensional structure from photographs. Using SfM, images can be combined to solve for surface topography (e.g., [20]). Advantages of SfM over traditional airborne stereo imaging is the ability to image steep slopes, to capture topography from a wide range of sensors without detailed flight planning and the ability to resolve very fine-scale features (Figure 3). Structure from motion also produces colored 3D models from the imagery, which can provide information about lithology and surficial characteristics (e.g., vegetation, moisture, texture, structure).

Figure 3. Example flight of a small unmanned aerial system (sUAS): in the Mecca Hills (**left**); and in Painted Canyon (**right**), acquiring low-altitude photos and to apply Structure from Motion along the San Andreas fault. The sUAS can image steep slopes to create topographic reconstructions with comparable horizontal and vertical resolution.

3. Results

GPS velocity measurements are useful for understanding how deformation is distributed across the California plate boundary. Grand inversions have been carried out to estimate earthquake slip and timing along hundreds of sections of fault throughout California [1]. Knowledge of these details of slip is important for informing estimates of earthquake hazard. Identifying the dominant fault zones is important for inferring details of the plate boundary at depth within the lithosphere.

Crustal blocks are surrounded by actively deforming zones in California (e.g., [6,66–68]). A simple exercise suggests that two zones of shear dominate in California. We plotted velocities in California relative to GPS stations near the SAF and in the Owens Valley (Figure 4). In those reference frames, GPS velocities to the west of the reference station move northward and stations to the east move southward. The inflection between northward and southward velocities can be traced as a zone of shear. These two zones join at the Gulf of California. Moving northward from there, one zone follows the San Jacinto fault, joining with the SAF to the north. The second zone follows the southernmost Coachella segment of the SAF and connects through the Eastern California Shear Zone and the Walker Lane. The zone of deformation is narrower, on the order of 10 km, along the more mature central SAF than it is to the south where it is about 40 km wide, or in the northern Walker Lane where it is about 50 km wide. These two zones likely represent deep lithospheric shear where the plate boundary transitions from the SAF system which is oblique to the overall plate boundary and therefore not oriented ideally within the overall stress field, to the more plate-boundary-parallel Eastern California Shear Zone. Dominant shear zones in California likely correlate with compliant zones between crustal blocks and may point to zones of lithospheric weaknesses. Measurement of surface deformation is necessary to drive models of lithospheric processes with heterogeneous mechanical properties.

In what follows, we sample locations along these plate boundary zones beginning at the northern end of the San Andreas branch of the Pacific–North American Plate boundary. We present key characteristics of the immature South Napa earthquake fault, the creeping section of the SAF, and the mature SAF in the Carrizo Plain. We discuss geomorphic modifications to the Landers earthquake rupture in the Eastern California Shear Zone. To the south, where these two zones merge, we discuss the locked San Jacinto fault, the intermittently creeping southern SAF, and the network of faults in the Salton Trough where these two major faults terminate.

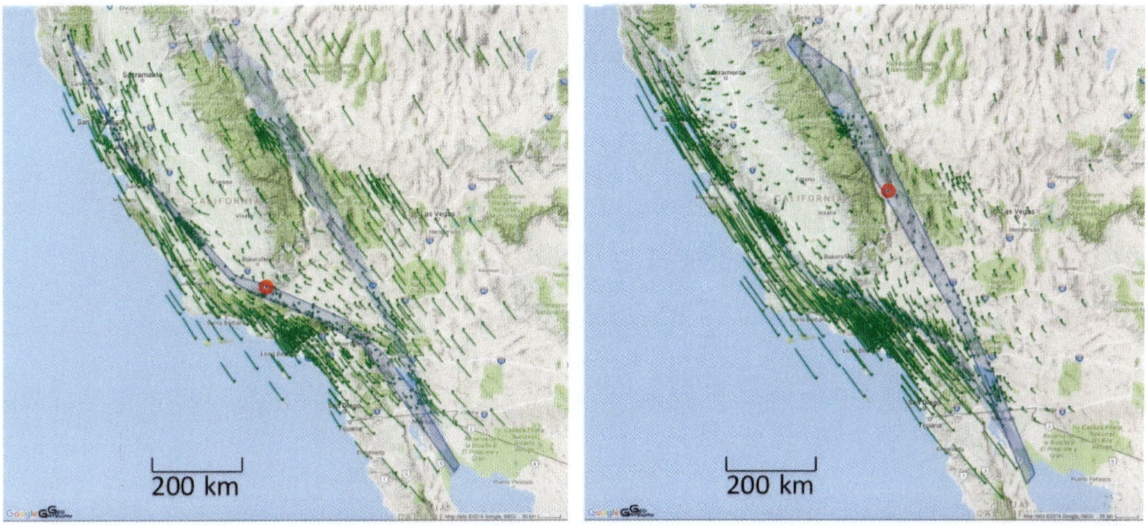

Figure 4. Global Positioning System velocity vectors are plotted relative to two different stations using GeoGateway [69,70]. Stations are marked by a small circle and a line with no arrowhead represents the velocity vector. Left panel shows velocities relative to station EDPP [71] near Tejon Pass noted by a red circle. Right panel shows velocities relative to station P093 in the Owens Valley, also noted by a red circle. The dividing region between northward and southward vectors was drawn in for: the San Andreas fault (**left**); and Eastern California Shear Zone/Walker Lane (**right**), and then plotted on each map.

3.1. South Napa Earthquake—Moderate Rupture through Several Mapped and Unmapped Traces

The Mw 6.0 South Napa earthquake of 24 August 2014 caused slip along several active fault strands within the West Napa fault zone. Field mapping identified 12.5 km of surface rupture and analysis of airborne UAVSAR data indicated more than ~30 km of surface deformation with a relatively complex distribution across a number of subparallel lineaments [43]. Postseismic slip exceeded coseismic slip along a portion of the main rupture trace, and maximum summed postseismic and coseismic slip approached 0.5 m [43]. A primary challenge after the earthquake was rapid and accurate surface rupture mapping, quantification of coseismic slip, and subsequent identification and monitoring of potentially damaging postseismic slip [72]. Analysis of UAVSAR data provides slip distributions along multiple fault traces, indicates connectivity and extent of secondary traces, and confirms that postseismic slip only occurred on the main trace of the rupture [40].

Of particular interest is the use of lidar, photography, and UAVSAR to map the fault rupture and image the fault zone. Figure 5 displays a section of the fault zone imaged using air photos, airborne lidar and terrestrial lidar. This highlights the challenge of mapping fine scale tectonic features and the need for topographic and imagery data with submeter and finer spatial resolution. Though vertical deformation over the coseismic interval was measureable using repeat airborne lidar, and some coarse estimations of horizontal measurement were made using repeat airborne lidar [72,73] the most precise (but spatially limited) quantifications of horizontal slip magnitudes were made during the postseismic interval using terrestrial lidar [65,74]. Figure 6 displays results from identifying geometric primitive objects from terrestrial lidar point clouds and tracking them through time. Because the earthquake was moderate in magnitude and in amount of surface rupture and deformation, this level of detection required data with centimeter level accuracy and precision acquired over multiple survey epochs. Larger earthquake deformation in contrast may be adequately quantified using somewhat coarser data.

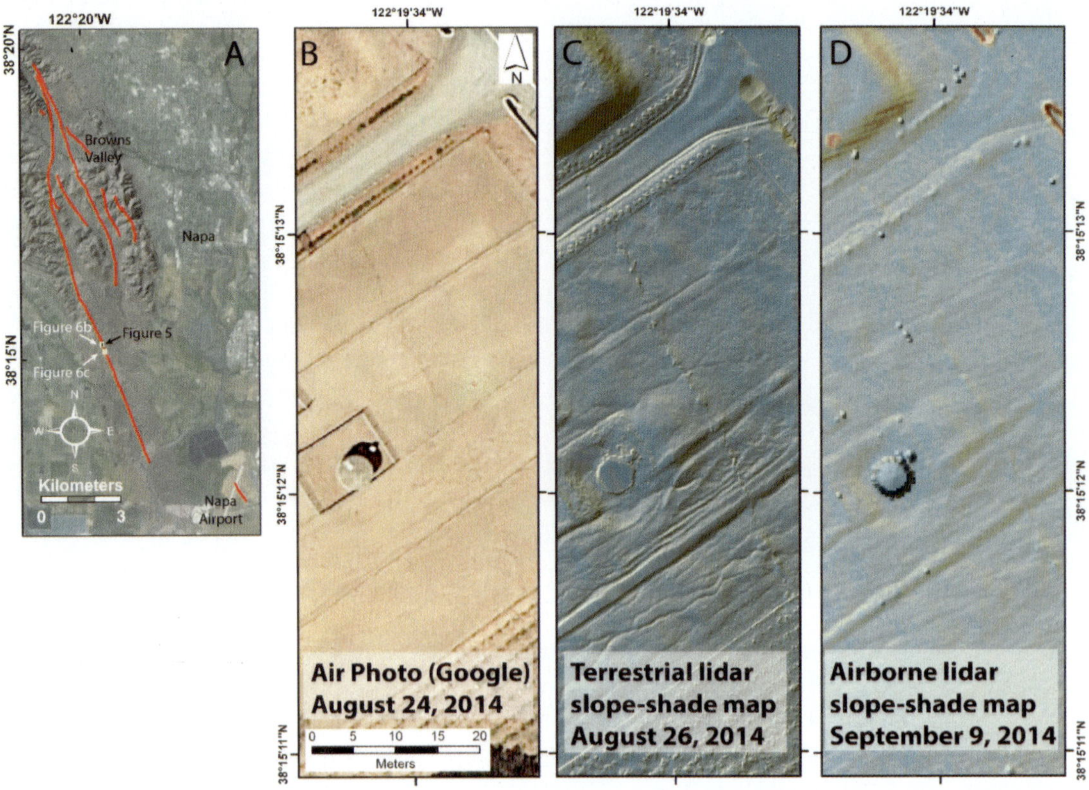

Figure 5. Images from 2014 South Napa Earthquake: (**A**) map of surface rupture with location of zoomed panel; (**B**) orthorectified air photos with ~30 cm pixel resolution collected the day of the earthquake by Google; (**C**) slope-shade map with 2.5 cm pixel resolution made from a digital elevation model generated from an ultra-high resolution 3D point cloud (>1000 points/m^2) from a terrestrial laser scanner; and (**D**) slope-shade map with 25 cm pixel resolution made from a digital elevation model generated from a high resolution (~20 points/m^2) 3D point cloud from an airborne laser scanner survey. Figure modified from [43]. In each panel, the rupture cuts the image from the lower right corner up towards the upper left corner.

As a tool to map the distribution of surface deformation, UAVSAR proved to be particularly valuable. UAVSAR interferograms were used iteratively with field observation to understand the distribution of surface deformation. Furthermore, UAVSAR provided the most robust means of generating full coseismic and postseismic slip distributions [40] (Figure 7). The wrapped interferogram shows considerable surface disruption, but the product is useful for identifying the location of fault rupture from discontinuities in the image. However, the severe surface disruption caused decorrelation in a zone about 4 km wide for the unwrapped interferogram. A combination of field measurements and displacement determined from the unwrapped interferogram were used to estimate fault slip.

The South Napa earthquake occurred within a low slip rate fault zone containing a network of what may be considered immature faults due to their complex connectivity and weak surface expression [75]. The surface rupture was located 2–2.5 km west of the part of the mapped West Napa fault trace thought to be most active [43]. Both faults lie within a series of nearly parallel faults in a 100 km wide plate boundary zone. It is interesting to note that South Napa rupture lies within the shear zone identified with GPS velocities (Figure 2). The shear zone connects to the south with the Calaveras fault and then joins the SAF at the north end of the creeping section of the SAF. The question remains how this system of parallel faults roots to the plate boundary at depth.

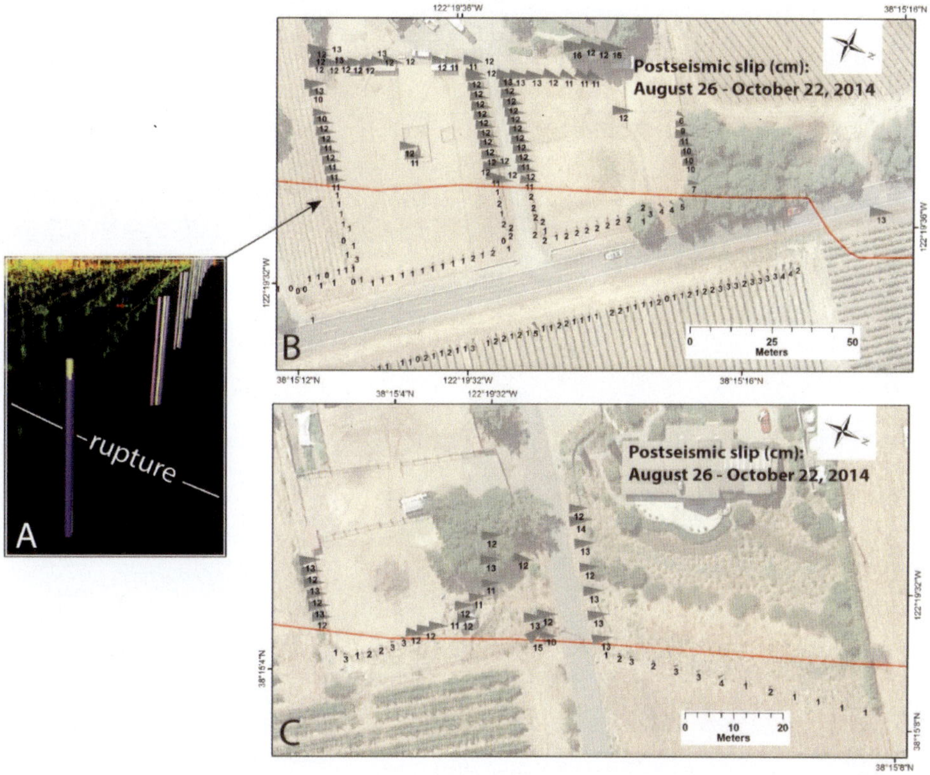

Figure 6. Postseismic deformation associated with 2014 South Napa earthquake as measured using objects in terrestrial lidar point cloud. (**A**) Cylinders modeled from fence posts in point clouds at three times post-earthquake, and 3D point cloud of vineyard rows in background. Note the progressive dextral offset relative to the fencepost in the foreground. (**B**,**C**) Display direction (grey arrow) and amount of translation of objects in the study area (as indicated by the size of the arrows and their labels), which are translation distances in cm. Red line indicates the generalized trace of the 2014 surface rupture. Base image is aerial photograph taken 24 August 2016 by Google. Figure modified from [67]. Study area locations are shown in Figure 5.

3.2. Creeping Section of the SAF

The 175-km long creeping segment of the SAF from Parkfield, California in the south to San Juan Bautista in the north is thought to prevent rupture propagation from the southern SAF to the northern SAF, effectively limiting the possible moment release on the greater San Andreas System. However, the possibility of significant coseismic moment release along the creeping segment remains a point of discussion in the literature [76,77] and is included in seismic hazard models, albeit with low probabilities [1]. As such, it remains critical to determine the distribution of relative plate motion across the SAF system along this fault section. This requires measuring and mapping the spatial and temporal distribution of aseismic slip in order to identify any slip deficit that might indicate potential for coseismic moment release. Aseismic slip rate varies along the creeping segment and also when measured over different apertures [78,79], but these measurements alone do not allow for proper slip partitioning onto mappable structures and/or distributed deformation [80]. Detailed fault mapping from field [81] and remote sensing (lidar) mapping [82,83] begin to explain the nature of the fault in a particularly complex zone of distributed faulting. Deformation may be accommodated along the creeping SAF as discrete fault offset only or with additional distributed deformation [80]. The question of whether plate boundary motion is accommodated as distributed deformation or accumulated as elastic strain or fault zone slip is fundamental to understanding rupture propagation and potential earthquake magnitude in central California. However, these studies do not constrain the motions or properly partition aseismic slip across multiple structures and/or measure distributed deformation between mappable structures. This can be addressed by combining UAVSAR, GPS, and collection of new geodetic imaging from sUAS and lidar platforms.

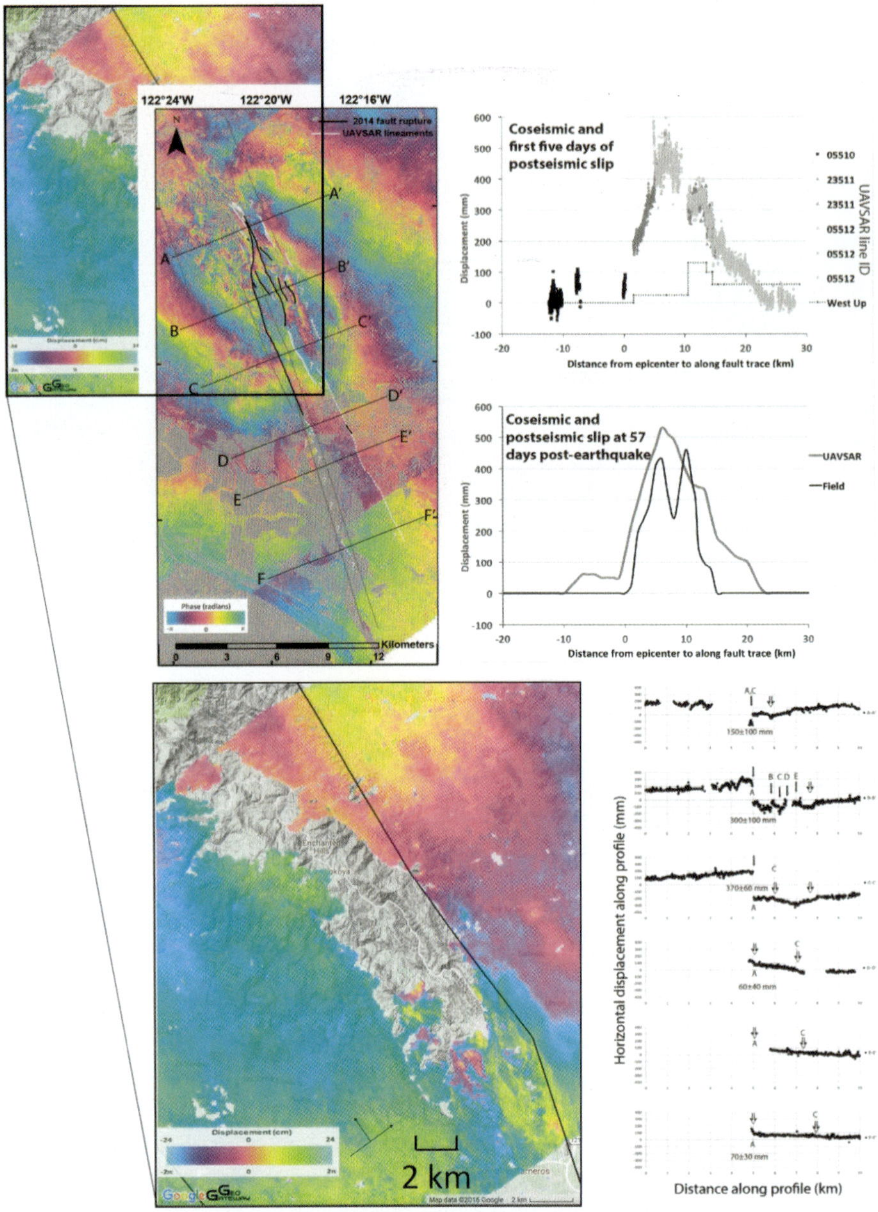

Figure 7. Deformation measurements from 2014 South Napa earthquake. Far upper left image places bottom image in context of top image, which shows interpreted UAVSAR interferograms with field-observed surface rupture in black and UAVSAR-interpreted deformation in white. Lines 05510, 23511, and 05512 are shown. The seams between images appear as discontinuities. Top right pane indicates event slip distribution from field study and UAVSAR analysis. Lower left panel shows unwrapped interferogram for line 05512 with decorrelated region. Mapped fault is shown as a black line from UCERF-3 [1]. Lower right panel displays horizontal displacements along selected profiles marked on top left image. Figure modified from [43].

Preliminary comparison of air and ground based lidar measurements from 2007 and 2016 respectively on the creeping section of the SAF constrain horizontal creep across the fault zone. At Mee Ranch along the creeping SAF, an alignment array with an aperture of 200 meters provided a near-field slip rate of 23.7–26.5 mm/year [79,81]. A reconnaissance-level application of lidar differencing is generally consistent with this result (Figure 8). The quantitative relationship between slope along a fault-parallel cross-section and local vertical difference gives the relative horizontal motion. The value of 31.2 ± 12 cm in nine years, gives a rate of 35.2 ± 11 mm/year for the past nine years, at the higher end of previously reported values, but over a greater width and with greater error bars than the alignment array data presented in [81]. An offset fence can be seen in the sUAS observations (Figure 8D). The main zone of creep is about 2 m wide as is evident from the deformation of the fence. This preliminary analysis motivates further development of analyses capable of producing creep and deformation measurements from topographic data.

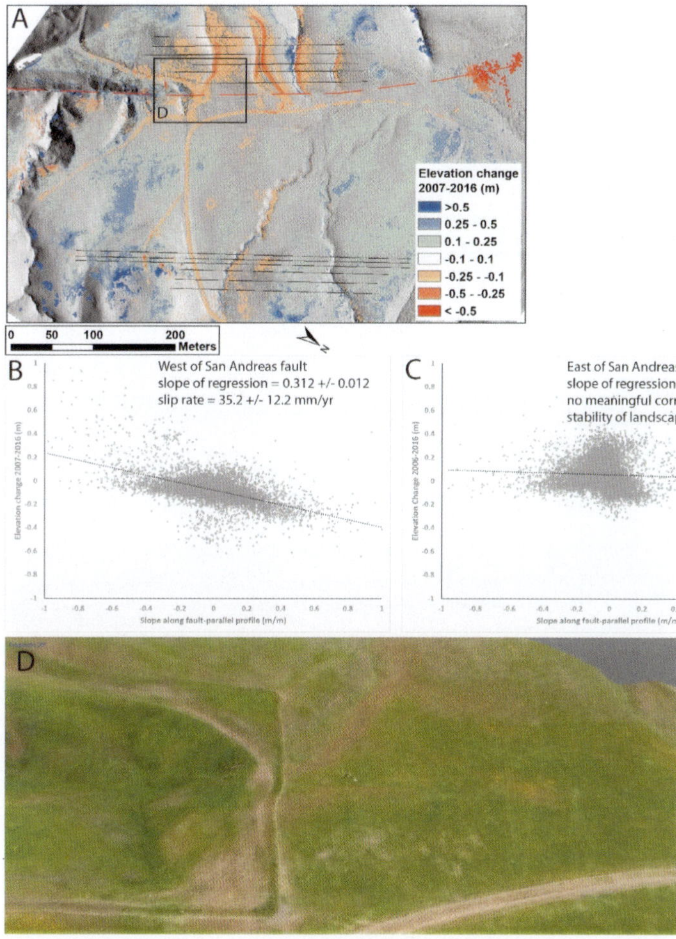

Figure 8. Rapid slip rate estimation from topographic change. (**A**) Map of elevation change from 2007 to 2016. This map displays vertical difference between 2007 airborne lidar and 2016 terrestrial lidar determined in a east-side-stable reference frame. Red indicates elevation decrease. Elevation decrease along south-facing slopes on west side of fault (top of figure) and elevation increase along north-facing slopes on west side of fault indicate dextral motion. (**B**,**C**) Plots of local fault-parallel slope versus elevation change on west and east sides of fault, respectively. These are taken from six fault parallel profiles on each side of fault as mapped in (**A**). For relatively smooth slopes and small slip magnitudes, the relationship between slope along a profile in the direction of fault motion and the vertical difference in elevation at that point at two times indicates the amount of horizontal movement. In this case, the slope of the regression line of slope versus elevation change along several fault-parallel profiles provides the amount of interseismic creep on the San Andreas fault from 2007 through 2016. In this case, the apparent horizontal offset is 0.312 ± 0.12 m over nine years, indicating a slip rate of 35.2 ± 11.2 mm/year, consistent with the higher end of previous estimates [81]. (**D**) 3D topographic model from sUAS-derived Structure from Motion photogrammetry. A fence offset by aseismic slip on the SAF is clearly visible just left of center. The fault cuts from left to right through the middle of the image.

UAVSAR observations over 1.73 years indicate a rate averaged between the cross and along fault difference profiles of about 32.5 ± 21 mm/year of horizontal motion if we assume that all of the motion is horizontal (Figure 9). The error is similarly large to the preliminary lidar analysis. The zone of deformation is about 200 m wide, however, suggesting additional off-fault deformation beyond the aperture of the alignment arrays, but within the aperture of the lidar analyses, with the possibility of additional off-fault deformation. The lack of UAVSAR data products for multiple look directions makes it difficult to separate horizontal from vertical motion here, which could also account for the possible differences among analytical results. GPS measurements show 27 ± 1 mm/year, 28 ± 1 mm/year, and 29 ± 1 mm/year for 1, 7, and 20 km GPS station spacing across the fault. The more recent lidar and UAVSAR measurements are more consistent with, but still larger than the farther field GPS results, than with the earlier alignment array measurements, leaving open the question whether measurement bias or time varying slip accounts for the discrepancy.

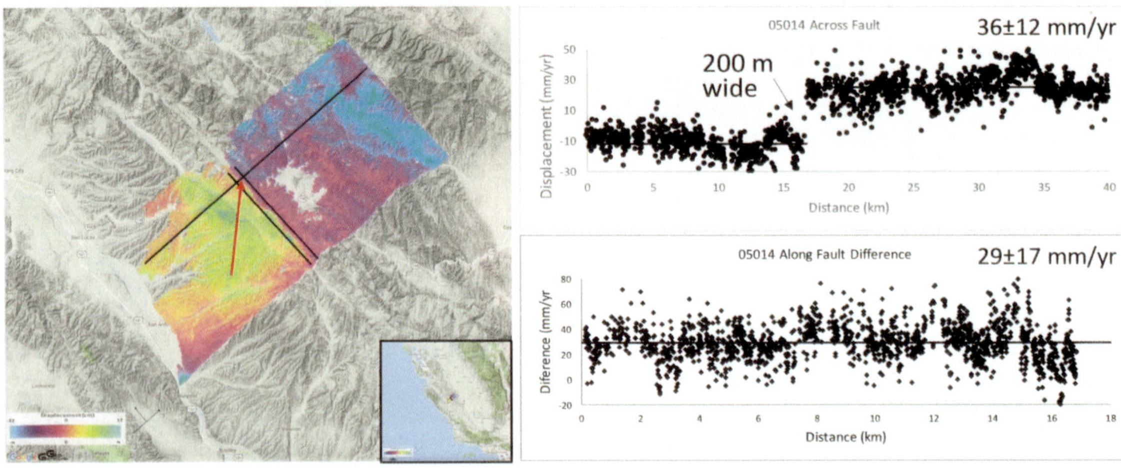

Figure 9. UAVSAR line SanAnd_05014_09009-001_10082-006_0630d_s01_L090HH_01 for data collected 26 February 2009 and 18 November 2010. The aircraft is flying northeast and looking left. The elevation angle between the ground points and airborne platform ranges from about 27°–65°. Top right: A profile line of sight ground range change across the fault is plotted and slip across the fault is estimated. Lower right: Two profiles on either side of the fault are differenced. The profiles are 1.25 km apart. Data and profiles were viewed and extracted from GeoGateway [69]. Principal investigator for this flight line is Paul Lundgren, JPL [84]. Red arrow points to location of image in Figure 8.

3.3. Carrizo Section San Andreas Fault—A Mature and Locked Fault

The Carrizo Plain area, situated in the southeastern California Coast Ranges (Figures 1 and 10), contains the premier geomorphic example of the San Andreas fault system in California. The SAF in the Carrizo Plain has the largest accumulated post-early Miocene offset and is the oldest reach of the entire active fault system (315 km and ~15 Ma; [85]). The area has long been regarded as a site of world-class examples of strike-slip faulting (e.g., [86–91]). The SAF follows the northeastern side of the plain and along it are offset, beheaded, and abandoned drainages that reflect recent strike-slip motion. That motion has been accommodated by great earthquakes; the most recent was the 1857 Fort Tejón earthquake [92–94]. Earlier offsets and paleoruptures are evident as well [88,93,95]. The geodetic measure of strain accumulation [96] is within error of the millennial scale fault slip rate of ~34 mm/year [88,97,98]. The record of faulting is of fundamental importance for testing models of earthquake physics and for constraining the significant earthquake hazard faced by southern California [1].

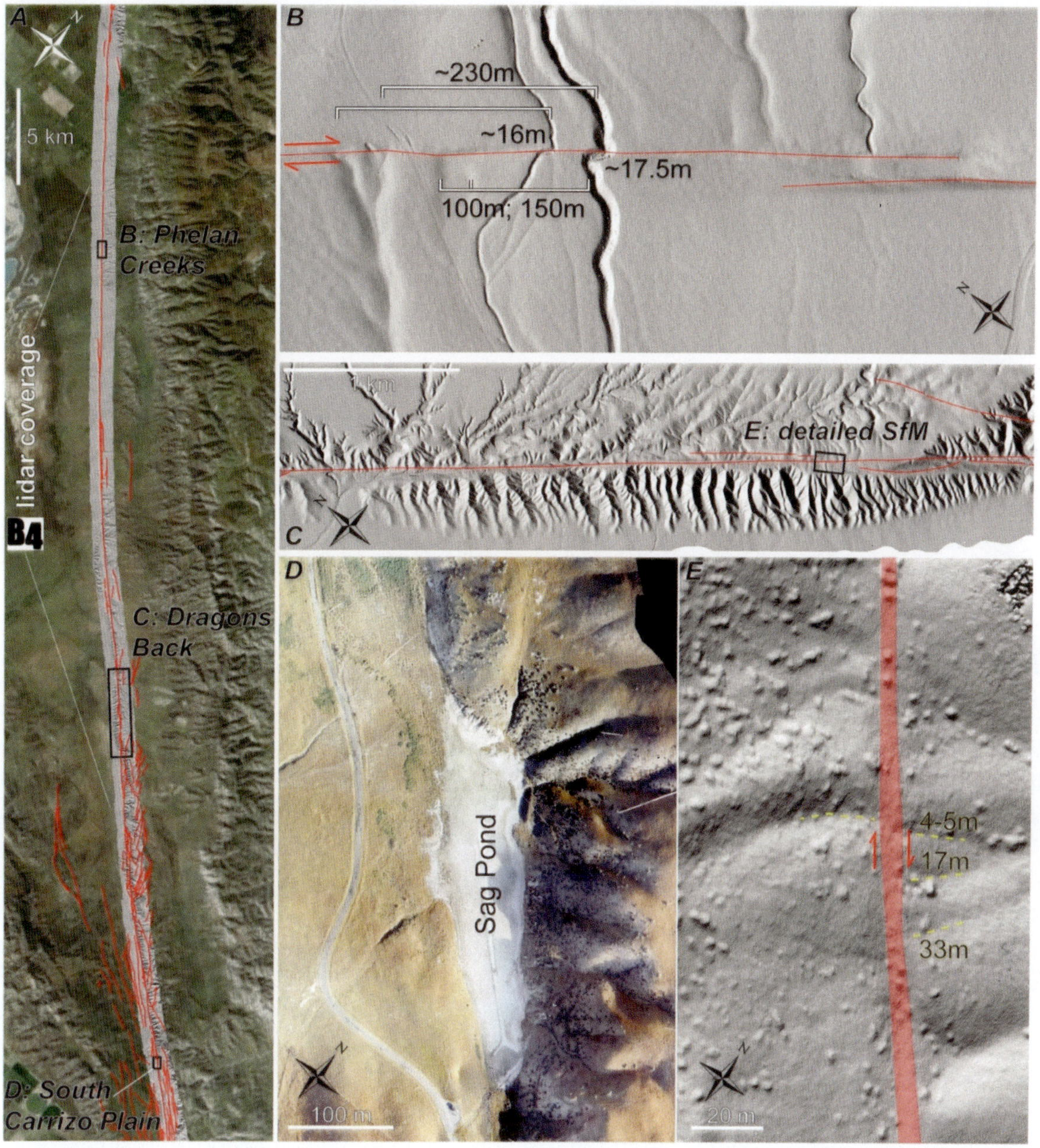

Figure 10. Lidar and Structure from Motion (SfM) for the Carrizo Plain: (**A**) B4 lidar [99] coverage strip along the San Andreas fault in the Carrizo Plain with US Geological Survey Quaternary Faults [100] (red); (**B**) B4 hillshade (0.5 m/pixel) depicts spectacular tectonic landforms at Phelan Creeks where millennial scale offsets are consistent with ~33 mm/year slip rate [98,101]; (**C**) lidar topography provides the basis for detailed geomorphic analysis of the Dragon's Back pressure ridge [90]; (**D**) SfM-derived orthophoto (6 cm/pixel) shows sag pond along the San Andreas fault in southern Carrizo Plain [91]; and (**E**) SfM-derived hillshade (0.2 m/pixel) provides basis for fine scale offset landform reconstruction (e.g., [98,102]).

Lidar topography and SfM have been transformative for active faulting studies along the SAF in the Carrizo Plain. Figure 10 illustrates a series of many possible examples. The SAF was laser scanned as part of the B4 project [99]. The data are invaluable for mapping the fault traces, reconstructing offsets, and inferring geomorphic response to rock uplift. Detailed mapping of the SAF trace is enabled with the fine resolution and high spatial accuracy of the lidar topography (e.g., [103]). The extent of lidar coverage from the 2005 B4 Lidar Project survey is 1–2 km wide with typical shot densities of 4 points/sq.

meter (Figure 10A). Millennial-scale offsets preserved at Phelan Creeks (Figure 10B) are consistent with 34 mm/year slip rates [98]. Reconstruction of the offsets and the complex, possibly climate-modulated incision, aggradation, abandonment, and offset sequences [101] are significantly aided with the high quality lidar topography. The Dragon's Back pressure ridge (Figure 10C) shows progressive landscape development associated with rock uplift and translation relative to a fixed "knuckle" along the SAF at a few hundred m depth [90]. The progression can be used in a substitution of space-for-time to explore surface process response to transient rock uplift and to assess the value of computed topographic metrics. Along with the high-resolution topography, the SfM provides cm/pixel resolution orthophotography, which acts as a valuable basis for site description for example at the South Carrizo sag pond site [91]. The sags are local depressions along the fault zone, developed from both structural depression and topographic juxtaposition. Finally, while the uniform quality (typically gridded at 0.5 to 1 m/pixel) and great extent of the B4 made it the main data source for offset reconstruction studies (see review [102]), the SfM provides even higher resolution topography (~0.1 m/pixel) for paleoseismic site investigations and more detailed offset reconstructions (e.g., [98]). Figure 10E shows an example of the SAF trace along the Dragon's Back where meter- to dekameter-scale offsets can be reconstructed. Most of the slip is confined to the fault trace and not distributed more broadly.

3.4. Landers Rupture: Geomorphic Modifications to Earthquake Surface Rupture

The Eastern California Shear Zone lies directly north and northeast of the southern Bend of the SAF. The 28 June 1992 Landers, California, earthquake (Mw 7.3) ruptured a northwest trending zone in the Mojave Desert [12] within the Eastern California Shear Zone. It formed a remarkable scarp along the Emerson Fault which has been repeatedly surveyed, allowing the original form of and modifications to the earthquake surface rupture to be described [20,104,105]. The rupture cut across a low relief and low gradient west-draining watershed along which there was limited evidence of pre-1992 faulting. The rupture was discontinuous with overlapping traces. At the Emerson Fault site, horizontal offsets were as much as 4 m, and 1–2 m high west-facing scarps formed.

Field observations and detailed topographic surveys document the modifications of the scarps and knickpoints. Early surveys were made with a total station with only hundreds of points to sample the ground [104]. In 2008, the site was scanned using a terrestrial laser scanner, producing more than 8 million ground points [105,106]. In 2012, a balloon-based system was used to fly a Nikon DSLR along the rupture and produced a high-quality SfM dataset of ~100 million points and a 2 cm digital elevation model [20,107] (Figure 11). Finally, in 2016, a sUAS was used to image a broad swath of the site with similar point densities as in 2012 (Figure 11).

Geomorphic modifications to the sharp surface rupture are notable. Some of the scarps are still steep and show modest smoothing from raindrop impact and eolian abrasion with small colluvial wedges at their bases. The most dramatic geomorphic response has been the incision by narrow channels as knickpoints cut headward across the faulted valleys. The initial forms of the valleys were broad with just a few decimeters of relief. After nearly 25 years, the meter-high knickpoints have been intricately incised. The incision signal has moved more than 20 m upstream in some of the valleys (Figure 11). Incision and headward propagation rate appear to depend on drainage basin hypsometry and substrate permeability. Assuming the pre-1992 rupture (approximately 10 ka; [108]) had a similar vertical component, consistent with the few degraded scarps observed, this style of narrow knickpoint incision must propagate headward and then channels widen and valley floor relief decreases attesting to the long time between events.

These topographic datasets provide invaluable information about the style, rates, and controls on landscape response to sharp perturbations such as surface rupture. A remaining challenge for their analysis will be alignment, scaling, and georeferencing of the successively acquired topographic datasets.

Figure 11. Structure from motion documentation of the degrading surface rupture of the 1992 Landers earthquake along the Emerson Fault in the Mojave Desert, California. The upper oblique image (view to the east northeast) is a portion of the model computed from the 2016 small unmanned aerial system survey (blue squares indicate focal plane positions and orientations for the photographs). The discontinuous and bending fault trace is evident. The image also provides context for the lower map. The 2012 data [20] are presented as a 2 cm per pixel hillshade map along which the overlapping traces at the left are evident. The incising valleys show sharp knick channels across the otherwise relatively smooth topography.

3.5. The Convergence of the San Andreas and Eastern California Shear Zones

The SAF and southern extension of the Landers rupture and Eastern California Shear zone meet at the Coachella segment of the SAF. It is also possible that SAF ruptures can jump to the San Jacinto fault a the Cajon Pass [109] consistent with geodetic evidence (Figure 4). The Eastern California Shear Zone extends into the southernmost SAF at the Coachella Segment [110]. The San Jacinto fault is more favorably oriented with the SAF to the north and the Coachella segment of the SAF and Eastern California Shear Zone are also favorably oriented. The deformation and extension of the San Jacinto and Coachella Segments of the SAF meet in the Imperial Valley appearing to form a single fault zone that extends into the Gulf of California.

The 2010 M7.2 El Mayor–Cucapah earthquake triggered slip on numerous faults north and northeast of the rupture between the Mexican border and the SAF [38,111,112]. These networks of faults connect the El Mayor–Cucapah rupture with the Elsinore, San Jacinto, and SAF. Postseismic deformation is greatest in the Salton Trough. Postseismic fault slip of ~10–40 cm was located mainly on the periphery of the main coseismic fault patches and additional deformation occurred further north [113]. A seismic swarm occurred in the Salton Trough in 2012 [114–116] and another one occurred further north in 2016. These are consistent with the postseismic deformation and suggest a connection to the Southern San Andreas Eastern California Shear Zone and a compliant crust. Long-term creep on the Coachella Segment [117] and triggered slip on the same segment following the Landers and El Mayor–Cucapah events also suggests that the fault is weak [118,119]. Additionally the Eureka Peak fault actively creeps following large earthquakes [7,120]. The San Jacinto fault shows no evidence of triggered slip or creep based on examination of seven years of UAVSAR observations implying that the fault is locked, stronger, and may produce larger earthquakes [1,38].

4. Measurement Requirements and Discussion

Across spatial scales, compliant zones of deformation are bounded by more rigid blocks. Distributed deformation occurs across the plate boundary, but narrow zones of shear extend from the Gulf of California and split to follow the San Andreas fault and the Eastern California Shear Zone. Distributed inelastic deformation is also associated with earthquakes and in most of the examples shallow slip is triggered along faults far from the main rupture. The interactions between surface process and deformation produce interpretable tectonic geomorphology and the resulting landscapes can be interrogated to establish time scales of events and relative process rates.

In this paper, we show observational evidence of localized and broad scale deformation ranging from meters to hundreds of kilometers. Understanding the full range of scales and separating plastic from elastic deformation is required to properly model lithospheric structure and assess potential rupture characteristics. The relative compliance of faults and lithospheric structures control the location of and fragmentation of the Pacific–North American plate boundary and rupture characteristics. The questions outlined in this paper can be addressed by a suite of targeted measurements and modeling of fault processes within the lithospheric structure (Table 1).

Table 1. Science objectives with needed measurements and resolution requirements.

Objective	Measurement	Technique	Resolution
Measure distribution of deformation across the plate boundary	Crustal Deformation	GPS	10 km spacing
			500 km × 1500 km field of view
			1 mm/year
		UAVSAR	10 m pixel spacing
			10 km × 10 km field of view
			1 cm/year
		Airborne lidar	1 m spacing
			10 km × 10 km field of view
			1 cm/year
Determine interactions between faults within the plate boundary	Fault Traces	GPS	10 km spacing
			50 km × 50 km field of view
			1 cm/day time dependent rates
		UAVSAR	≤10 m posting
			5 km × 5 km field of view
			1 cm
		Airborne lidar	2 km × 2 km field of view
			≤1 m
		Structure from motion (SfM) from ground or near surface	1 m posting
			2 km × 2 km field of view
Assess what fraction of strain accumulation is released aseismically	Fault zone landforms	UAVSAR	≤10 m posting
			5 km × 5 km field of view
			1 cm
		Ground and airborne lidar	2 km × 2 km field of view
			≤1 m
		Structure from motion (SfM) from ground or near surface	1 m posting
			2 km × 2 km field of view

4.1. Connecting Plate Boundary Motions to Surface Faults and Their Associated Deformation

Understanding how faults interact across the plate boundary requires measurement of crustal deformation at the level of individual faults. GPS measurements elucidate the total plate rate as well as block motions and zones of rapid deformation. The approximately 10 km station spacing in California at 1 mm/year rates is useful for focusing attention on areas of rapid deformation and for determining slip partitioning between faults. The high temporal sampling of GPS is particularly important for measuring and modeling postseismic deformation and processes.

The 10 km scale of GPS velocity measurements does not provide details about how slip is distributed across individual fault zones, nor does it provide detail of slip variations along strike. UAVSAR measurements fill in that gap in measurement. The 7 m pixel spacing of UAVSAR provides details of slip near fault zones. In the near-field UAVSAR accuracy is on the order of 1 cm. Over a few km, due to error introduced by tropospheric water vapor the accuracy can degrade to up to 3 cm. These measurements are important for detecting shallow slip that does not influence the farther field. When combined with GPS measurements, a much better assessment can be made of how deeply shallow slip extends and the extent of off fault deformation. Though UAVSAR measurements are not as accurate as GPS they have proven useful for identifying localized slip across faults [38–40], including faults triggered by distant earthquakes [72,112]. In general, significant triggered slip is several cm, outside the error of the UAVSAR measurements. UAVSAR measurements can decorrelate in areas most disrupted by an earthquake, leaving a gap in areas that may be of substantial interest.

Topographic differencing provides a means of identifying surface ruptures when UAVSAR fails. Optical SfM topographic data have the additional value of allowing for image change detection or for simply providing imagery for studying surface disruption, and surface soil, rock, or vegetation characteristics. One-meter resolution is adequate for measuring offset features and local surface disruption. Topographic differencing applied to landslides (e.g., [121]) and earthquake displacements (e.g., [59]) requires sufficiently dense sampling to measure submeter offsets. In addition, temporal repeat and response must be sufficient to capture events of interest which can be sub-daily for immediate postseismic deformation and yearly for long-term deformation.

4.2. Fault Rupture and Slip throughout Multiple Earthquake Cycles

Building a history of fault rupture and slip characteristics throughout multiple earthquake cycles is necessary for determining how faults interact within the plate boundary. Seismology provides an excellent means of understanding fault ruptures. Computer simulations (e.g., [122]) provide a means of building up long catalogs of possible earthquake histories to interrogate for statistical relations between earthquakes and fault network geometry. Paleoseismology provides information about fault rupture, slip, and timing of events at points along fault segments. Measurements that encompass entire fault traces show rupture characteristics, fault zone complexity from geomorphic features such as branches or jumps, and time-dependent slip properties.

InSAR and topographic differencing along fault ruptures allows for unprecedented observation of surface rupture characteristics and triggered slip on other faults. Rapid response following an event and then repeated measurements following the event yield time-dependent deformation from which models of postseismic processes can be constructed. Airborne UAVSAR provides higher resolution than spaceborne InSAR with configurable flight paths. Spaceborne InSAR has the advantage of producing systematic repeated measurements. When fused with GPS position time series, a high resolution spatio-temporal observation of crustal deformation can be constructed. GPS measurements provide time-dependent characterization of faults and the crustal deformation following earthquakes and provide potential for detecting transient slip behavior or the propagation of deformation associated with earthquakes along fault traces.

Topographic models draped with color imagery can be used to elucidate details about fault traces and past ruptures. Identification of meter-scale features requires high accuracy georeferenced images for mapping. From these measurements offset landforms can be reconstructed. Decimeter scale sampling is required to identify and interpret post-offset modifications for meter to dekameter scale offsets as well as for landscape elements sensitive to repeated earthquake displacements. Critical breaks in landscape length scale depend on surface process types and rates, degree of soil mantle and rock outcrop, etc. While hillslope lengths and drainage density may be sufficiently resolved at dekameter length scales (see review [123]), some sensitive elements are much finer (e.g., places like Carrizo Plain). These include hilltop curvature (e.g., [124]), channel head locations and shallow landslide behavior ([90] for example).

4.3. *Implications for Earthquake Hazard*

Geodetic imaging contributes to assessment of earthquake hazard, providing estimates of rupture potential and fault interactions. Determining the partitioning of seismic and aseismic motion enables a more accurate assessment of potential moment release from earthquakes. Evidence is accumulating that a non-trivial component of strain accumulation is relieved aseismically. The measurements discussed here can provide an accounting of these aseismic contributions to the plate boundary motion. Fault afterslip is pervasive in California following large earthquakes (e.g., [120,125,126]) and can contribute substantial slip to a fault rupture. GPS measurements and rapid repeated measurement from UAVSAR and topographic differencing can aid in measurement and assessment of afterslip. Decimeter level estimate of afterslip for large earthquakes would improve estimates of the fraction of aseismic versus seismic slip on faults.

Aseismic deformation can also occur as triggered slip on other faults, discussed earlier, or can be plastically distributed across a fault zone. Such fault zone distributed deformation is suggested along the SAF for the 1857 Fort Tejon earthquake [127] and possibly for the creeping section of the SAF. Spatial sampling of 1 m or finer across fault zones will clarify the significance of off-fault plastic deformation of the crust.

5. Conclusions

Multiscale measurements of spatio-temporal patterns of deformation and landforms within plate boundary systems (Figure 12) are necessary for improved understanding of earthquakes and tectonic deformation. Accurate assessment of rupture characteristics and seismic moment release through seismic cycles requires an accounting of deformation ranging from meter-scale fault zones to 100-km-scale plate boundaries. Several geodetic imaging techniques fill in different temporal and spatial scales of lithospheric processes: GPS, InSAR, lidar, and Structure from Motion. In California, the South Napa earthquake slipped several kilometers from the main mapped fault trace and produced substantial afterslip on the main rupture. The creeping section of the San Andreas fault, while weak, may exhibit more distributed off-fault deformation than the stronger Carrizo segment of the fault. Farther south, the Coachella segment of the San Andreas fault may be weak and connect the Eastern California Shear Zone and the Gulf of California, while the stronger San Jacinto fault connects the Gulf of California with the San Andreas fault further north.

At the finest scale, fault creep, coseismic and postseismic deformation, as well as triggered slip of nearby structures can be measured using topographic and UAVSAR data, as exemplified by work done following the south Napa earthquake, the El Mayor–Cucapah earthquake, and along the creeping San Andreas fault. Landforms and deformation patterns at coarser scale, such as those preserved along the Carrizo Plain San Andreas fault and the San Jacinto fault can indicate fault behavior over longer timescales. Fine scale landscape changes following earthquakes can be detected using topographic measurements from laser scanners or optical sensors. Determining fault slip rates at the mm/year level from GPS can enable determination of partitioning between faults. Measurements across and within fault zones will clarify how fault ruptures have occurred previously, may propagate in the future, and illuminate what fraction of total strain accumulation is released aseismically.

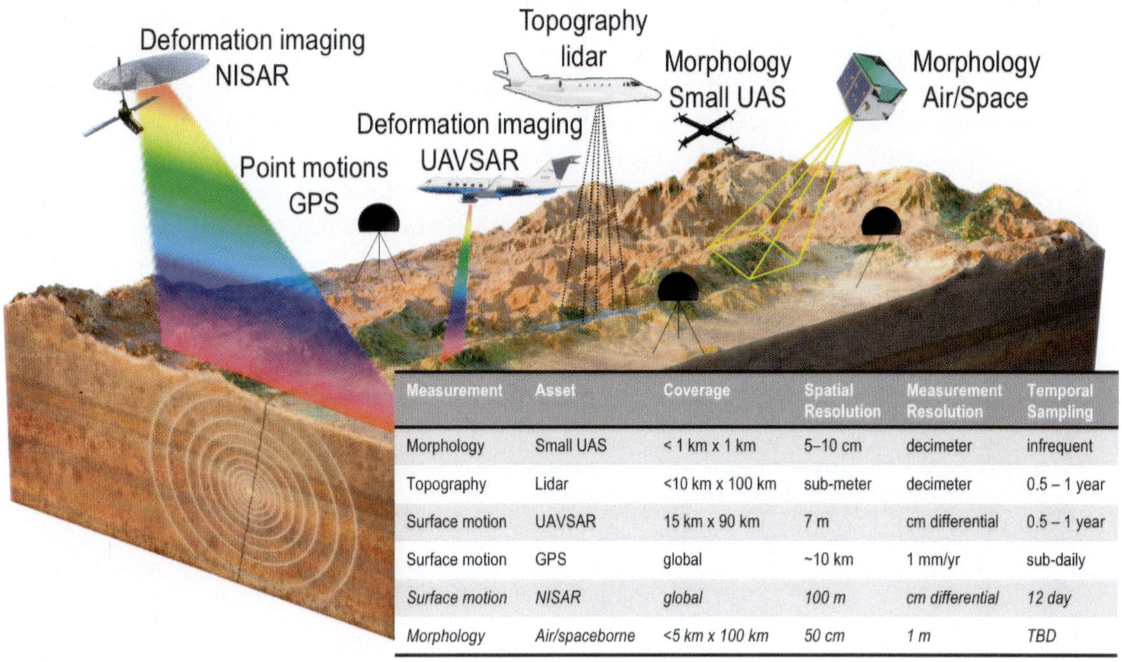

Measurement	Asset	Coverage	Spatial Resolution	Measurement Resolution	Temporal Sampling
Morphology	Small UAS	< 1 km x 1 km	5–10 cm	decimeter	infrequent
Topography	Lidar	<10 km x 100 km	sub-meter	decimeter	0.5 – 1 year
Surface motion	UAVSAR	15 km x 90 km	7 m	cm differential	0.5 – 1 year
Surface motion	GPS	global	~10 km	1 mm/yr	sub-daily
Surface motion	*NISAR*	*global*	*100 m*	*cm differential*	*12 day*
Morphology	*Air/spaceborne*	*<5 km x 100 km*	*50 cm*	*1 m*	*TBD*

Figure 12. Measurements and assets applied to understanding plate boundary and fault processes with approximate coverage and resolution. The last two measurements are planned or notional and are italicized. Base art created by Chuck Carter. Figure modified from [128].

Acknowledgments: Portions of this work were carried out at the Jet Propulsion Laboratory, California Institute of Technology under contract with NASA. Sinan Akciz assisted in the imagery collection for the sag pond in Figure 9. Jessica Sutton assisted with SfM computations for Figures 9 and 10. Alexandra Pickering helped with terrestrial lidar data acquisition and processing for South Napa and the creeping SAF. The airborne lidar from the South Napa Earthquake was funded by the California Department of Water Resources, the US Geological Survey, the California Geological Survey, and PEER-GEER, coordinated by Kenneth Hudnut. Discussions with C. Crosby were helpful for the topography sections. DeLong was funded by the National Earthquake Hazards Reduction Program of the U.S. Geological Survey. Any use of trade, firm, or product names is for descriptive purposes only and does not imply endorsement by the U.S. Government.

Author Contributions: The authors of this paper contributed evenly throughout the paper. A. Donnellan led discussion of UAVSAR and GPS measurements. R. Arrowsmith and S. DeLong focused on the topographic and geomorphic discussion. R. Arrowsmith focused on the Carrizo Plain and Landers regions. S. DeLong focused on the South Napa and Creeping sections of the SAF. A. Donnellan focused on the broader picture of crustal deformation.

Conflicts of Interest: The authors declare no conflict of interest.

References

1. Field, E.H.; Arrowsmith, R.J.; Biasi, G.P.; Bird, P.; Dawson, T.E.; Felzer, K.R.; Jackson, D.D.; Johnson, K.M.; Jordan, T.H.; Madden, C. Uniform California Earthquake Rupture Forecast, version 3 (UCERF3)—The time-independent model. *Bull. Seismol. Soc. Am.* **2014**, *104*, 1122–1180. [CrossRef]
2. Rundle, J.B.; Turcotte, D.L.; Shcherbakov, R.; Klein, W.; Sammis, C. Statistical physics approach to understanding the multiscale dynamics of earthquake fault systems. *Rev. Geophys.* **2003**, *41*. [CrossRef]
3. Council, N.R. *Landscapes on the Edge: New Horizons for Research on Earth's Surface*; National Academies Press: Washington, DC, USA, 2010.
4. Sauber, J.; Thatcher, W.; Solomon, S.C.; Lisowski, M. Geodetic slip rate for the eastern California shear zone and the recurrence time of Mojave Desert earthquakes. *Nature* **1994**, *367*, 264–266. [CrossRef]
5. Wesnousky, S.G. The San Andreas and Walker Lane fault systems, western North America: Transpression, transtension, cumulative slip and the structural evolution of a major transform plate boundary. *J. Struct. Geol.* **2005**, *27*, 1505–1512. [CrossRef]

6. Meade, B.J.; Hager, B.H. Block models of crustal motion in southern California constrained by GPS measurements. *J. Geophys. Res. Solid Earth* **2005**, *110*, 1–19. [CrossRef]
7. Spinler, J.C.; Bennett, R.A.; Anderson, M.L.; McGill, S.F.; Hreinsdóttir, S.; McCallister, A. Present-day strain accumulation and slip rates associated with southern San Andreas and eastern California shear zone faults. *J. Geophys. Res. Solid Earth* **2010**, *115*, B11407. [CrossRef]
8. Rockwell, T.; Loughman, C.; Merifield, P. Late Quaternary rate of slip along the San Jacinto fault zone near Anza, southern California. *J. Geophys. Res. Solid Earth* **1990**, *95*, 8593–8605. [CrossRef]
9. Fletcher, K.E.; Rockwell, T.K.; Sharp, W.D. Late Quaternary slip rate of the southern Elsinore fault, Southern California: Dating offset alluvial fans via 230Th/U on pedogenic carbonate. *J. Geophys. Res. Earth Surf.* **2011**, *116*. [CrossRef]
10. Stein, R.S.; Barka, A.A.; Dieterich, J.H. Progressive failure on the North Anatolian fault since 1939 by earthquake stress triggering. *Geophys. J. Int.* **1997**, *128*, 594–604. [CrossRef]
11. Barka, A.; Akyüz, H.S.; Altunel, E.; Sunal, G.; Cakir, Z.; Dikbas, A.; Yerli, B.; Armijo, R.; Meyer, B.; De Chabalier, J.B. The surface rupture and slip distribution of the 17 August 1999 Izmit earthquake (M 7.4), North Anatolian fault. *Bull. Seismol. Soc. Am.* **2002**, *92*, 43–60. [CrossRef]
12. Sieh, K.; Jones, L.; Hauksson, E.; Hudnut, K.; Eberhart-Phillips, D.; Heaton, T.; Hough, S.; Hutton, K.; Kanamori, H.; Lilje, A. Near-field investigations of the Landers earthquake sequence, April to July 1992. *Science* **1993**, *260*, 171–176. [CrossRef] [PubMed]
13. Heflin, M.; Dong, D.; Donnellan, A.; Hurst, K.; Jefferson, D.; Watkins, M.; Webb, F.; Zumberge, J.; Dauger, D.; Lyzenga, G. Rate change observed at JPLM after the Northridge earthquake. *Geophys. Res. Lett.* **1998**, *25*, 93–96. [CrossRef]
14. Wells, D.L.; Coppersmith, K.J. New empirical relationships among magnitude, rupture length, rupture width, rupture area, and surface displacement. *Bull. Seismol. Soc. Am.* **1994**, *84*, 974–1002.
15. Aki, K. Characterization of barriers on an earthquake fault. *J. Geophys. Res. Solid Earth* **1979**, *84*, 6140–6148. [CrossRef]
16. Harris, R.A.; Day, S.M. Dynamics of fault interaction: Parallel strike-slip faults. *J. Geophys. Res. Solid Earth* **1993**, *98*, 4461–4472. [CrossRef]
17. Harding, T.P. Seismic characteristics and identification of negative flower structures, positive flower structures, and positive structural inversion. *AAPG Bull.* **1985**, *69*, 582–600.
18. Cowie, P. A.; Scholz, C. H. Physical explanation for the displacement-length relationship of faults using a post-yield fracture mechanics model. *J. Struct. Geol.* **1992**, *14*, 1133–1148. [CrossRef]
19. Cochran, E. S.; Li, Y.-G.; Shearer, P. M.; Barbot, S.; Fialko, Y.; Vidale, J. E. Seismic and geodetic evidence for extensive, long-lived fault damage zones. *Geology* **2009**, *37*, 315–318. [CrossRef]
20. Johnson, K.; Nissen, E.; Saripalli, S.; Arrowsmith, J.R.; McGarey, P.; Scharer, K.; Williams, P.; Blisniuk, K. Rapid mapping of ultrafine fault zone topography with structure from motion. *Geosphere* **2014**, *10*, 969–986. [CrossRef]
21. Hudnut, K.W.; Bock, Y.; Galetzka, J.E.; Webb, F.H.; Young, W.H. The southern California integrated GPS network (SCIGN). In Proceedings of the 10th FIG International Symposium on Deformation Measurements, Orange, CA, USA, 19–22 March 2001; pp. 19–22.
22. Donnellan, A.; Hager, B.H.; King, R.W.; Herring, T.A. Geodetic measurement of deformation in the Ventura Basin region, Southern California. *J. Geophys. Res. Solid Earth* **1993**, *98*, 21727–21739.
23. Feigl, K.L.; Agnew, D.C.; Bock, Y.; Dong, D.; Donnellan, A.; Hager, B.H.; Herring, T.A.; Jackson, D.D.; Jordan, T.H.; King, R.W. Space geodetic measurement of crustal deformation in central and southern California, 1984–1992. *J. Geophys. Res. Solid Earth* **1993**, *98*, 21677–21712. [CrossRef]
24. Donnellan, A.; Hager, B.H.; King, R.W. Discrepancy between geological and geodetic deformation rates in the Ventura Basin. *Nature* **1993**, *366*, 333–336. [CrossRef]
25. Hudnut, K.W.; Shen, Z.; Murray, M.; McClusky, S.; King, R.; Herring, T.; Hager, B.; Feng, Y.; Fang, P.; Donnellan, A. Co-seismic displacements of the 1994 Northridge, California, earthquake. *Bull. Seismol. Soc. Am.* **1996**, *86*, S19–S36.
26. Silver, P.G.; Bock, Y.; Agnew, D.C.; Henyey, T.; Linde, A.T.; McEvilly, T.V.; Minster, J.-B.; Romanowicz, B.A.; Sachs, I.S.; Smith, R.B.; et al. A plate boundary observatory. *IRIS Newsl.* **1999**, *16*, 3. Available online: http://ftp.iris.edu/news/IRISnewsletter/EE.Fall98.web/plate.html (accessed on 14 March 2017).

27. Herring, T.A.; Melbourne, T.I.; Murray, M.H.; Floyd, M.A.; Szeliga, W.M.; King, R.W.; Phillips, D.A.; Puskas, C.M.; Santillan, M.; Wang, L. Plate Boundary Observatory and Related Networks: GPS Data Analysis Methods and Geodetic Products. *Rev. Geophys.* **2016**, *54*, 759–808. [CrossRef]
28. Argus, D.F.; Heflin, M.B. Plate motion and crustal deformation estimated with geodetic data from the Global Positioning System. *Geophys. Res. Lett.* **1995**, *22*, 1973–1976. [CrossRef]
29. Argus, D.F.; Fu, Y.; Landerer, F.W. Seasonal variation in total water storage in California inferred from GPS observations of vertical land motion. *Geophys. Res. Lett.* **2014**, *41*, 1971–1980.
30. Smith-Konter, B.R.; Sandwell, D.T.; Shearer, P. Locking depths estimated from geodesy and seismology along the San Andreas Fault System: Implications for seismic moment release. *J. Geophys. Res. Solid Earth* **2011**, *116*. [CrossRef]
31. Bürgmann, R.; Rosen, P.A.; Fielding, E.J. Synthetic aperture radar interferometry to measure Earth's surface topography and its deformation. *Annu. Rev. Earth Planet. Sci.* **2000**, *28*, 169–209.
32. Jones, C.E. A practical guide to synthetic aperture radar for engineering and environmental geologists. In *Applied Geology in California, Special Publication 26*; Anderson, R.L., Ferriz, H., Eds.; Association of Environmental and Engineering Geologists, Star Publishing Company: Redwood City, CA, USA, 2016.
33. Massonnet, D.; Rossi, M.; Carmona, C.; Adragna, F.; Peltzer, G.; Feigl, K.; Rabaute, T. The displacement field of the Landers earthquake mapped by radar interferometry. *Nature* **1993**, *364*, 138–142. [CrossRef]
34. Rosen, P.A.; Kim, Y.; Hensley, S.; Shaffer, S.; Veilleux, L.; Hoffman, J.; Chuang, C.-L.; Chakraborty, M.; Sagi, V.R.; Satish, R. An L-and S-band SAR mission concept for Earth science and applications. In Proceedings of the EUSAR 2014 10th European Conference on Synthetic Aperture Radar, Berlin, Germany, 3–5 June 2014; pp. 1–4.
35. Hensley, S.; Wheeler, K.; Sadowy, G.; Jones, C.; Shaffer, S.; Zebker, H.; Miller, T.; Heavey, B.; Chuang, E.; Chao, R. The UAVSAR instrument: Description and first results. In Proceedings of the 2008 IEEE Radar Conference, Rome, Italy, 26–30 May 2008; pp. 1–6.
36. Hensley, S.; Zebker, H.; Jones, C.; Michel, T.; Muellerschoen, R.; Chapman, B. First deformation results using the NASA/JPL UAVSAR instrument. In Proceedings of the APSAR 2009 2nd Asian-Pacific Conference on Synthetic Aperture Radar, Xian, China, 26–30 October 2009; pp. 1051–1055.
37. Rosen, P.A.; Hensley, S.; Wheeler, K.; Sadowy, G.; Miller, T.; Shaffer, S.; Muellerschoen, R.; Jones, C.; Zebker, H.; Madsen, S. UAVSAR: A new NASA airborne SAR system for science and technology research. In Proceedings of the 2006 IEEE Conference on Radar, Verona, NY, USA, 24–27 April 2006; p. 8.
38. Donnellan, A.; Parker, J.; Hensley, S.; Pierce, M.; Wang, J.; Rundle, J. UAVSAR observations of triggered slip on the Imperial, Superstition Hills, and East Elmore Ranch Faults associated with the 2010 M 7.2 El Mayor-Cucapah earthquake. *Geochem. Geophys. Geosyst.* **2014**, *15*, 815–829. [CrossRef]
39. Donnellan, A.; Grant Ludwig, L.; Parker, J.W.; Rundle, J.B.; Wang, J.; Pierce, M.; Blewitt, G.; Hensley, S. Potential for a large earthquake near Los Angeles inferred from the 2014 La Habra earthquake. *Earth Space Sci.* **2015**, *2*, 378–385. [CrossRef] [PubMed]
40. DeLong, S.B.; Donnellan, A.; Ponti, D.J.; Rubin, R.S.; Lienkaemper, J.J.; Prentice, C.S.; Dawson, T.E.; Seitz, G.; Schwartz, D.P.; Hudnut, K.W. Tearing the terroir: Details and implications of surface rupture and deformation from the 24 August 2014 M6. 0 South Napa earthquake, California. *Earth Space Sci.* **2016**, *3*, 416–430. [CrossRef]
41. Willett, S.D.; Brandon, M.T. On steady states in mountain belts. *Geology* **2002**, *30*, 175–178. [CrossRef]
42. Kirby, E.; Whipple, K.X. Expression of active tectonics in erosional landscapes. *J. Struct. Geol.* **2012**, *44*, 54–75. [CrossRef]
43. Gallen, S.F.; Wegmann, K.W.; Bohnenstiehl, D.R. Miocene rejuvenation of topographic relief in the southern Appalachians. *GSA Today* **2013**, *23*, 4–10.
44. Arrowsmith, J.R. *High Resolution Topography and Active Faulting*; Grutzner, C., Choi, J.H., Edwards, P., Kim, Y.S., Eds.; Geological Society of Korea and the Korean Institute of Geoscience and Mineral Resources: Busan, Korea, 2014; Volume 5.
45. Donnellan, A.; Arrowsmith, R.; Ben-Zion, Y.; Rundle, J.; Grant Ludwig, L.; Glasscoe, M.; Ansar, A.; Parker, J.; De Jong, E.; Lundgren, P.; et al. *Connecting Plate Boundary Processes to Earthquake Faults Using Geodetic and Topographic Imaging*. Available online: http://surveygizmoresponseuploads.s3.amazonaws.com/fileuploads/15647/2604456/105-57af35cccc6e67b33cf5caed44ad11e6_DonnellanAndrea.pdf (accessed 17 March 2017).

46. Milliner, C.W.; Dolan, J.F.; Hollingsworth, J.; Leprince, S.; Ayoub, F.; Sammis, C.G. Quantifying near-field and off-fault deformation patterns of the 1992 Mw 7.3 Landers earthquake. *Geochem. Geophys. Geosyst.* **2015**, *16*, 1577–1598. [CrossRef]
47. Farr, T.G.; Kobrick, M. Shuttle Radar Topography Mission produces a wealth of data. *Eos Trans. Am. Geophys. Union* **2000**, *81*, 583–585. [CrossRef]
48. Reuter, H.I.; Neison, A.; Strobl, P.; Mehl, W.; Jarvis, A. A first assessment of ASTER GDEM tiles for absolute accuracy, relative accuracy and terrain parameters. In Proceedings of the 2009 IEEE International Geoscience and Remote Sensing Symposium, Cape Town, South Africa, 12–17 July 2009; Volume 5, pp. V240–V243.
49. Tachikawa, T.; Hato, M.; Kaku, M.; Iwasaki, A. Characteristics of ASTER GDEM version 2. In Proceedings of the 2011 IEEE International Geoscience and Remote Sensing Symposium (IGARSS), Vancouver, BC, Canada, 24–29 July 2011; pp. 3657–3660.
50. Ryan, W.B.; Carbotte, S.M.; Coplan, J.O.; O'Hara, S.; Melkonian, A.; Arko, R.; Weissel, R.A.; Ferrini, V.; Goodwillie, A.; Nitsche, F. Global multi-resolution topography synthesis. *Geochem. Geophys. Geosyst.* **2009**, *10*. [CrossRef]
51. Krieger, G.; Moreira, A.; Fiedler, H.; Hajnsek, I.; Werner, M.; Younis, M.; Zink, M. TanDEM-X: A satellite formation for high-resolution SAR interferometry. *IEEE Trans. Geosci. Remote Sens.* **2007**, *45*, 3317–3341. [CrossRef]
52. Crosby, C.J.; Arrowsmith, R.; Nandigam, V.; Baru, C. *Online Access and Processing of LiDAR Topography Data*; Cambridge University Press: Cambridge, UK, 2011.
53. Passalacqua, P.; Belmont, P.; Staley, D.M.; Simley, J.D.; Arrowsmith, J.R.; Bode, C.A.; Crosby, C.; DeLong, S.B.; Glenn, N.F.; Kelly, S.A. Analyzing high resolution topography for advancing the understanding of mass and energy transfer through landscapes: A review. *Earth Sci. Rev.* **2015**, *148*, 174–193. [CrossRef]
54. Prentice, C.S.; Crosby, C.J.; Whitehill, C.S.; Arrowsmith, J.R.; Furlong, K.P.; Phillips, D.A. Illuminating Northern California's active faults. *Eos Trans. Am. Geophys. Union* **2009**, *90*, 55.
55. Prentice, C.S. Coastal Marine Terraces of Northern California: A Datum for Estimating Pleistocene Slip Rate across the Northern San Andreas Fault. In Proceedings of the 2008 Joint Meeting of the Geological Society of America, Soil Science Society of America, American Society of Agronomy, Crop Science Society of America, Gulf Coast Association of Geological Societies with the Gulf Coast Section of SEPM, Houston, TX, USA, 5–9 October 2008.
56. Snyder, G.I.; Sugarbaker, L.J.; Jason, A.L.; Maune, D.F. *National Requirements for Enhanced Elevation Data*; Open File Report; US Geological Survey: Reston, VA, USA, 2013; Volume 1237, p. 371.
57. Oskin, M.E.; Arrowsmith, J.R.; Corona, A.H.; Elliott, A.J.; Fletcher, J.M.; Fielding, E.J.; Gold, P.O.; Garcia, J.J.G.; Hudnut, K.W.; Liu-Zeng, J. Near-field deformation from the El Mayor–Cucapah earthquake revealed by differential LIDAR. *Science* **2012**, *335*, 702–705. [CrossRef] [PubMed]
58. Glennie, C.L.; Hinojosa-Corona, A.; Nissen, E.; Kusari, A.; Oskin, M.E.; Arrowsmith, J.R.; Borsa, A. Optimization of legacy lidar data sets for measuring near-field earthquake displacements. *Geophys. Res. Lett.* **2014**, *41*, 3494–3501. [CrossRef]
59. Nissen, E.; Maruyama, T.; Arrowsmith, J.R.; Elliott, J.R.; Krishnan, A.K.; Oskin, M.E.; Saripalli, S. Coseismic fault zone deformation revealed with differential LiDAR: Examples from Japanese Mw$\sim$7 intraplate earthquakes. *Earth Planet. Sci. Lett.* **2014**, *405*, 244–256. [CrossRef]
60. Duffy, B.; Quigley, M.; Barrell, D.J.; Van Dissen, R.; Stahl, T.; Leprince, S.; McInnes, C.; Bilderback, E. Fault kinematics and surface deformation across a releasing bend during the 2010 Mw 7.1 Darfield, New Zealand, earthquake revealed by differential LiDAR and cadastral surveying. *Geol. Soc. Am. Bull.* **2013**, *125*, 420–431.
61. Brooks, B.A.; Glennie, C.; Hudnut, K.W.; Ericksen, T.; Hauser, D. Mobile laser scanning applied to the earth sciences. *Eos Trans. Am. Geophys. Union* **2013**, *94*, 313–315. [CrossRef]
62. Kayen, R.; Pack, R.T.; Bay, J.; Sugimoto, S.; Tanaka, H. Terrestrial-LIDAR visualization of surface and structural deformations of the 2004 Niigata Ken Chuetsu, Japan, earthquake. *Earthq. Spectra* **2006**, *22*, 147–162.
63. Gold, P.O.; Oskin, M.E.; Elliott, A.J.; Hinojosa-Corona, A.; Taylor, M.H.; Kreylos, O.; Cowgill, E. Coseismic slip variation assessed from terrestrial LiDAR scans of the El Mayor–Cucapah surface rupture. *Earth Planet. Sci. Lett.* **2013**, *366*, 151–162.
64. Wilkinson, M.; McCaffrey, K.J.W.; Roberts, G.; Cowie, P.A.; Phillips, R.J.; Michetti, A.M.; Vittori, E.; Guerrieri, L.; Blumetti, A.M.; Bubeck, A. Partitioned postseismic deformation associated with the 2009 Mw 6.3 L'Aquila earthquake surface rupture measured using a terrestrial laser scanner. *Geophys. Res. Lett.* **2010**, *37*. [CrossRef]

65. DeLong, S.B.; Lienkaemper, J.J.; Pickering, A.J.; Avdievitch, N.N. Rates and patterns of surface deformation from laser scanning following the South Napa earthquake, California. *Geosphere* **2015**, *11*. [CrossRef]
66. McClusky, S.C.; Bjornstad, S.C.; Hager, B.H.; King, R.W.; Meade, B.J.; Miller, M.M.; Monastero, F.C.; Souter, B.J. Present day kinematics of the eastern California shear zone from a geodetically constrained block model. *Geophys. Res. Lett.* **2001**, *28*, 3369–3372. [CrossRef]
67. Dixon, T.H.; Miller, M.; Farina, F.; Wang, H.; Johnson, D. Present-day motion of the Sierra Nevada block and some tectonic implications for the Basin and Range province, North American Cordillera. *Tectonics* **2000**, *19*, 1–24. [CrossRef]
68. Hearn, E.H.; Humphreys, E.D. Kinematics of the southern Walker Lane Belt and motion of the Sierra Nevada block, California. *J. Geophys. Res. Solid Earth* **1998**, *103*, 27033–27049.
69. GeoGateway Tools for Modeling, Analysis, and Response. Available online: http://geo-gateway.org (accessed on 17 March 2017).
70. Donnellan, A.; Parker, J.; Glasscoe, M.; Granat, R.; Pierce, M.; Wang, J.; Ma, Y.; Ludwig, L.G.; Rundle, J. GeoGateway: A system for analysis of UAVSAR data products. In Proceedings of the 2016 IEEE International Geoscience and Remote Sensing Symposium (IGARSS), Beijing, China, 10–15 July 2016; pp. 210–213.
71. Jet Propulsion Laboratory GPS Time Series. Available online: http://sideshow.jpl.nasa.gov/post/series.html (accessed on 17 March 2017).
72. Hudnut, K.W.; Brocher, T.M.; Prentice, C.S.; Boatwright, J.; Brooks, B.A.; Aagaard, B.T.; Blair, J.L.; Fletcher, J.B.; Erdem, J.E.; Wicks, C.W. *Key Recovery Factors for the August 24, 2014, South Napa Earthquake*; Open File Report; US Geological Survey: Reston, VA, USA, 2014; Volume 1249, p. 51.
73. Lydaa, A.W.; Zhanga, X.; Glenniea, C.L.; Hudnutb, K.; Brooksc, B.A. Airborne Light Detection and Ranging (LIDAR) Derived Deformation from the Mw 6.0 24 August, 2014 South Napa Earthquake Estimated by Two and Three Dimensional Point Cloud Change Detection Techniques. In Proceedings of the International Archives of the Photogrammetry, Remote Sensing and Spatial Information Sciences, Prague, Czech Republic, 12–19 July 2016; pp. 35–42.
74. DeLong, S.B. 3D Point Cloud Data from Laser Scanning along the 2014 South Napa Earthquake Surface Rupture, California, USA. Available online: https://www.sciencebase.gov/catalog/item/57f26aebe4b0bc0bebfff915 (accessed on 14 March 2017).
75. Graymer, R.W.; Brabb, E.E.; Jones, D.J.; Barnes, J.; Nicholson, R.S.; Stamski, R.E. *Geologic Map and Map Database of Eastern Sonoma and Western Napa Counties, California*; Geological Survey: Reston, VA, USA, 2007.
76. Jolivet, R.; Simons, M.; Agram, P.S.; Duputel, Z.; Shen, Z.-K. Aseismic slip and seismogenic coupling along the central San Andreas Fault. *Geophys. Res. Lett.* **2015**, *42*, 297–306.
77. Maurer, J.; Johnson, K. Fault coupling and potential for earthquakes on the creeping section of the central San Andreas Fault. *J. Geophys. Res. Solid Earth* **2014**, *119*, 4414–4428. [CrossRef]
78. Burford, R.O.; Harsh, P.W. Slip on the San Andreas fault in central California from alinement array surveys. *Bull. Seismol. Soc. Am.* **1980**, *70*, 1233–1261.
79. Lisowski, M.; Prescott, W.H. Short-range distance measurements along the San Andreas fault system in central California, 1975 to 1979. *Bull. Seismol. Soc. Am.* **1981**, *71*, 1607–1624.
80. Titus, S.J.; DeMets, C.; Tikoff, B. Thirty-five-year creep rates for the creeping segment of the San Andreas fault and the effects of the 2004 Parkfield earthquake: Constraints from alignment arrays, continuous global positioning system, and creepmeters. *Bull. Seismol. Soc. Am.* **2006**, *96*, S250–S268.
81. Rymer, M.J.; Lisowski, M.; Burford, R.O. Structural explanation for low creep rates on the San Andreas fault near Monarch Peak, central California. *Bull. Seismol. Soc. Am.* **1984**, *74*, 925–931.
82. DeLong, S.B.; Hilley, G.E.; Rymer, M.J.; Prentice, C. Fault zone structure from topography: Signatures of en echelon fault slip at Mustang Ridge on the San Andreas fault, Monterey County, California. *Tectonics* **2010**, *29*. [CrossRef]
83. Hilley, G.E.; DeLong, S.; Prentice, C.; Blisniuk, K.; Arrowsmith, J.R. Morphologic dating of fault scarps using airborne laser swath mapping (ALSM) data. *Geophys. Res. Lett.* **2010**, *37*, L04301.
84. Liu, Z.; Lundgren, P.; Fielding, E.J.; Hensley, S. Imaging fault slip variation along the central San Andreas fault from satellite, airborne InSAR and GPS. Available online: http://abstractsearch.agu.org/meetings/2011/FM/T43I-06.html (accessed on 17 March 2017).
85. Page, B.M. Evolution and complexities of the transform system in California, USA. *Ann. Tecton.* **1990**, *4*, 53–69.

86. Lawson, A.C.; Reid, H.F. *The California Earthquake of April 18, 1906: Report of the State Earthquake Investigation Commission*; Carnegie Institution of Washington: Washington, DC, USA, 1908.
87. Wood, H.O.; Buwalda, J.P. Horizontal displacement along the San Andreas fault in the Carrizo Plain, California. *Bull. Geol. Soc. Am.* **1931**, *42*, 298–299.
88. Sieh, K.E.; Jahns, R.H. Holocene activity of the San Andreas fault at Wallace creek, California. *Geol. Soc. Am. Bull.* **1984**, *95*, 883–896.
89. Wallace, R.E. *The San Andreas Fault System, California*; US Government Printing Office: Washington, DC, USA, 1990.
90. Hilley, G.E.; Arrowsmith, J.R. Geomorphic response to uplift along the Dragon's Back pressure ridge, Carrizo Plain, California. *Geology* **2008**, *36*, 367–370.
91. Akçiz, S.O.; Arrowsmith, J.R. New views on the evolution of the San Andreas fault zone in central California and the Carrizo Plain. *Field Guid.* **2013**, *32*, 1–12.
92. Sieh, K.E. Slip along the San Andreas fault associated with the great 1857 earthquake. *Bull. Seismol. Soc. Am.* **1978**, *68*, 1421–1448.
93. Zielke, O.; Arrowsmith, J.R.; Ludwig, L.G.; Akçiz, S.O. Slip in the 1857 and earlier large earthquakes along the Carrizo Plain, San Andreas fault. *Science* **2010**, *327*, 1119–1122.
94. Zielke, O.; Arrowsmith, J.R.; Ludwig, L.G.; Akciz, S.O. High-Resolution Topography-Derived Offsets along the 1857 Fort Tejon Earthquake Rupture Trace, San Andreas Fault. *Bull. Seismol. Soc. Am.* **2012**, *102*, 1135–1154.
95. Akçiz, S.O.; Ludwig, L.G.; Arrowsmith, J.R.; Zielke, O. Century-long average time intervals between earthquake ruptures of the San Andreas fault in the Carrizo Plain, California. *Geology* **2010**, *38*, 787–790.
96. Schmalzle, G.; Dixon, T.; Malservisi, R.; Govers, R. Strain accumulation across the Carrizo segment of the San Andreas Fault, California: Impact of laterally varying crustal properties. *J. Geophys. Res. Solid Earth* **2006**, *111*, B05403. [CrossRef]
97. Noriega, G.R.; Arrowsmith, J.R.; Grant, L.B.; Young, J.J. Stream channel offset and late Holocene slip rate of the San Andreas fault at the Van Matre ranch site, Carrizo Plain, California. *Bull. Seismol. Soc. Am.* **2006**, *96*, 33–47.
98. Salisbury, J.B. Coupling Tectonic Geomorphology and Paleoseismology for Understanding of Earthquake Recurrence. Ph.D. Dissertation, Arizona State University, Tempe, AZ, USA, 2016.
99. Bevis, M.; Hudnut, K.; Sanchez, R.; Toth, C.; Grejner-Brzezinska, D.; Kendrick, E.; Caccamise, D.; Raleigh, D.; Zhou, H.; Shan, S.; et al. The B4 Project: Scanning the San Andreas and San Jacinto Fault Zones. Available online: http://abstractsearch.agu.org/meetings/2005/FM/H34B-01.html (accessed on 17 March 2017).
100. U.S. Geological Survey and California Geological Survey, Quaternary Fault and Fold Database for the United States. Available online: https://earthquake.usgs.gov/hazards/qfaults/ (accessed 17 March 2017).
101. Sims, J.D. Stream channel offset and abandonment and a 200-year average recurrence interval of earthquakes on the San Andreas fault at Phelan Creeks, Carrizo Plain, California. In Proceedings of the Workshop on Paleoseismology, Marshall, CA, USA, 18–22 September 1994; pp. 170–172.
102. Zielke, O.; Klinger, Y.; Arrowsmith, J.R. Fault slip and earthquake recurrence along strike-slip faults—Contributions of high-resolution geomorphic data. *Tectonophysics* **2015**, *638*, 43–62.
103. Arrowsmith, J.R.; Zielke, O. Tectonic geomorphology of the San Andreas Fault zone from high resolution topography: An example from the Cholame segment. *Geomorphology* **2009**, *113*, 70–81. [CrossRef]
104. Arrowsmith, J.R.; Rhodes, D.D. Original forms and initial modifications of the Galway Lake Road scarp formed along the Emerson fault during the 28 June 1992 Landers, California, earthquake. *Bull. Seismol. Soc. Am.* **1994**, *84*, 511–527.
105. Haddad, D.E.; Akçiz, S.O.; Arrowsmith, J.R.; Rhodes, D.D.; Oldow, J.S.; Zielke, O.; Toké, N.A.; Haddad, A.G.; Mauer, J.; Shilpakar, P. Applications of airborne and terrestrial laser scanning to paleoseismology. *Geosphere* **2012**, *8*, 771–786.
106. Landers Surface Rupture Terrestrial Laser Scanning Dataset. Available online: https://doi.org/10.5069/G91Z4298 (accessed on 17 March 2017).
107. Listing of /v1/AUTH_opentopography/hosted_data/SfM_GalwayLakeRd_1.23.2014/. Available online: https://cloud.sdsc.edu/v1/AUTH_opentopography/hosted_data/SfM_GalwayLakeRd_1.23.2014/ (accessed on 17 March 2017).
108. Rubin, C.M.; Sieh, K. Long dormancy, low slip rate, and similar slip-per-event for the Emerson fault, eastern California shear zone. *J. Geophys. Res. Solid Earth* **1997**, *102*, 15319–15333.
109. Ludwig, L.G.; Brune, J.N.; Anooshehpoor, A.; Purvance, M.D.; Brune, R.J.; Lozos, J.C. Reconciling precariously balanced rocks (PBRs) with large earthquakes on the San Andreas fault system. *Seismol. Res. Lett.* **2015**. [CrossRef]

110. Thatcher, W.; Savage, J.C.; Simpson, R.W. The Eastern California Shear Zone as the northward extension of the southern San Andreas fault. *J. Geophys. Res. Solid Earth* **2016**. [CrossRef]
111. Hauksson, E.; Stock, J.; Hutton, K.; Yang, W.; Vidal-Villegas, J.A.; Kanamori, H. The 2010 Mw 7.2 El Mayor-Cucapah Earthquake Sequence, Baja California, Mexico and Southernmost California, USA: Active Seismotectonics along the Mexican Pacific Margin. *Pure Appl. Geophys.* **2011**, *168*, 1255–1277.
112. Rymer, M.J.; Treiman, J.A.; Kendrick, K.J.; Lienkaemper, J.J.; Weldon, R.J.; Bilham, R.; Wei, M.; Fielding, E.J.; Hernandez, J.L.; Olson, B.P. *Triggered Surface Slips in Southern California Associated with the 2010 El Mayor-Cucapah, Baja California, Mexico, Earthquake*; US Geological Survey: Reston, VA, USA, 2011.
113. Gonzalez-Ortega, A.; Fialko, Y.; Sandwell, D.; Alejandro Nava-Pichardo, F.; Fletcher, J.; Gonzalez-Garcia, J.; Lipovsky, B.; Floyd, M.; Funning, G. El Mayor-Cucapah (Mw 7.2) earthquake: Early near-field postseismic deformation from InSAR and GPS observations. *J. Geophys. Res. Solid Earth* **2014**, *119*, 1482–1497.
114. Hauksson, E.; Stock, J.; Bilham, R.; Boese, M.; Chen, X.; Fielding, E.J.; Galetzka, J.; Hudnut, K.W.; Hutton, K.; Jones, L.M. Report on the August 2012 Brawley earthquake swarm in Imperial Valley, Southern California. *Seismol. Res. Lett.* **2013**, *84*, 177–189.
115. Lohman, R.B.; McGuire, J.J. Earthquake swarms driven by aseismic creep in the Salton Trough, California. *J. Geophys. Res. Solid Earth* **2007**, *112*, B04405.
116. Wei, S.; Avouac, J.-P.; Hudnut, K.W.; Donnellan, A.; Parker, J.W.; Graves, R.W.; Helmberger, D.; Fielding, E.; Liu, Z.; Cappa, F. The 2012 Brawley swarm triggered by injection-induced aseismic slip. *Earth Planet. Sci. Lett.* **2015**, *422*, 115–125.
117. Sieh, K.; Williams, P.L. Behavior of the southernmost San Andreas fault during the past 300 years. *J. Geophys. Res.* **1990**. [CrossRef]
118. Rymer, M.J. Triggered surface slips in the Coachella Valley area associated with the 1992 Joshua Tree and Landers, California, earthquakes. *Bull. Seismol. Soc. Am.* **2000**, *90*, 832–848. [CrossRef]
119. Vincent, P. Aseismic slip events along the southern San Andreas fault system captured by radar interferometry. In Proceedings of the 3rd Conference on Tectonic Problems of the San Andreas Faul System, Stanford, CA, USA, 6–8 September 2000; Bokelmann, G., Kovach, R.L., Eds.; 2000; p. 193.
120. Behr, J.; Bilham, R.; Bodin, P.; Gross, S. Eureka peak fault afterslip following the 28 June 1992 Landers earthquake. *Bull. Seismol. Soc. Am.* **1994**, *84*, 826–834.
121. DeLong, S.B.; Prentice, C.S.; Hilley, G.E.; Ebert, Y. Multitemporal ALSM change detection, sediment delivery, and process mapping at an active earthflow. *Earth Surf. Process. Landf.* **2012**, *37*, 262–272. [CrossRef]
122. Rundle, P.B.; Rundle, J.B.; Tiampo, K.F.; Donnellan, A.; Turcotte, D.L. Virtual California: Fault model, frictional parameters, applications. *Pure Appl. Geophys.* **2006**, *163*, 1819–1846. [CrossRef]
123. Grieve, S.W.; Mudd, S.M.; Milodowski, D.T.; Clubb, F.J.; Furbish, D.J. How does grid-resolution modulate the topographic expression of geomorphic processes? *Earth Surf. Dyn.* **2016**, *4*, 627–653.
124. Hurst, M.D.; Mudd, S.M.; Attal, M.; Hilley, G. Hillslopes record the growth and decay of landscapes. *Science* **2013**, *341*, 868–871. [CrossRef]
125. Donnellan, A.; Lyzenga, G.A. GPS observations of fault afterslip and upper crustal deformation following the Northridge earthquake. *J. Geophys. Res. Solid Earth* **1998**, *103*, 21285–21297.
126. Azúa, B.M.; DeMets, C.; Masterlark, T. Strong interseismic coupling, fault afterslip, and viscoelastic flow before and after the Oct. 9, 1995 Colima-Jalisco earthquake: Continuous GPS measurements from Colima, Mexico. *Geophys. Res. Lett.* **2002**, *29*. [CrossRef]
127. Grant, L.B.; Donnellan, A. 1855 and 1991 surveys of the San Andreas fault: Implications for fault mechanics. *Bull. Seismol. Soc. Am.* **1994**, *84*, 241–246.
128. Davis, J.L.; Kellogg, L.H.; Arrowsmith, J.R.; Buffett, B.A.; Constable, C.G.; Donnellan, A.; Ivins, E.R.; Mattioli, G.S.; Owen, S.E.; Pritchard, M.E.; et al. Challenges and Opportunities for Research in ESI (CORE). Available online: https://smd-prod.s3.amazonaws.com/science-green/s3fs-public/atoms/files/CORE2016.pdf (accessed on 14 March 2017).

3

The Value of Original Natural Stone in the Context of Architectural Heritage

Dolores Pereira [1,*,†] **and Brian Marker** [2,†]

1 Department of Geology, University of Salamanca, Plaza de la Merced s/n, Salamanca 37008, Spain
2 Independent consultant, 40 Kingsdown Avenue, London W13 9PT, UK; brian@amarker.freeserve.co.uk
* Correspondence: mdp@usal.es
† These authors contributed equally to this work.

Academic Editor: Carlos Alves

Abstract: Natural stone is required for maintenance and repair of historically, archaeologically and culturally important structures to prevent deterioration due to weathering and use. Those that have national or World Heritage significance are historically and culturally important. Sometimes, severely damaged buildings may require extensive or complete restoration. The choice of stone should be appropriate in physical and aesthetic terms, ideally from the original source. Researchers, architects, those who specify contracts and other stakeholders need reliable information, but access to, and awareness of, important sources such as manuscripts, publications and collections of samples is often limited. Easier access to information is needed when planning and commissioning works that require the replacement of stone. Making important information more widely available and promoting awareness of the need to maintain adequate supplies of natural stone of suitable quality is a task for both national and international organizations including the Heritage Stone Task Group of IUGS and IAEG. This paper provides some illustrative examples and pointers towards some recent major publications, as well as describing current Heritage Stone Resource and European Union initiatives, all the while requesting further participation from colleagues in this field.

Keywords: natural stone; architectural heritage; building; Global Heritage Stone Resource

---

1. Introduction

This paper reviews aspects of natural stones that are required for maintenance and repair of historically, archaeologically and culturally important structures that have national or World Heritage significance. Proper repair and maintenance to prevent deterioration of structures due to weathering and use is essential but, from time to time, important buildings are severely damaged by events such as fires, earthquakes or military actions to the extent that they may be extensively or wholly restored [1,2].

In either situation, the choice of stone should be appropriate in terms of physical performance and aesthetic appearance. Researchers, architects, those who specify contracts and other stakeholders need reliable information to achieve that. However, access to, and awareness of, important sources is often limited. Sources include publications, manuscripts, reference collections of rock samples and technical data in commercial websites and brochures. It is important to make important information more readily known and available as well as promoting awareness of the need to maintain adequate supplies of natural stone of suitable quality. That is a task for both national and international organizations.

There is space here only to provide some illustrative examples and pointers towards some recent major publications, and to describe current Heritage Stone Resource and European Union initiatives, all the while requesting further participation from colleagues.

2. Historical and Cultural Significance

Natural stone has been used as a construction material since ancient times (e.g., since about 4700 BC in Egypt). Initially, these expensive materials were used for temples, tombs, palaces, civic buildings and major infrastructure as well as decoration and sculptures. Selection of stone was guided partly by suitability but was also influenced by personal prestige and social and mystic beliefs. For instance, green "marbles" (now identified as serpentinites) were related to joy in Egypt. Also, the traditional royal colour was blue until Queen Cleopatra had the walls of her banqueting hall in Alexandria adorned with red porphyry. Roman Emperors continued to use purplish-red stone encouraging wealthy people to surround themselves with beautiful objects made of porphyry to mark the importance of the owner [3]. These porphyries continued to be widely associated with nobility and Imperial monuments [4]. Several works have dealt with Egyptian rocks in general and porphyries in particular [5,6]. Another example is the mineral variety onyx which was regarded as being associated with instinct and intuition and was associated with business and management capabilities in ancient Greece and Rome [7].

The European Renaissance revived interest in European Classical culture. Royal Families became increasingly interested in stone as an expression of power and status in those classical styles. That led to an interest in preserving or initiating quarries to supply desirable materials for their projects. In particular, Carlos II of Spain was closely involved in securing his preferred stones for details (e.g., pavements, columns) in his palace. Similarly, in Sweden, the monarchy showed a close interest in natural stones, both for constructions and for gifts to hand to international dignitaries. A porphyry, discovered in Sweden in the 1730s, closely resembled the Egyptian red porphyry and became so desired that the Royal Family owned the porphyry-works from 1818 to 1856 [3].

3. Some Reference Collections of Specimens

Collections of samples of natural stones are an important source of information especially if they are accompanied by information on buildings and structures in which they have been used, technical properties and aesthetic characteristics. Unfortunately, that has rarely been the case but, even so, there are important exceptions.

One was developed for the Spanish Royal Family when, after a severe fire at their palace in 1734, they decided to build a new palace using as many "marbles and ornamental stones" as possible in construction [8]. They favoured stones from Spain but, if materials did not suit the King's aesthetic tastes, "marbles" were sought elsewhere, especially from Italy because that was the source of contemporary artistic and architectural inspiration. So, once again, the Roman preference for red stone became fashionable. Assembly of a set of samples was begun in 1740 and continued through the reigns of Kings Fernando VI, Carlos III and Carlos IV until 1759. The set is now part of the national heritage collections kept at the Royal Palace in Madrid (Figure 1). There are eight boxes with 40 compartments in each. Each compartment contains a sample of "marble" together with its location reference. Additional information is provided on a document included in each box, such as origin, distance from the quarry to the palace, (in Spanish leguas = leagues: originally, this was the distance that a person could walk in an hour) and abundance. Most samples came from Andalucía, Aragón and Catalonia. The collection is sometimes loaned for exhibitions such as that organized by the Colouste Gulbenkian foundation in Lisbon in 2014. It remains of great importance in establishing the selection of natural stones required for repair and maintenance by the royal authorities and has wider significance, but may not be well known outside Spain.

A systematic search of national quarries was undertaken for a specific stone, the *diáspero sanguine,* (similar to Rosso Diaspro Marble quarried in Sicily), that was referred to in archive documents as having been quarried in Andalucía, for the construction of the palace. The architect in charge was sent to Andalucía, visiting numerous quarries from 1761 to 1772 before finding a perfect match [9].

Figure 1. Showcase of marbles from the Spanish Royal Palace, Box number 3, inventory number 1013958.

An inventory of natural stones used for construction and for royal gifts such as vases in Baroque and Renaissance styles, known as "Jewels of the Dolphin" (the "Dolphin" being the French *Grand Dauphin* Louis de Borbón) survives. This is held by the Prado Museum in Madrid [10].

Another particularly valuable collection was made by Faustino Corsi [11], a lawyer who lived in Rome (1771–1846). He became a reputable judge but had artistic hobbies. He was passionate about the decorative stones that had been used in Ancient Rome and built up a collection of over a thousand polished specimens. He added descriptions, geological classifications, local names used by the marble cutters of contemporary Rome and ancient and contemporary names used by authors. In 1825, he published the *Catalogo ragionato d'una collezione di pietre di decorazione* [12] and, in 1827, he described each stone in his *Supplemento* [13]. In addition to those descriptions, Corsi drew attention to exceptional examples of stones that could be seen in the historic buildings of Rome. The collection is now held by the Oxford University Museum of Natural History [14] (Figure 2). It remains the most complete set of individual stones that were used in Roman times and for traditional architecture in Italy, related to examples of buildings and structures where they are best seen, but this, otherwise, exemplary work considered only polished decorative stones.

Figure 2. Drawers with Corsi's collection, by permission of Monica Price, Head of Earth Collections at Oxford University Museum of Natural History.

Significant collections can be seen in other important museums such as the Louvre in Paris [8] and others are held by geological surveys and universities.

4. Written Sources and Terminology

There are many written sources of material on natural stones spanning several centuries but these are often not readily accessible. These include old manuscripts held in archives of libraries and museums where access may be restricted because of fragility. Also, the locations may not be widely known. Relevant literature has been aimed at various audiences (e.g., geoscientists, architects, historians and the general public) which may not be aware of publications aimed at the other sectors.

Another factor is language [15]. For instance, many important quarries and stones have been described in the regions of the Urals, Karelia, the Caucasus and around St Petersburg but are available only in Russian and, therefore, have not been widely disseminated outside Russian speaking areas. This is particularly important in relation to St Petersburg where a variety of stones from Scandinavia were imported for major buildings and art works [16,17].

Problems also arise through translations of terms. An example is "freestone" which, in English, means a stone used in masonry that can be readily carved with a chisel. The word "freemason" probably derives from that term as "one who is capable of carving freestone" [18] (in contrast to "roughstone", hard stones worked by the hard hewers or "rough masons"). The French term for "freestone" is "Pierre de Taille" or, in Spanish, "Piedra Franca". "Piedra Franca" literally means "honest stone", meaning stone that is easy to cut and carve. In Spain, this is applied to all stones of sedimentary origin that have uniform colour and ease of cutting/carving in order to obtain dimension stone blocks while, in England, the term has been generally restricted to limestones with those characteristics.

The earliest references to "freestone" are found in medieval documents when builders were choosing the best materials to construct magnificent cathedrals and castles. England had extensive suitable stone resources but, for historical reasons—the Norman invasion of England in 1066AD and subsequent expansion in major ecclesiastical and military building—a large volume of stone was imported from Caen in Normandy. Caen limestone was widely used for major buildings in Western Europe and more widely (e.g., Church of Saint-Étienne, Church of La Trinité, both Abbeys in Caen, Canterbury Cathedral, Reading Abbey, the Tower of London, the clock tower at Westminster Palace, and the Old South Church in Boston).

While the original French "Pierre de Taille" referred to the limestones extracted at Caen, over time, the meaning in adjacent countries, such as Spain, expanded to include other natural stones which had similar colours and were easy to carve. They included many sandstones, one of the most important of which was the Villamayor sandstone from Salamanca. However, other historically important natural stones are also known as "Piedra Franca" including calcarenite from Santa Pudia (Granada), limestone from Alava, and sandstones from Jaen and Cádiz [19]. All of these were used in construction of the Spanish architectural heritage and, superficially, share superficially similar exterior characteristics: several are known as "golden stones".

Very few papers have been published on individual freestones (or Pierre de Taille or Piedra Franca) and most of those are in relatively obscure journals or languages other than English.

Another issue is the use of the same name for more than one type of stone. Much stone is traded internationally and stone from developing countries often has the attraction of a lower price but this may not have a guaranteed quality characterization. Instances are known of stone being sold under European local names even though it comes from far away. The European Commission has recently launched a public consultation on the protection of names of specific important stones [20]. This protection would seek to make illegal within the EU to apply these names to other products, either man-made or imported from other geographical areas inside or outside the Union. This protection would be similar to that which already exists for agricultural products. In the case of natural stones, this protection would also be related to potential economic, social and cultural benefits. Several contributions regarding preservation of individual natural stones were made by European stone producers including Portland Stone, Carrara Marble, Bath Stone, and natural stone from Bourgogne.

5. The Heritage Stone Task Group (HSTG) Initiative

The HSTG was convened by the International Union of Geological Sciences at the 33th International Geological Congress (IGC), held in Oslo in 2008, in association with the International Association of Engineering Geology and the Environment (IAEG) Commission C-10 on Building Stones and Ornamental Rocks. The Group developed the concept of Global Heritage Stone Resource (GHSR) with the aims of improving awareness of the need for continuing supplies of important stone and protection of sources. This new formal international geological designation is being applied to important types of natural stone that have been widely used and/or have widespread cultural and architectural recognition [21].

This activity has already stimulated numerous presentations on specific natural stones from many parts of the world at international meetings (e.g., EGU 2013, EGU 2014, IAEG 2014, EGU 2015) and most have now been published. This has helped to increase awareness in the geosciences community of stones that were previously only appreciated nationally. Cooper *et al.* [21] set out the aim and objectives of the HSTG which included a protocol for describing natural stones that are proposed as candidates for GHSR status in terms of their occurrence, technical and aesthetic characteristics and examples of major heritage works in which they have been used. The first use of the protocol, for Portland Stone (United Kingdom), was published by [22] establishing the model that is being followed in subsequent proposals. A series of HSTG international events and publications are ongoing and include the European Geosciences Union General Assembly at Vienna and 35th International Geological Congress at Cape Town, both in 2016.

The first important volume dedicated only to GHSR candidate stones was published by the Geological Society of London as Special Publication 417. This volume was edited by Pereira *et al.* [23], and contained full descriptions of four natural stones from Spain, two from Sweden, one from Norway, two from Portugal, two from Slovenia, one from Brazil, one from Russia and five from Italy. The first edition of "Geoscience of the Built Environment", published in 2012 [24], also contained reports from HSTG activities on the Syracuse Limestone [25] and Verde Macael, a serpentinite from Andalucía, Spain. In this latter article, Navarro *et al.* described difficulties when dealing commercially with serpentinites which are commonly described as marbles which may lead some users to select stone incorrectly. Verde Macael had already been presented in different volumes to be recognized as a heritage stone [26].

The second major publication fully dedicated to heritage stones was a special issue of the journal Episodes, edited by Pereira, Kramar and Cooper [27], all of them board members of the HSTG at the time. This contained full descriptions of ten natural stones used in the architectural heritage: two from Sweden, two from Belgium, two from Spain, one from Russia, two from the United Kingdom and one from Australia.

Subsequently, Pereira and Pratt (in press) edited seven papers on stones from Spain, Brazil, Italy, Slovenia and Egypt based on papers presented at the European Geoscience Union (EGU) 2015 meeting. This issue also contains an introduction by Pereira and Marker highlighting the aesthetic and structural problems caused by the use of inappropriate stones in repair and maintenance [28].

6. Physical and Aesthetic Properties

Another aspect of raising awareness of appropriate stone for repair, maintenance and restoration of historic structures concerns the availability of reliable technical data. Replacement stone should ideally be of the same type as the original, or the nearest possible equivalent. Use of inappropriate stone can result in significant damage to the heritage. In order choose the correct stone, both the original stone and the potential replacements should be fully characterized. The stone industry has an important role in providing accurate technical and aesthetic information on their products.

There is extensive literature on the variety, assessment, durability and resistance to weathering of natural stones which tend to be very individual in their behaviour. For instance, a serpentinite from Galicia (Verde Pirineo [29,30] behaves in a different way that a serpentinite from Andalucía (Verde Macael, [31]), or Rajasthan (Rajashtan Green, [32]) or Italy (green stones of the central Alps, [33]).

The Geological Society of London has strongly supported publication of many papers related to weathering and conservation of natural stones. In 2002, Siegesmund *et al.* [34] edited a volume dealing with weathering of natural building stones in different contexts, different weathering processes affecting the buildings, the fabric dependence of physical properties, biodeterioration, quality assessment, conservation of stones, and environmental conditions. Přikryl & Smith [35] edited a book where authors analysed the causes, consequences, prevention and solution of several stone decay problems. Smith *et al.* [36] compiled information on limestones and the challenges for the preservation of heritage built with them. Přikryl & Tőrők [37] referred to natural stone resources that have been used for the construction of historical monuments. One of the major topics in this book was the availability of certain stone types that were used for monuments. Gómez-Heras *et al.* [38] presented a study on Oxford Stone and the consequences of its use in Oxford's historic buildings. In the same issue, Pereira *et al.* provided a paper on serpentinites from Galicia, Spain, reporting the unacceptable characteristics of these rocks for construction purposes, in contrast to some other serpentinites from elsewhere. Three tuffs from Italy were described by [39] to validate their use as replacement in some historical building. Figueiredo *et al.* [40] presented two Portuguese stones and Alloca *et al.* described the Vitulano marbles, which are important for the preservation of cultural heritage of Southern Italy [41].

JoAnn Cassar *et al.* edited a volume dealing with characterization and performance of natural stone in historic buildings [42]. Most of the papers in that book presented specific natural stones that had been used in architectural heritage, identifying the lithology and other characteristics of the stone, specific buildings where it had been used, and the provenance in parts of Spain, France, Belgium, Italy, and the United Kingdom.

7. Conclusions

Worldwide, the archaeological and historic stone built heritage has deep cultural significance. This has been recognised by national and World Heritage designation of important structures. These structures must be maintained, repaired and in some cases restored. However, there are many instances of inappropriate use of natural stone that does not match the original partly due to cost but also to lack of knowledge of the importance of using the same or very similar materials to those in the original fabrics to avoid physical or aesthetic damage. Barriers to progress include limited awareness of existing sources of information scattered across the technical and other literature, manuscripts and in historic collections of stone samples. Easier access to information is needed by researchers, constructors, funding organizations, architects and other stakeholders when planning and commissioning works that require the replacement of stone.

Increasing numbers of contributions are being published, partly through the Heritage Stone Task Group of the IUGS and IAEG, drawing wider attention to heritage stones, but these works are found in the geological literature while many relevant stakeholders are not geoscientists. The HSTG has developed a protocol for description of natural stones that is drawing together technical material and information on the major heritage structures in which the stones have been used. This may also help in stimulating awareness. Because stone is internationally traded, it is important to have proper technical descriptions of stone so that only appropriate types are used. It is significant that the EU is considering controlling the names that can be used to clarify what is "appropriate stone". It is increasingly important to summarize key information and to make this more readily available to both geological and non-geologist stakeholders to help to properly preserve the built heritage [43].

Geoscientists should discuss how this can be achieved and are invited to engage with the Heritage Stone Task Group.

Acknowledgments: Thanks to the editor of this special volume, Carlos Alves, for inviting us to contribute to the issue. Brian R. Pratt made a critical revision of the first version of the manuscript, for which he is greatly acknowledged. Monica Price, from the Oxford University Museum of Natural History, showed me the Corsi collection, triggering the idea of this review. Lidia Catarino called attention to the exhibition of the Spanish treasures at the Colouste Gulbenkian Foundation in Lisbon. The Spanish National Heritage office provided a high-resolution picture of the stone collection from the Spanish Royal Palace. The authors are grateful for the continuous support of the International Union of Geological Sciences (IUGS) to Heritage Stone Task Group (HSTG) activities, and the International Geoscience Program (IGCP) 637 project for partially funding some of the HSTG meetings.

Author Contributions: Both authors contributed equally to this paper.

Conflicts of Interest: The authors declare no conflict of interest.

References

1. Pereira, D.; Marker, B. Repair and maintenance of natural stone in historical structures: The potential role of the IUGS Global Heritage Stone initiative. *Geosci. Can.* **2016**, accepted.
2. List of World Heritage in Danger. Available online: http://whc.unesco.org/en/danger/ (accessed on 15 December 2015).
3. Wikström, A.; Pereira, D.; Lundqvist, T.; Cooper, B. The Dala (Älvdalen) porphyries from Sweden. *Episodes* **2015**, *38*, 79–84.
4. Bulak, A. "Porphyries" from Russia and Sweden Used in St Petersburg and Russian "porphyry" Used in Paris: Misuse of a Geological Term for some Possible Candidate as a Global Heritage Stone Resource. *Episodes* **2015**, *38*, 114–117.
5. Maxfield, V.; Peacock, D. *The Roman Imperial Quarries—Survey and Excavation at Mons Porphyrites 1994–1998. Volume 1—Topography and Quarries*; Egypt Exploration Society: London, UK, 2001; p. 339.
6. Harrell, J.A.; Storemyr, P. Ancient Egyptian quarries—An illustrated overview. In *QuarryScapes: Ancient Stone Quarry Landscapes in the Eastern Mediterranean*; Abu-Jaber, N., Bloxam, E.G., Degryse, P., Heldal, T., Eds.; Geological Survey of Norway Special Publication; Geological Survey of Norway: Trondheim, Norway, 2009; Volume 12, pp. 7–50.
7. Hamida, D. Onyx from Ain Smara (North-Eastern Algeria): A natural stone to be considered as a Global Heritage Stone. In Proceedings of the 2015 GSA Annual Meeting, Baltimore, MD, USA, 1–4 November 2015.
8. Tárraga Baldó, M.L. Mármoles y Rocas Ornamentales en la decoración del Palacio Real de Madrid. *Arch. Esp. Arte* **2009**, *82*, 367–392. [CrossRef]
9. Falcón Márquez, T. *Documentos Sobre Jaspes Utilizados en el Escorial y en el Palacio Real de Madrid, Procedentes de la Cuenca Minera y su Entorno*; I Congreso Nacional Cuenca Minera de Riotinto; Biblioteca Capitular y Colombina: Sevilla, Spain, 1988.
10. Angulo Iñiguez, D. *Catálogo de las Alhajas del Delfín*; Museo del Prado: Madrid, Spain, 1989; p. 267.
11. Corsi, F. *Delle Pietre Antiche Libri Quattro*; Da' Torchj di Giuseppe Salviucci e Figlio: Roma, Italy, 1828.
12. Corsi, F. *Catalogo Ragionato D'una Collezione di Pietre di Decorazione*; Da' Torchj del Salviucci: Roma, Italy, 1825.
13. Corsi, F. *Supplemento al Catalogo Ragionato D'una Collezione di Pietre di Decorazione Formata in Roma Dall' Avvocato Faustino Corsi acquistata Dall' Onorevole Signor Stefano Jarrett Inglese e Posseduta Dall' Universita di Oxford*; Da' Torchj del Salviucci: Roma, Italy, 1827.
14. Corsi Collection of Decorative Stones. Available online: http://www.oum.ox.ac.uk/corsi/ (accessed on 15 December 2015).
15. Pereira, D.; Cooper, B. Building stone as part of a World Heritage site: "Piedra Pajarilla" granite and the city of Salamanca (Spain). In *Stone in Historic Buildings: Characterization and Performance*; Geological Society Special Publications: London, UK, 2014; Volume 391, pp. 7–16.
16. Bulak, A. *From the Granite Vase to the Door of the Palace. Aelvdalen Porphyry Manufacture, and Its Vases in St Petersburg*; St Petersburg State University: St Petersburg, Russia, 2015; p. 128.
17. Gavrilenko, V. (Russian Pedagogical University, St. Petersburg, Russian Federation). Personal communication, 2015.
18. Schlosser, J. La Piedra Franca. Una Historia Masónica. Edición La Fraternidad: Tel Aviv, Israel, 2002. Available online: http://www.librospdf.net/libro/la-piedra-franca-una-historia-masonica/50930/ (accessed on 28 December 2015).

19. Pereira, D.; Navarro, R.; Baltuille, J.M. "Piedra Franca": The same name for many different natural stones. *Geophys. Res. Abstr.* **2014**, *16*, EGU2014-31.
20. The EU single market. Available online: http://ec.europa.eu/internal_market/consultations/2014/geo-indications-non-agri/index_en.htm (accessed on 17 December 2015).
21. Cooper, B.; Marker, B.; Pereira, D.; Schouenborg, B. Establishment of the "Heritage Stone Task Group" (HSTG). *Episodes* **2013**, *36*, 8–10.
22. Hughes, T.; Lott, G.K.; Poultney, M.J.; Cooper, B.J. Portland Stone: A nomination for "Global Heritage Stone Resource" from the United Kingdom. *Episodes* **2013**, *36*, 221–226.
23. Pereira, D.; Marker, B.; Kramar, S.; Cooper, B.; Schouenborg, B. *Global Heritage Stone: Towards International Recognition of Building and Ornamental Stones*; Geological Society Special Publications: London, UK, 2015; Volume 417, p. 275.
24. Special Issue "Geoscience of the Built Environment". Available online: http://www.mdpi.com/journal/geosciences/special_issues/built_environ (accessed on 28 December 2015).
25. Giuffrida, A.; Ciliberto, E. Syracuse limestone: From the past a prospect for contemporary buildings. *Geosciences* **2013**, *3*, 159–175. [CrossRef]
26. Navarro, R.; Pereira, D.; Rodríguez-Navarro, C.; Sebastián, E. The Sierra Nevada serpentinites: The serpentinites most used in Spanish heritage buildings. In *Global Heritage Stone: Towards International Recognition of Building and Ornamental Stones*; Pereira, D., Marker, B., Kramar, S., Cooper, B., Schouenborg, B., Eds.; Geological Society Special Publications: London, UK, 2015; Volume 417, pp. 101–108.
27. Pereira, D.; Kramar, S.; Cooper, B. Global Heritage Stone Resource: An Update. *Episodes* **2015**, *38*, 78.
28. Pereira, D.; Pratt, B.R. Heritage Stones of the world. *Geosci. Can.* **2016**, *43*, 3–4.
29. Pereira, D.; Peinado, M.; Yenes, M.; Monterrubio, S.; Nespereira, J.; Blanco, J.A. *Serpentinites from Cabo Ortegal (Galicia, Spain): A Search for Correct Use as Ornamental Stones*; Geological Society Special Publication: London, UK, 2010; Volume 333, pp. 81–85.
30. Pereira, D. A report on serpentinites in the context of heritage stone resources. *Episodes* **2012**, *35*, 478–480.
31. Navarro, R.; Pereira, D.; Gimeno, A.; del Barrio, S. Verde Macael: A serpentinite wrongly referred to as a marble. *Geosciences* **2013**, *3*, 102–113. [CrossRef]
32. Pereira, D.; Peinado, M.; Blanco, J.A. Misuse of natural stone for construction and the consequences in buildings. Case of study of serpentinites. *J. Mater. Civ. Eng.* **2013**, *25*, 1563–1567. [CrossRef]
33. Cavallo, A. The "green stones" of Valtellina and Valchiavenna (central Alps, northern Italy). *Geophys. Res. Abstr.* **2015**, *17*, EGU2015–12319.
34. Siegesmund, S.; Weiss, T.; Vollbrecht, A. *Natural Stone, Weathering Phenomena, Conservation Strategies and Case Studies*; Geological Society Special Publications: London, UK, 2002; p. 205.
35. Přikryl, R.; Smith, B.J. *Building Stone Decay: From Diagnosis to Conservation*; Geological Society Special Publications: London, UK, 2007; p. 271.
36. Smith, B.J.; Gomez-Heras, M.; Viles, H.A.; Cassar, J. *Limestone in the Built Environment: Present-Day Challenges for the Preservation of the Past*; Geological Society Special Publications: London, UK, 2010; Volume 331, p. 249.
37. Přikryl, R.; Tőrők, A. *Natural Stone Resources for Historical Monuments*; Geological Society Special Publications: London, UK, 2010; Volume 333, p. 231.
38. Gomez-Heras, M.; Smith, B.J.; Viles, H.V. Oxford stone revisited: Causes and consequences of diversity in building limestone used in the historic centre of Oxford, England. In *Natural Stone Resources for Historical Monuments*; Přikryl, R., Tőrők, A., Eds.; Geological Society Special Publications: London, UK, 2010; Volume 333, pp. 101–110.
39. Nijland, T.G.; van Hees, R.P.J.; Bolondi, L. Evaluation of three Italian tuffs (Neapolitan Yellow Tuff, Tufo Romano and Tufo Etrusco) as compatible replacement stone for Römer tuff in Dutch built cultural heritage. In *Natural Stone Resources for Historical Monuments*; Přikryl, R., Tőrők, A., Eds.; Geological Society Special Publications: London, UK, 2010; Volume 333, pp. 119–128.
40. Figueiredo, C.; Folha, R.; Maurício, A.; Alves, C.; Aires-Barros, L. Contribution to the technological characterization of two widely used Portuguese dimension stones: The "Semi-rijo" and "Moca Creme" stones. In *Natural Stone Resources for Historical Monuments*; Přikryl, R., Tőrők, A., Eds.; Geological Society Special Publications: London, UK, 2010; Volume 333, pp. 153–164.

41. Allocca, F.; Calcaterra, D.; Calicchio, G.; Cappelletti, P.; Colella, A.; Langella, A.; de' Gennaro, M. Ornamental stones in the cultural heritage of Campania region (southern Italy): The Vitulano marbles. In *Natural Stone Resources for Historical Monuments*; Přikryl, R., Tőrők, A., Eds.; Geological Society Special Publications: London, UK, 2010; Volume 333, pp. 219–231.
42. Cassar, J.; Winter, M.G.; Marker, B.R.; Walton, N.R.G.; Entwisle, D.C.; Bromhead, E.N.; Smith, J.W.N. *Introduction to Stone in Historic Buildings: Characterization and Performance*; Geological Society Special Publications: London, UK, 2013; Volume 391, p. 200.
43. Thomas, I. *Quarrying Industry in Wales. A History*; National Stone Centre: Middleton, UK, 2015; p. 224.

4

Assessment of the Evolution of a Landslide using Digital Photogrammetry and LiDAR Techniques in the Alpujarras Region (Granada, Southeastern Spain)

Tomás Fernández [1,2,*], José Luis Pérez [1,2], Carlos Colomo [1], Javier Cardenal [1,2], Jorge Delgado [1], José Antonio Palenzuela [3], Clemente Irigaray [3] and José Chacón [3]

1. Department of Cartographic, Geodetic and Photogrammetric Engineering, University of Jaén, Campus de las Lagunillas s/n, 23071 Jaén, Spain; jlperez@ujaen.es (J.L.P); cmcj0002@red.ujaen.es (C.C.); jcardena@ujaen.es (J.C.); jdelgado@ujaen.es (J.D.)
2. Centre for Advanced Studies in Earth Sciences (CEACTierra), University of Jaén, Campus de las Lagunillas s/n, 23071 Jaén, Spain
3. Department of Civil Engineering, University of Granada, Campus de Fuentenueva s/n, 18071 Granada, Spain; jpalbae@ugr.es (J.A.P.), clemente@ugr.es (C.I.); jchacon@ugr.es (J.Ch.)

* Correspondence: tfernan@ujaen.es

Abstract: In this work a detailed analysis of the temporal evolution of the Almegíjar landslide is presented. It is a rock slide located in the Alpujarras region (Granada, Spain) that has developed over the last 30 years. Six datasets and photogrammetric flights corresponding to the years 1956, 1984, 1992, 2001, 2008, and 2010 were surveyed. The more recent flight of 2010 combined an aerial digital camera and a LiDAR sensor and was oriented by means of in-flight data and tie points. This 2010 flight allowed for the generation of a reliable and high-precision Digital Terrain Model (DTM). The other flights were oriented using second-order ground control points transferred from the 2010 flight, and the corresponding DTMs were prepared by automatic matching and subsequent editing from the stereoscopic models. After comparing the DTMs of different dates, it has been observed that the landslide was triggered after 1984 and since then has evolved in an irregular pattern with periods of variable activity. On average, the ground surface dropped more than 8 m in depleted zones and rose nearly 4 m in the accumulation zones, with a velocity catalogued as very slow (about 15–30 cm/year) over a time span corresponding to a degree VIII of diachroneity. The total volume of the mobilized mass of this large contemporary slide was about 300×10^3 m^3.

Keywords: landslide evolution; DTM; digital photogrammetry; LiDAR; Alpujarras region

---

1. Introduction

This article is an extended and improved version of the conference paper titled "Digital photogrammetry and LiDAR techniques to study the evolution of a landslide" [1], presented at the Eighth International Conference on Geo-information for Disaster Management (Gi4DM), celebrated in Enschede (The Netherlands) in December 2012.

The application of remote sensing techniques to natural hazards and landslide research has been steadily expanding in the last two decades [2,3], with different approaches from the optical spectrum to Synthetic Aperture Radar (SAR) techniques [4,5]. Among these techniques, those based on photogrammetry and Light Detection and Ranging (LiDAR) are adequate for middle- to high-resolution studies.

Regarding photogrammetric techniques, it should be taken into account that the only current approach that enables reconstruction of the landslide kinematics at sufficiently significant temporal and spatial resolution is the use of aerial photographs from historical flights [6,7] and, alternatively, but with less accuracy, the use of historical cartography. Therefore the photogrammetric techniques have been widely employed [1,7–22], sometimes combined with LiDAR [1,23,24], Global Navigation Satellite Systems, GNSS [25,26], digitization of old maps [24], and other techniques such as the electric resistivity tomography [27]. Recently, the use of Unmanned Aerial Vehicles (UAV) has extended the use of photogrammetric techniques to very high resolution studies [28–39].

Most studies based on photogrammetric techniques follow a similar methodology, starting with the orientation of the digital aerial images using photo-triangulation techniques by means of bundle block adjustment [1,7,10,11,13–16,19–23,26,27]. Then it is possible to use a reduced number of ground control points (GCPs), which are generally established by means of GNSS techniques, discarding in each survey any control points that are not adequately identified in the corresponding images.

Once the stereoscopic models are oriented, Digital Elevation Models (DEMs)—Digital Terrain Models (DTMs) or Digital Surface Models (DSMs)—and orthophotographs are generated by image matching techniques. After this step, almost all these studies perform different quantitative analyses, including those to prepare differential DEMs between surveys, profiles, and volumetric calculations. Also, in some studies, displacement vectors between single points are determined [10,11,13–18,21,26,29,30,32,36–38] or correlation algorithms are used with a much larger number of points [20,31,33]. From these analyses, useful qualitative features concerning their evolution are also established [1,7,13–16,20,23,24,27,32].

The quality or positional accuracy of these analyses is easily assessed in most of the aforementioned studies by means of the root mean square error (RMS) at the residuals of the control and/or check points [19,22,29,32,33,37,39,40], including the error propagation to DEMs and their differentials [7,23,38].

On the other hand, LiDAR techniques supply reliable and high-density point clouds from which true Digital Terrain Models (DTMs) can be generated [41–52], as the processes of classification and filtering of ground points work better in LiDAR than in photogrammetric approaches. In this way, DTMs from LiDAR data corresponding to different dates, together with those from digital photogrammetry, may be compared in order to determine vertical ground displacements or calculate volumes [23,24,47,48,51,52]. Also, these DTMs, both photogrammetry and LiDAR-derived, provide the starting steps for geomorphological studies based on different parameters (slope, aspect, roughness, etc.), enabling landforms and landslides to be identified [42,45,46,49–52]. In some cases, data from different surveys are compared in order to assess the ground surface evolution [46]. In addition, accuracy analysis of LiDAR data and models are accomplished [47,48,52].

The methodology of this study starts from a combined camera and LiDAR flight oriented with in-flight information (onboard GNSS and inertial data), which provides a common reference system for both the photogrammetric and the high-resolution LiDAR digital models. Taking this flight as a reference, the other photogrammetric flights were oriented by means of second-order ground control points that were transferred from the reference flight. Once these historical flights were oriented, the DTMs corresponding to each survey were obtained by outdating the reference DSM (from LiDAR data) by means of stereoscopic editing in a photogrammetric workstation. Then differential DTMs and subsequent profiles and volumetric changes were computed. It is a rather efficient methodology that considerably reduces the number of field-surveyed ground control points and allows for the geo-referencing of all data in a common reference system.

2. Geographical and Geological Setting

The study zone is located in the Alpujarras region, a mountainous area in the Sierra Nevada range (Granada province, Southern Spain), near the Mediterranean Sea (Figure 1). The region is widely affected by slope instability processes, giving rise to abundant rock falls, slides, and debris flows resulting from a combination of abrupt relief and metamorphic geological units of the Internal Zone of the Betic Cordillera, rather susceptible to landslides [53–59].

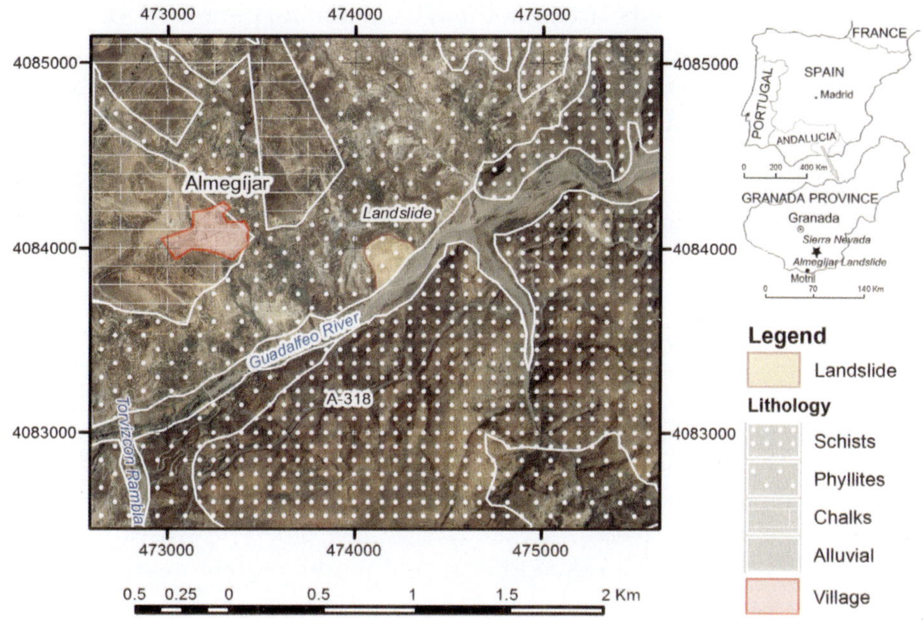

Figure 1. Geographical and geological setting. Coordinates are in ETRS89/UTM30.

The main triggering factor in the region is the combination of torrential rainfalls, the morphological slope, and a river channel evolution controlled by semiarid environmental conditions which gives place to a "rambla" segment in the Guadalfeo River. This channel has a planar bottom in which ephemeral streams alternate with variable flooding controlled by extraordinary intense storms or the quick melting of Sierra Nevada snow during spring rising of temperatures.

The landslide studied is a good example of the many landslides resulting from the slope evolution on the intensely eroded northern margin of the Guadalfeo River [59]. It is settled near the village of Almegíjar (Granada, Spain), but clearly outside the urban area and affecting only terrain free of any buildings or infrastructure. Given the fresh morphological features and the sliding material, it has been classified as a translational rock slide, with a clear upper scarp and a planar crown of 20 m height. The landslide mass is 250 m long, 300 m wide, and 140 m high (Figure 2). Also, some secondary minor scarps are visible, besides rock fall and debris flow in the frontal part of the mass [1,60].

Figure 2. The Almegíjar landslide.

The landslide is considered to be active, although alternating periods of very slow speed with more active periods during ephemeral floods of the "rambla" [53,58,59]. Also, according to [60], the slide reactivation in the period 2009–2010 showed a bulging in the lower sector of the displaced mass, following a mechanism described by [61].

The mass of the Almegíjar slide is composed of an Alpujarride unit of quartzites and quartzphyllites of Permian to Triassic age [62]. Nevertheless, the slide is part of an unstable but dormant slope showing an upward extension of the instability expressed by the appearance of large rock blocks and open cracks affecting the overlying marble unit, with water flows in its bottom infiltrating the mass of phyllite rock.

3. Materials and Methods

Digital photogrammetry and LiDAR techniques represent a significant advance in the study of ground surface evolution derived from landslides and other processes. The results using photogrammetric workstations have offered better and more accurate interpretations of changes on the terrain surface over the past decade [8,13–16].

Thus, landslide inventories and their geomorphic features (scarps, crown, lateral flanks, cracks, etc.) may be prepared by means of 3D digital stereoplotting. In this sense, not only single landslides but also the identification of unstable slopes in a region may be analyzed with these techniques, starting from high-resolution DEMs and 3D inventories [22,44,47]. In addition, if data from different surveys are available, it becomes possible to study landslide evolution. The overlaying of landslide inventories in Geographical Information Systems (GIS) allows the study of landslide evolution in a given region and the identification of reactivated landslide areas as well as the growth or generation of new landslides. Even landslide multitemporal landslides can be obtained from accurate DEMs of different epochs [44,47]. Thus it is possible to determine landslide activity, differentiating between active, dormant, and relict landslides. The concept of landslide diachroneity [63] distinguishes landslides considering the time span of the active periods and the initial triggering data.

For the analysis of single landslides, two approaches are available: the first is based on a comparison between DEMs from different flights, calculating distances and volumes [1,7–11,13–26,29–32,36–38,48,52]; the second approach consists of a comparison between points and other elements identified without any ambiguity between the different surveys, enabling the determination of 3D displacement vectors for accurate identification of the landslide kinematics [10,11,13–18,21,26,29,30,32,36–38]. In this paper the quantitative analysis of ground changes will be approached by comparing different DTMs.

The starting point is a dataset made up by photogrammetric flights and LiDAR data. The methodology, established by the research team based on previous results [1,19], includes a set of tasks summarized in the following steps (Figure 3):

1. Dataset compilation and digitization of images from historical flights.
2. Definition of reference system and orientation of reference data.
3. Orientation of historical photogrammetric flights.
4. Building and editing DTMs.
5. Comparison between models and calculations.

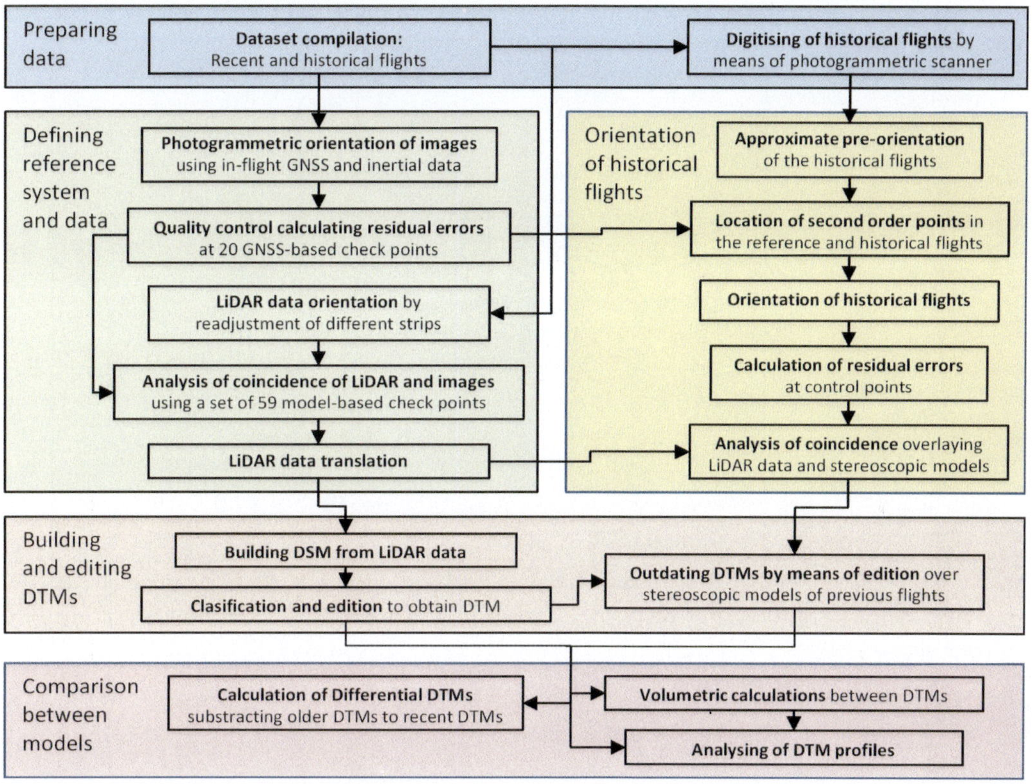

Figure 3. Flow chart of the methodology.

3.1. Dataset Compilation and Digitization of Images from Historical Flights

First, a photogrammetric flight corresponding to the year 2010 was available, combined with a digital camera (Z/I DMC) and a LiDAR sensor (Leica ALS50-II), as well as attached GNSS/IMU systems for direct orientation. The Z/I digital camera offers a resolution of 0.20 m and has four spectral bands, three in the visible spectrum (RGB) and one in the near-infrared (NIR). The LiDAR instrument produces a ground resolution of about 1–1.5 points per m^2. A second photogrammetric flight of 2008 with the same properties of the first one was also available but without LiDAR dataset.

Besides the highly accurate and recent flights, historical images from aerial film cameras were also available, corresponding to four photogrammetric flights of the years 1956, 1984, 1992, and 2001. The first one was the so-called "American flight", as it was undertaken by the U.S. government in 1956, and it was panchromatic at a middle scale (1:33,000). The second one (1984) was a panchromatic flight made by the National Geographic Institute of Spain (IGN) at scale of 1:30,000. The other two (1992 and 2001) were also panchromatic flights at a scale of 1:20,000, made by the Andalusian Cartographic and Statistical Institute (IECA). The main flight features are indicated in Table 1 and their distributions are shown in Figure 4.

Table 1. Properties of the image datasets.

Date	Bands	Format	Pixel [1]	GSD [2]	Camera	Focal Distance [3]	LiDAR
1956	Panchromatic	Film	15	0.60	-	151.42	No
1984	Panchromatic	Film	20	0.80	UAG-II	153.03	No
1992	Panchromatic	Film	15	0.30	UAG-II	153.03	No
2001	Panchromatic	Film	15	0.30	Wild UAO-S	152.75	No
2008	RGB-NIR	Digital	12	0.20	Z/I DMC	120.00	No
2010	RGB-NIR	Digital	12	0.20	Z/I DMC	120.00	Yes

[1] Pixel size is in microns; [2] GSD (ground sample distance) is in meters; [3] Focal distance is in mm.

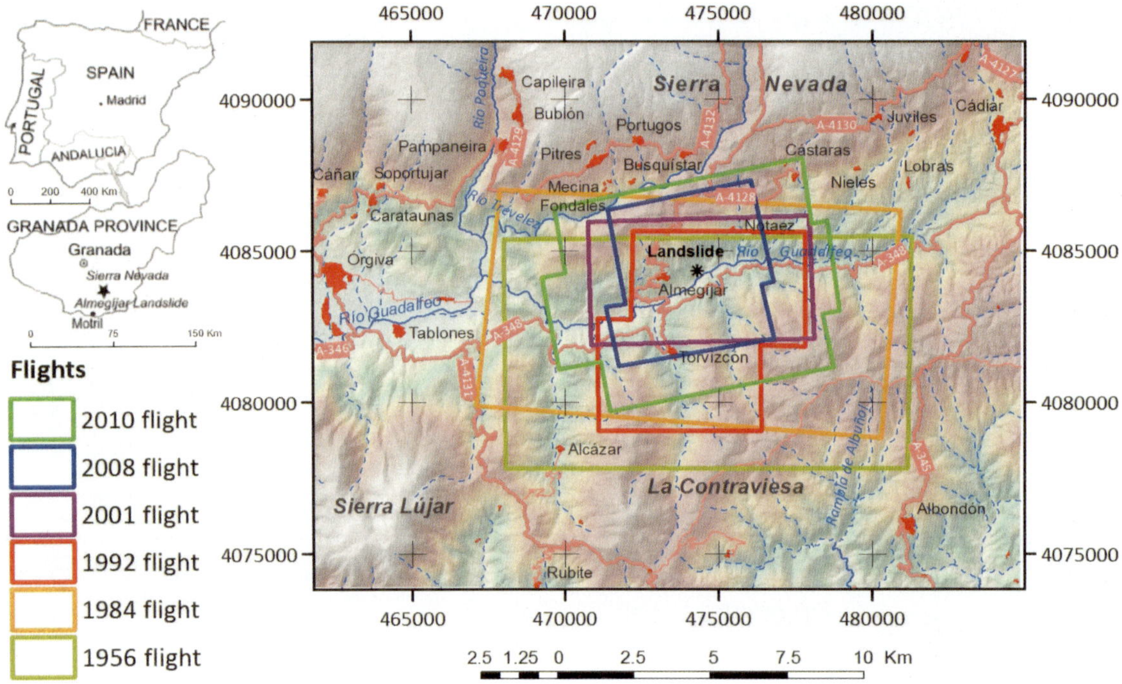

Figure 4. Distribution of available flights. Coordinates are in ETRS89/UTM30.

These historical flights were digitized with a precision photogrammetric scanner Vexcel Ultrascan 5000 at a resolution of 15 microns, which, in turn, implies a ground sample distance (GSD) of 0.60 m in the 1956 flight and 0.30 m in the 1992 and 2001 flights. The 1984 flight was already supplied in digital format after a scanning at 20 microns that provided a GSD of 0.80 m.

The period considered (c. 50 years) is long enough to study short-term changes in the relief and, more precisely, to analyze the landslide evolution in the study zone.

3.2. Definition of Reference System and Orientation of Reference Data

As a starting point for the entire process, the 2010 flight (camera and LiDAR) was used. This flight was oriented by means of in-flight orientation systems (direct orientation). As this dataset was considered to be of the highest quality, both in image resolution and point cloud accuracy, it has been proposed as the reference data and the previous flights have been registered with respect to it.

Thus, it is important to ensure the adjustment and minimal data discrepancies between the photogrammetric flight and the LiDAR point cloud of this 2010 reference flight. The discrepancies between them are originated as a consequence of the independent orientation systems of these sensors—despite the simultaneous recording of data—and thus its different positioning, which may prompt a lack of geometrical coincidence in the data compiled. This raises the need for an optimal adjustment between the two types of data [64].

In accordance with the above, the reference data are compiled and refined by three steps:

1. Photogrammetric orientation using in-flight GNSS and inertial data to ensure the geometrical quality of the whole photogrammetric block. This procedure was performed by means of a photogrammetric workstation and Socet Set 5.2 software (BAE Systems Plc., London, UK) [65] by measuring about 200 tie points and readjusting the orientation parameters. The RMS errors of the resulting model, based on the residual of 20 check points surveyed with GNSS (GNSS-based check points), are 0.258 m in XY and 0.100 m in Z (Table 2).
2. LiDAR data orientation, by readjustment of different strips, which is employed to obtain a single homogeneous point cloud, even in overlapping zones. This task was carried out with the TerraMatch module of TerraSolid software (Terrasolid Ltd., Helsinki, Finland) [66].

3. Analysis of height coincidence of both datasets. For this analysis an internal control process of the LiDAR point cloud was made using check points (model-based check points) taken from the refined photogrammetric model of the step 1. These points were selected taking into account the following:

   - Points on stable and flat surfaces, avoiding high slopes or roughness zones, were selected.
   - The points had to be well distributed throughout the study area and the number high enough to get an appropriate redundancy and significant results.

Table 2. Orientation errors.

Date	Photographs Number	GCP/CHK Number	Tie Points Number	Errors				
				RMS Pixel	GCP/CHK RMS [1]		Propagation [1]	
					RMS_{XY}	RMS_Z	PE_{XY}	PE_Z
2010	36	20	195	0.529	0.258	0.100	-	-
2008	9	8 xyz	82	0.246	0.145	0.042	0.296	0.108
2001	3	8 xyz	14	0.363	0.187	0.213	0.319	0.235
1992	4	7 xyz	21	0.423	0.255	0.067	0.363	0.120
1984	3	8 xyz	15	0.621	0.517	0.234	0.578	0.254
1956	3	8 xyz.4 z	14	1.038	0.566	0.269	0.622	0.287

[1] Root Mean Square (RMS) and propagation errors are in meters. For 2010, flight errors are calculated in check points (CHK). For the remaining flights, errors correspond to the residual calculated in ground control points (GCPs). GCP types: full (xyz) and height (z).

Taking into account the good adjustment of the image orientation process, with mean values of about 0.10 m in GNSS-based check points, what really was tested in Step 3 were the errors of the LiDAR data. The vertical accuracy of aerial LiDAR data is well documented in many works [47–49,51,52] although the values present a great dispersion—from 0.16 to 3.26 m [47]—depending on different factors (flight altitude, field of view, beam angle respect to the vertical, coverage reflectance, vegetation, etc.). Furthermore, according to [67], the vertical error in LIDAR increases by 0.1 m per 1000 m of flight altitude, so it reaches values of about 0.10–0.20 m in conventional flights. In this sense, some works, in which the point density is usually higher than 1 point/m^2, establish a vertical accuracy of 0.10–0.20 m [48,49,52,68]; but in other works, with point densities lower than 1 point/m^2, the accuracy is about 0.50 m [47,50]. Thus, the higher the flight is, the lower the point accuracy and density, and therefore the vertical errors and uncertainties usually increase with the flight altitude. Meanwhile, in [47] the vertical discrepancies—expressed as RMS—between two LIDAR surveys in a close time interval were of 0.28 m, measured in stable zones (out of a mask drawn with the landslide inventory); however, the calculated uncertainties from comparison of LiDAR data with GNSS-measured points ranges from 0.50 m in leaf-off conditions and 0.75 m in leaf-on conditions. Bearing in mind these previous data, a reference value of accuracy of 0.20 m has been established for the present work, according the flight altitude and LiDAR configuration.

Nevertheless, a set of 59 model-based check points was extracted in a wide area surrounding the studied landslide to test the adjustment of LIDAR data and the photogrammetric model. The distribution of these points, all of them located outside of the landslide limits, is shown in Figure 5. The discrepancies or errors between the datasets range from 0 to near 1 m, with a mean positive value of 0.529 m, as can be observed in Table 3 and Figure 5a,c. Considering the good adjustment of the photogrammetric orientation process, this analysis allows us to infer a systematic error upwards in LiDAR data, corresponding to a translation of about 0.5 m, due to weak absolute georeferencing of the whole LiDAR block. This discrepancy could be eliminated by a simple translation downwards in height of the LiDAR data equal to the mean error. Meanwhile, the standard deviation (0.260 m) was of the same order of expected accuracy of LiDAR [48,49,52,67,68] and remained constant after correction. The discrepancies in the points after correction are also shown in Table 5 and Figure 5b,c. Thus, in the new situation, a refined adjustment has been reached with a zero mean error and a standard deviation around the considered accuracy of LiDAR data.

Table 3. Statistics of errors between LiDAR and photogrammetric data of the 2010 flight.

Statistics of Errors	Before Correction [1]	After Correction [1]
Mean	0.529	0.000
Standard deviation	0.259	0.259
Maximum	0.939	0.410
Minimum	0.045	−0.484
Median	0.564	0.035

[1] Errors are in meters.

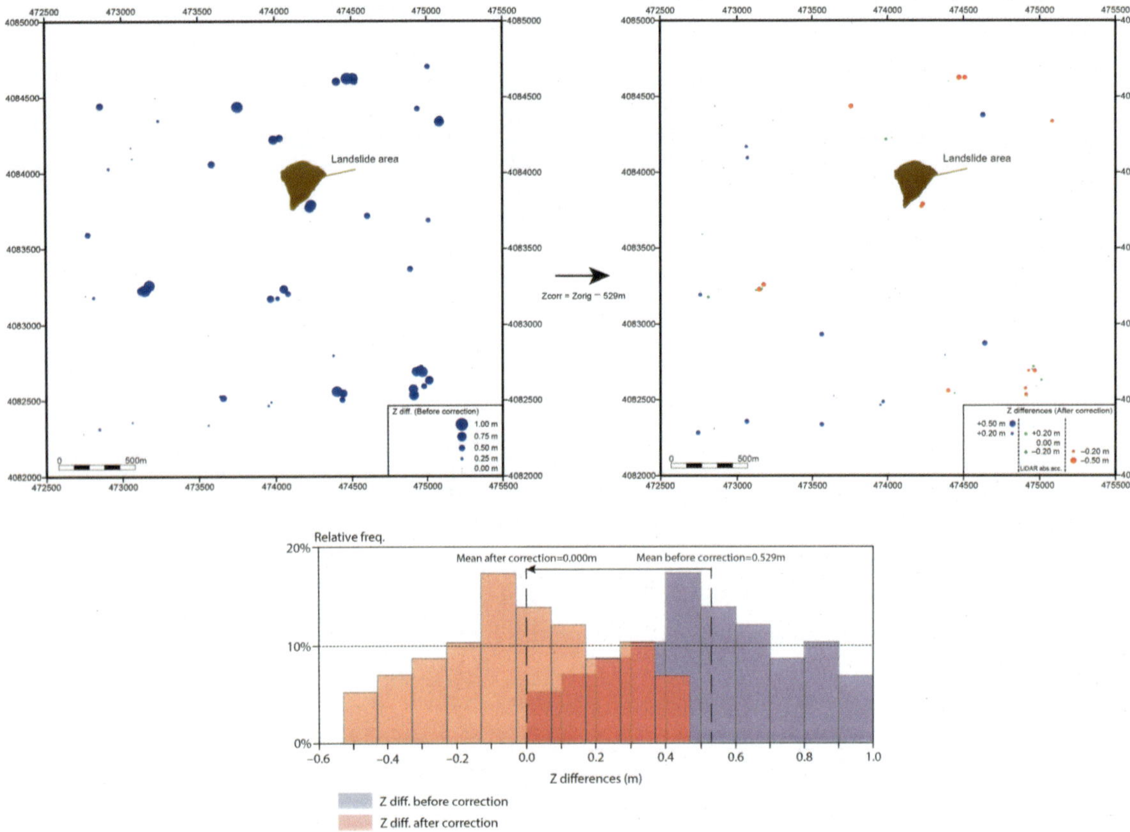

Figure 5. Distribution of errors or discrepancies in height between 2010 LiDAR and photogrammetric data: (**a**) before the adjustment; (**b**) after the adjustment; (**c**) histograms.

3.3. Orientation of Historical Photogrammetric Flights

Once the 2010 reference flight was re-oriented and the agreement between the photogrammetric and LiDAR data was ensured, the historical flights were oriented with respect to the 2010 reference system. The processing of several historical flights corresponding to a broad time span is rather difficult as the field ground control points (GCPs) must be stable, accessible, and unequivocally recognizable, not only in images from the time period considered but also in the field, where the points have to be measured with differential GNSS observations. Thus, this paper proposes the use of GCPs transferred from the reference photogrammetric flight (second-order GCPs) recognizable in flights corresponding to each survey. As the GCPs are directly located on the photogrammetric flights (the one to be oriented and the reference flight), it will be possible to certify in real time whether the point is observable and unequivocally recognizable on both flights. In addition, with a zenithal perspective the observer would have access to the entire study area, without limitation in accessibility or observation. Furthermore, with this procedure the number of required GCPs measured in the field is optimized, as well as the field work and processing time.

The main constraint of this methodology is a theoretical reduction in the accuracy of the coordinates in the second-order GCPs with regard to the accuracy of the first-order GCPs due to error propagation. Nevertheless, as the 2010 photogrammetric flight with higher resolution and a RMS lower than the pixel size was always taken as the reference, the accuracy in the GCPs transferred in this way is proved to be better than the resolution of the historical flights.

The process of transferring points from the reference flight (more recent) to the previous flights to be oriented is made iteratively following the next steps (Figure 3):

1. Approximate pre-orientation of the flight to be oriented in the reference system.
2. Once the two photogrammetric flights are pre-oriented in the same reference system, a set of several common points in both flights are located and uploaded simultaneously in the digital photogrammetric workstation.
3. All these points located and measured in the reference flight are considered second-order GCPs for the flight to be oriented (Figure 6). This process includes full GCPs (known X,Y,Z), horizontal GCPs (known X,Y), and height GCPs (known Z). Table 2 shows the number and type of the second-order GCPs used in the orientation of the historical flights.
4. The flight to be oriented is adjusted and the process is repeated from step 2 until a final proper orientation is attained, that is when the difference between coordinates of GCPs in both datasets was lower than the image resolution of the flight to be oriented.

Figure 6. Example of transferred second-order control point (GCP).

In Table 2, the results of the process of orientation of historical flights are presented together with data from the reference flight. The orientation errors of the 2008 flight are satisfactory but the errors increase in the older flights. Thus, XY errors range from 0.15 m (2008), about 0.20 m (2001 and 1992), 0.50 m (1984), to near 0.60 m (1956). Regarding to heights, Z errors are between 0.04 m for 2008 flight, 0.07 m for 1992 flight and 0.20–0.30 m for 2001, 1984 and 1956 flights. But, as the historical flights were oriented using GCPs obtained from the reference 2010 flight, the precision of any measurement in those flights is influenced by the error propagation from the 2010 flight. Therefore the final precision of these flights can be calculated according to the following expression:

$$Prop_{YEAR} = (RMS_{2010}^2 + RMS_{YEAR}^2)^{0.5} \quad (1)$$

Thus, the propagated error increases in the following way: XY errors range between 0.30–0.36 m for the 2008, 2001, and 1992 flights to near 0.60 m for the 1984 and 1956 flights; and Z errors are between 0.11–0.12 m for 2008 and 1992 flights to 0.24–0.30 m for 2001, 1984, and 1956 flights.

After applying this procedure, the spatial correspondence between the two compared flights is checked out under the premise of finding a total correspondence in stable zones where no modifications were detected between the two surveys. The comparison was made by overlaying the reference 2010 LiDAR data and the stereomodels of the historical flights with the stereoscopic viewing tools of the Socet Set and the photogrammetric workstation. If the orientation was correct, then DTM contours and points derived from LiDAR data had to be adjusted to the ground in the stable areas of the study zone. If significant discrepancies were found, then an orientation re-adjustment would be necessary. In the case studies, the stereoviewing display comparison showed a good coincidence between elevation and stereoscopic models in all the flights, and therefore no readjustments of the flight orientation were necessary.

For this comparison, different types of color compositions were applied in recent flights: true color (Blue-Green-Red) and false color (Green-Red-NIR), which was possible in recent flights (2008 and 2010), since they incorporated the near infrared (NIR) band. With these compositions the different ground features were more visible and identifiable in addition to the element over them, such as vegetation, which is very sensible to detection with near infrared.

3.4. Building and Editing DTMs

A Digital Surface Model (DSM) was obtained from the 2010 LiDAR data. Then a classification was made between ground-points and non-ground-points (vegetation, buildings, artifacts, etc.). This classification process was performed with the software TerraSolid (module TerraScan). Once this classification was finished, the ground surface corresponding to the reference survey was defined by those filtered ground-points, thus obtaining the Digital Terrain Model (DTM).

From the initial 4,240,499 points in the 2010 LiDAR point cloud, 1,943,545 were classified as ground-points while 2,296,954 were non-ground points. Then, the results of this classification were displayed in the stereo photogrammetric workstation together with the reference photogrammetric flight, and edited using the stereoscopic tools supplied by Socet Set. This editing process consisted basically on correcting wrong classified points and stereoplotting breaklines and some typical ground features, such as scarps, water streams, slope changes, etc., which allowed for a more realistic and precise DTM generation.

After preparing a reference DTM from LiDAR data classification, the models corresponding to the historical flights are obtained by editing and outdating the 2010 reference model. Other authors [7,20,21,23,24,26] have generated DEMs corresponding to each survey by a process of image correlation independent for each flight. However, in this work another approach has been adopted for the following two reasons. First, the accuracy of models is consistent with the resolution and the radiometric quality of the aerial images, which in the older flights has led to a lower accuracy in the models. A second reason is the need to build the model again by matching, which in addition requires the editing of all the surveys in the study zone, no matter if there have been any changes in the terrain or not, thereby giving rise to inevitable false ground modifications in the comparison of models corresponding to different surveys and noise in the detection of changes.

For these reasons, an outdating of the reference model—of a higher quality and geometric resolution than those models that would be obtained by a matching technique in each case—is proposed. This outdating process consists of uploading the reference model overlaid to the historical flights in the photogrammetric workstation. Then, by means of the stereoscopic viewing tools is possible to analyze the lack of coincidence between the two datasets. When that lack of coincidence occurs, the 2010 DTM is edited in that zone to adapt the terrain to the ground surface in the stereomodel of the historical flight. In the case of changes affecting large surfaces, it is necessary to launch a matching process exclusively for that area. In ground areas where no change is detected, the model is not modified, thus reducing the editing process time as well as avoiding noise in the analysis of changes and maintaining a better overall quality of the model provided by the higher positional accuracy of the 2010 data.

After finishing the outdating of the reference model to the immediately previous historical survey under analysis, this resulting model was used as a reference model in the preceding survey, repeating the process in all cases until the 1956 DTM was obtained. In Figure 7, the DTMs generated in the unstable area are shown, allowing a visual observation of significant changes between these six dates considered.

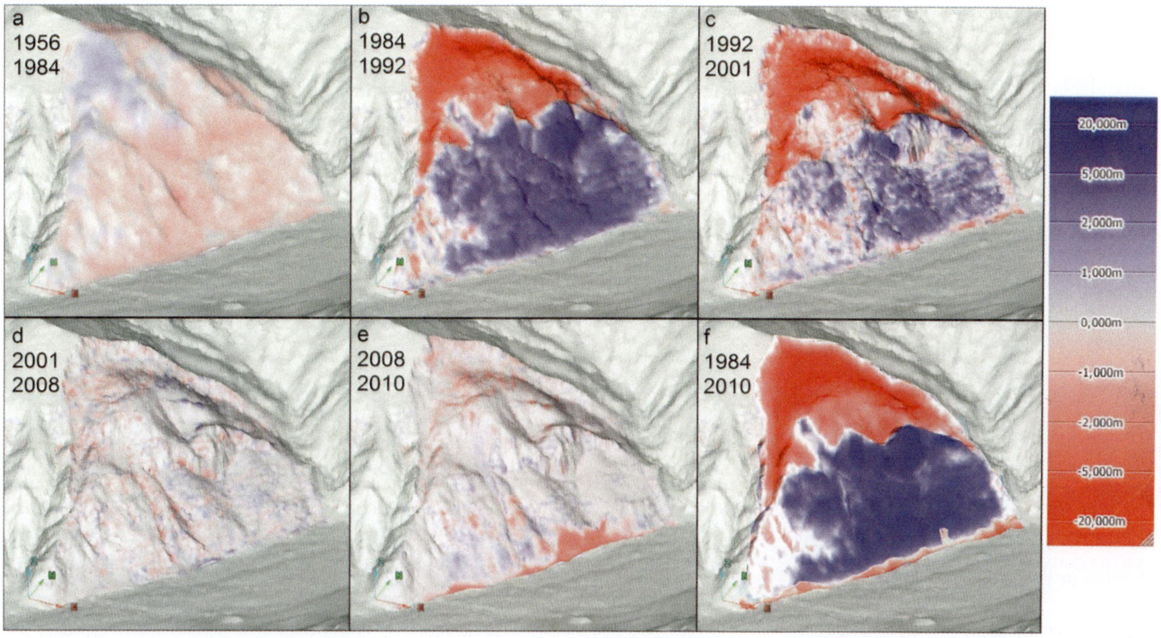

Figure 7. Differential DTMs (Digital Terrain Model). Depletion areas are in red and accumulation areas are in blue: (**a**) 1956–1984; (**b**) 1984–1992; (**c**) 1992–2001; (**d**) 2001–2008; (**e**) 2008–2010; (**f**) 1984–2010.

3.5. Comparison between Models and Calculations

The comparison between models was carried out from different perspectives:

1. Vertical displacements by subtracting DTMs (differential DTMs) (Figure 7). They may be negative or positive, depending on whether the model compared with the reference lies below or above, allowing the identification of areas of mass depletion or accumulation, respectively.
2. Volumetric calculations between models. As in the previous case, the volumes can be negative (depletion areas) or positive (accumulation areas).
3. Analysis of DTMs profiles through the landslides permitting a better observation of terrain displacements (Figure 8). With enough profiles, it is possible to attain an overview of the landslide kinematics, for instance if it is planar or there are rotations around vertical or horizontal axis indicating rotational sliding.

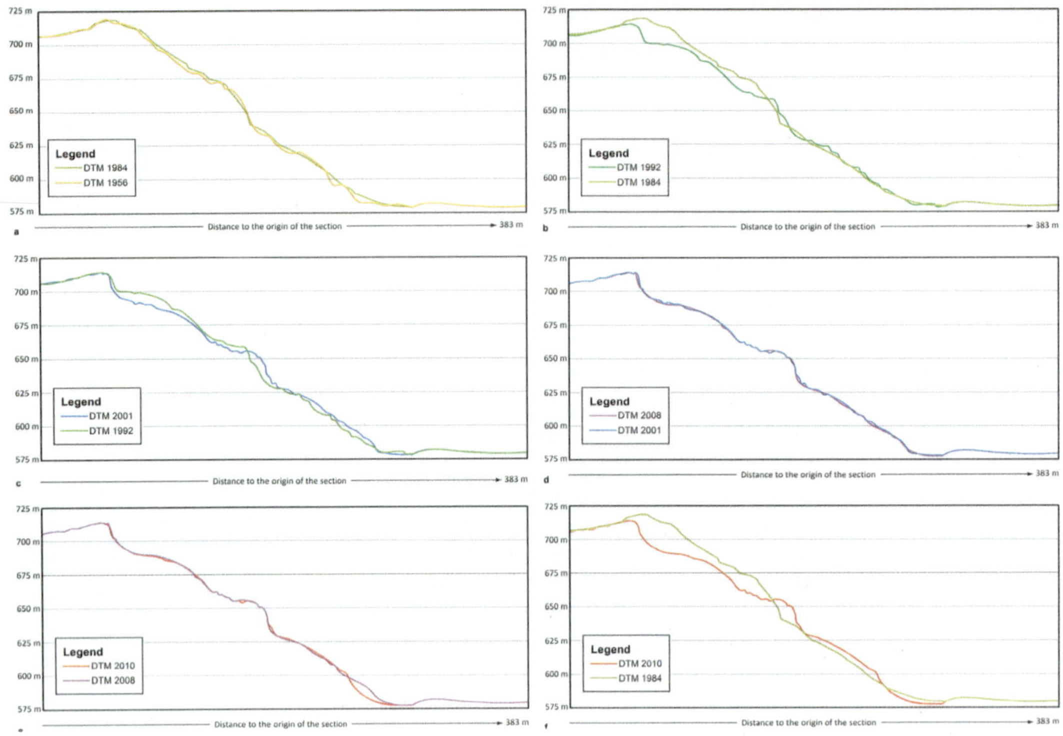

Figure 8. Cross sections or profiles of the DTMs: (**a**) 1956–1984; (**b**) 1984–1992; (**c**) 1992–2001; (**d**) 2001–2008; (**e**) 2008–2010; (**f**) 1984–2010.

These calculations and analyses have been made by means of the Maptek I-Site Studio 6.0 software (Maptek Pty. Ltd., Adelaide, Australia) [69] and ArcGIS 10.2 (Esri, Redlands CA, USA) [70]. The errors and uncertainties—usually expressed as root mean square (RMS) and standard deviation (SD), respectively—of these procedures have to be estimated in order to validate the results of these calculations. In the case of the 2010 flight, the value of standard deviation (0.26 m) is assumed to be vertical uncertainty or the minimum level of detection [52,71,72]. Meanwhile, from previous works [7,23,32] it has been observed that the vertical errors and uncertainties in DEMs obtained by means of matching from photogrammetric oriented flights are about two or three times higher than the orientation residual errors in Z. Therefore the vertical uncertainties of DTMs obtained in this work range from 0.27 to 0.30 m for 2008 and 1992 flights, about 0.60 m for 2001 and 1984 flights, to over 0.70 m for the 1956 flight (Table 4).

Table 4. Estimated uncertainties of DTMs, differential models, and volumetric calculations.

Date	Propagated Error GCPs [1]	Uncertainty in DTMs [1]	Interval	Differential Models [1]	Volumes [2]		
					DEP	ACC	WAS
1956	0.287	0.718	1956–1984	0.958	33,548	22,266	40,265
1984	0.254	0.635	1984–1992	0.702	25,735	17,350	31,037
1992	0.120	0.300	1992–2001	0.660	18,374	15,782	24,222
2001	0.235	0.588	2001–2008	0.647	23,933	23,755	33,720
2008	0.108	0.270	2008–2010	0.375	16,024	14,278	21,462
2010	-	0.260	1956–2010	0.763	26,558	20,600	33,611
-	-	-	1984–2010	0.686	22,717	16,343	27,985

[1] Errors in GCPs and the estimated uncertainty in DTMs (Digital Terrain Model) and differential models are in meters.
[2] The estimated uncertainty in volumetric calculations is in m^3. Volumes: DEP: Depletion; ACC: Accumulation; WAS: Waste.

Furthermore, the vertical uncertainties of the differential DEMs are estimated as [7,52,71,72]:

$$\text{Uncert.}_{\text{YEAR1-YEAR2}} = (\text{Uncert.}_{\text{YEAR1}}{}^2 + \text{Uncert.}_{\text{YEAR2}}{}^2)^{0.5} \quad (2)$$

In this case, the uncertainties are lower than 0.40 m for the 2008–2010 period, 0.65–0.70 m for the 2001–2008, 1992–2001, and 1984–1992 periods, and close to 1 m for the 1956–1984 period (Table 4).

In addition, the uncertainties of the volumetric calculations are estimated by multiplying the vertical uncertainties by the areas of depletion and accumulation [52]. The uncertainty of the waste material (difference between depletion and accumulation volumes) is calculated as a propagated error (square root of the sum of squares of depletion and accumulation volumetric uncertainties). These last ones range from 21×10^3 m^3 for the 2008–2010 period, 31–33 $\times 10^3$ m^3 for the 2001–2008, 1992–2001, and 1984–1992 periods, to about 40×10^3 m^3 for the 1956–1984 period (Table 4).

4. Results

The analysis of orthophotographs and models since 1956 of the area located in the Guadalfeo River, near Almegíjar village, has allowed for the identification of a landslide as well as its characteristics and evolution. The landslide does not appear in the aerial photographs of the 1956 and 1984 flights, so it was triggered after 1984, although located in a broadly unstable zone, with evidence of older activity upslope. Thus in the stereoscopic models of later dates (1992, 2001, 2008, and 2010), the landslide area presents some clear features that include a clean scarp, steps and terraces, minor scarps, and an accumulation zone invading the river channel. The involved mass reaches about 300×10^3 m^3 according to the volumetric calculations, so the landslide can be defined as moderately large. Other features identified through aerial photographs, but also field work, are a large extent of debris flows and the development of pervasive erosion in rills and gullies on the weathered rock-mass surface, even before 1984 and at the start of the main landslide.

The differential models between the 1956 and 2010 surveys (the whole period considered) show two clearly differentiated areas, the depletion area at the upper part of the mass and the accumulation area in the lower part as shown in Figures 7 and 8. As the differential models are calculated by subtracting the older models to the recent ones, negative differences in elevation and volume can be observed in the depleted area, and positive differences are recorded in the accumulation zone. In the Table 5, the proportion of areas of depletion, accumulation, and below the uncertainty is shown. If uncertainty is not taken into account, the areas of depletion and accumulation are practically compensated for (near 50%), although in some periods (1992–2001 and 2008–2010) depletion areas predominate slightly. When the uncertainty areas are introduced, some periods show a high proportion of areas below the uncertainty, such as 1956–1992 and 2001–2008 (less than 10%) or even 2008–2010 (16%). In these cases, the depleted and accumulated areas are compensated for except in the 2008–2010 period where the depleted area predominates. Meanwhile, other periods such as 1984–1992, 1992–2001, and the global periods show a higher proportion of area depleted and accumulated (over 70%), usually compensated for, except in the 1992–2001 period, when depleted area also predominates.

Table 5. Areas resulting in from the comparisons between DTMs.

Period	Without Uncertainty				With Uncertainty					
	Areas [1]		Proportion [2]		Areas [1]			Proportion [2]		
	DEP	ACC	DEP	ACC	DEP	ACC	UNC	DEP	ACC	UNC
1956–1984	37,846	35,859	51.35	48.65	4738	4048	64,919	6.43	5.49	88.08
198–1992	44,024	29,681	59.73	40.27	33,555	19,950	20,200	45.53	27.07	27.41
199–2001	39,650	34,055	53.80	46.20	28,985	22,764	21,955	39.33	30.89	29.79
200–2008	36,990	36,715	50.19	49.81	5955	5425	62,325	8.08	7.36	84.56
200–2010	38,975	34,730	52.88	47.12	12,232	7448	54,026	16.60	10.10	73.30
195–2010	45,887	27,818	62.26	37.74	37,523	29,523	6659	50.91	40.06	9.03
198–2010	41,509	32,196	56.32	43.68	37,770	28,533	7403	51.24	38.71	10.04

[1] Areas are in m^2. [2] Proportions are in %. Volumes and areas: DEP: Depletion; ACC: Accumulation; UNC: Uncertainty.

Significant vertical displacements are recorded in the landslide area, as shown in Table 6, with mean values higher than an order of magnitude with respect to the estimated uncertainties (Table 4). Thus, in the depleted area, there is an average descent of 8.5 m of the ground surface, with maximum values around 28 m. In addition, in the accumulation area, vertical displacements related to the elevation of the ground surface show an average value of 3.9 m, with maximum peaks of almost 28 m. Considering that the landslide was triggered after 1984, the annual displacement rates reach at least 0.32 m/year and 0.15 m/year in the depleted and accumulation areas, respectively. Consequently, the landslide velocity can be estimated on the order of some few cm/year so the slide can be catalogued as "very slow" if it were considered continuous and uniform in time [73,74].

The analysis of the results of all five time periods considered (Table 6), between the six available flights, reveals that the vertical mean displacements are practically null in the first period of 1956–1984 (0.09 m and 0.08 of terrain elevation in the depletion and accumulation areas), higher in the second period of 1984–1992 (3.70 m of decrease in the depletion area and 3.11 m of elevation in the accumulation area), lower in the third period of 1992–2001 (2.87 m and 1.78 m, respectively), and far lower in the last two periods of 2001–2008 (0.30 and 0.27 m) and 2008–2010 (0.43 and 0.23 m). These partial values are significantly higher than the uncertainties of Table 4, except for the periods of 1956–1984 and 2001–2008 (clearly lower than uncertainties) and the period 2008–2010 (similar to uncertainties). More informative is the comparison of annual rates, with values clearly above average in the period 1984–1992 (0.46 m/year of depletion and 0.39 m/year of accumulation); values near average in the period 1992–2001 (0.32 and 0.20 m/year respectively), and slightly lower in 2008–2010 (0.22 and 0.12 m/year); finally, rate values are clearly lower in the period 2001–2008 (0.04 and 0.04 m/year) and practically null in the period 1956–1984.

Table 6. Vertical displacements resulting from differential DTMs.

Period	Displacements [1]								Rates [2]	
	Depletion Area				Accumulation Area				DEP	ACC
	MIN	MAX	ME	SD	MIN	MAX	ME	SD	ME	ME
1956–1984	−3.39	2.91	−0.09	1.95	−2.83	2.37	−0.08	1.66	0.00	0.00
1984–1992	−12.40	2.57	−3.70	3.94	−2.17	24.75	3.11	4.21	−0.46	0.39
1992–2001	−10.89	2.35	−2.87	2.43	−1.98	19.09	1.78	2.22	−0.32	0.20
2001–2008	−6.14	0.65	−0.30	0.36	−0.83	7.95	0.27	0.41	−0.04	0.04
2008–2010	−5.95	1.02	−0.43	0.34	−1.34	6.14	0.23	0.65	−0.22	0.12
1956–2010	−27.75	3.12	−8.49	6.29	−3.53	27.60	3.94	4.82	−0.16	0.07
1984–2010	−25.67	3.03	−8.41	6.18	−3.23	25.70	3.89	4.74	−0.32	0.15

[1] Displacements are in meters. [2] Displacement rates are in m/year. Rates: DEP: Depletion; ACC: Accumulation. MIN: Minimum; MAX: Maximum; ME: Mean; SD: Standard deviation.

The volume assessment in Table 7 shows a depletion of more than 290 × 10^3 m^3 of mass in the upper part of the slide, but the accumulated mass in the lower part near the river channel is about 120 × 10^3 m^3. This implies mass losses of nearly 175 × 10^3 m^3, probably evacuated by the river flow. These values are slightly modified to 280 × 10^3 m^3, over 110 × 10^3 m^3 and 167 × 10^3 m^3 if a starting date of 1984 is considered. By periods, the values are very low in the first period of 1956–1984 (24 × 10^3 m^3 of depleted material, 17 × 10^3 m^3 of accumulated material and 8 × 10^3 m^3 of waste material), clearly higher in the second period of 1984–1992 (169 × 10^3, 53 × 10^3 and 116 × 10^3 m^3, respectively), lower in the third period of 1992–2001 (86 × 10^3, 46 × 10^3 and 40 × 10^3 m^3), and far lower in the last two periods of 2001–2008 (9,2 × 10^3, 8,7 × 10^3 and 0,5 × 10^3 m^3) and 2008–2010 (18 × 10^3, 7 × 10^3 and 11 × 10^3 m^3). The values corresponding to the periods 1984–1992 and 1992–2001 are clearly higher than the uncertainties in Table 4, but the values of the periods 1956–1984 and 2001–2008 are lower than the uncertainties. The values of the period 2008–2010 are on the same order as the uncertainties, higher in depleted material and lower in accumulated and waste materials.

The annual rates considering the landslide starting after 1984 show values over 10.7 × 10^3 m^3/year of depletion, near 4.3 × 10^3 m^3/year of accumulation, and about 6.4 × 10^3 m^3/year of mass waste. By periods, the annual rates are highest in the interval 1984–1992 (21 × 10^3 m^3/year of depleted material, almost 7 × 10^3 m^3/year of accumulated material and 15 × 10^3 m^3/year of waste material), followed by the intervals of 1992–2001 and 2008–2010, with rates near to those of the whole period (9.2–9.5 × 10^3, 3.6–5.1 × 10^3 and 4.4–5.6 × 10^3 m^3/year, respectively), and much lower in 2001–2008 (1.3 × 10^3, 1.2 × 10^3 and 0.1 × 10^3 m^3/year) and 1956–1992, when the mass losses are practically null.

Table 7. Volumetric calculations between DTMs.

Period	Volumes [1]			Rates [2]		
	DEP	ACC	WAS	DEP	ACC	WAS
1956–1984	−24,503	16,803	−7700	−875	600	−275
1984–1992	−168,806	52,765	−116,041	−21,101	6596	−14,505
1992–2001	−85,784	46,182	−39,602	−9532	5131	−4400
2001–2008	−9252	8705	−547	−1322	1244	−78
2008–2010	−18,389	7124	−11,265	−9195	3562	−5632
1956–2010	−292,333	117,707	−174,626	−5414	2180	−3234
1984–2010	−280,231	112,776	−167,455	−10,778	4338	−6441

[1] Volumes are in m^3; [2] Volume rates are in m^3/year. Volumes and rates: DEP: Depletion; ACC: Accumulation; WAS: Waste.

5. Discussion

As the landslide does not appear in aerial photographs from the 1956 and 1984 flights, the origin of the landslide had to be after 1984, probably in 1989, the rainiest year in this period [75]. However, due to the uncertainty about this date, the calculations presented in Tables 6 and 7 are made assuming the landslide origin in 1984, so the rate values should be considered as minimum values for the period 1984–1992.

The development of recent features, such as a clean scarp, steps and terraces, minor scarps, and an accumulation zone invading the river channel, can be clearly observed in the orthophotographs and models from 1992 to 2010, which evidences significant activity in the last 30 years. As mentioned before, other features identified in the orthoimages of 1956 and 1984 such as debris flows and the development of rills and gullies allow the zone to be described as unstable, with some evidence of older activity upslope.

Furthermore, landslides and these other features evidence a relevant geomorphic activity in the study area, confirmed by large mass losses associated with seasonal "rambla" flash-flooding events. Thus, the recorded negative displacements or volume changes (ground decrease or mass erosion) are larger than the positive changes (ground elevation or mass accumulation), so it can be deduced that the resulting mass losses are due to their evacuation by river flows. The great volume of mass waste is consistent with the significant geomorphic activity and the high indexes of tectonic activity along the southern flank of the Sierra Nevada, where the Almegíjar landslide is set [58]. This currently visible activity is related to the overall tectonic evolution of the region since Late Neogene and through the Holocene [76].

The parameters related to the landslide and determined from the calculation and comparison of models, longitudinal sections, and other evidence (Tables 5–7 and Figures 7 and 8) allow for the assessment of its dimensions and morphology, as well as the identification of trends concerning age, velocity, activity, diachroneity, and intensity following published classifications [63,73,74,77–81].

Thus the affected area is about 75 × 10^3 m^2, with a balanced distribution of depletion and accumulation areas (near 50%, although with a certain predominance of the depletion areas) that leads to a movement of rockslide type without development of flow mechanism, in which the material can expand downslope. In this sense, according to [60], the landslide, reactivated in the period 2009–2010, shows a strain shortening along the slope with an extension perpendicular to this axis, resulting in a bulging in the lower sector of the displaced mass, following a mechanism described by [61]. The morphology of the landslide scarp and the profiles suggest a planar sliding surface and a mass thickness ranging from 40 to 70 m, partially eroded and affected by debris flow of the weathered rock-mass surface. According to it, the movement can be classified as a rock translational or planar slide [74,77] without evidence of a rotational sliding component in the models and sections. The mobilized (depleted) mass has a volume of about 300×10^3 m^3, which leads us to categorize the landslide as "moderately large" [63,78,80]. Part of the mobilized mass is now displaced downslope, while another part invaded the river channel and was evacuated by the water flow, which has also limited the development of the accumulation areas.

The age is catalogued as "contemporary" [63] and its velocity—in average terms—can be considered as "very slow" [73,74], with vertical displacements of 0.32 and 0.15 m/year in the depleted and in the accumulated area, respectively, over the whole period of 26 years (1984–2010). However, as discussed below, this velocity is not uniform along this period and the landslide has shown periods of different rates of movement, with inactive and active phases, where the velocity ranges from "extremely slow" to "very slow." Thus, more difficult is the assessment of the landslide activity and diachroneity, because the results do not provide a detailed pattern of active periods, which is particularly difficult in these regions where the triggering factors, mainly the rainfall, do not follow a regular or seasonal pattern. However, the analysis of the differential changes between DTMs for the five time periods considered, and some additional data from terrestrial laser scanner (TLS) techniques [60], make it possible to assess certain activity features of interest.

The style of activity [81] is "single," corresponding to the displacement of a well-delimited mass in this single planar slide. On the other hand, the distribution of the activity may be considered "in advance" and also "confined," as the failure plane—shown only at the slide crown of the main scarp—is below the mass at slide toe. Regarding the state of activity [81], the movement can be considered active, although alternating periods of low activity with more active periods during ephemeral floods of the "rambla" [53,58,59].

Furthermore, the results described in the previous section show some periods of activity with displacements rates—measured in the vertical direction—on the order of decimeters by year. Regarding them, the accuracy of the technique has to be taken into account. The accuracy has been established in this work from errors of the flight orientation at check points that allow the estimation of the uncertainty of DTMs and differential DTMs. This uncertainty ranges from 0.36 m in the most favorable case (2008–2010) to 0.95 m in the least favorable (1956–1984), with the other cases between 0.60–0.70 m. Thus the vertical displacements measured in the periods 1984–1992 and 1992–2001, clearly higher than uncertainty, can be considered significant, but those measured in the periods 1956–1984 and 2001–2008 (next to zero values) are considered non-significant. The displacements calculated in the short period of 2008–2010—slightly higher than uncertainty—are at the limit of accuracy of the analysis, although additional data confirm the activity in this period, as will be discussed below. The volumetric accuracy and calculations agree with this, and therefore the depleted, accumulated, and waste volumes in the periods 1984–1992 and 1992–2001 are significant (much larger than the uncertainty), while the other periods (1956–1984, 2001–2008, and 2008–2010) are not significant (lower or similar to the uncertainty).

With this premise, the movement started after 1984 with the maximum rates observed (0.46 m/year of descent in the head or depleted area and 0.39 m/year in the foot or accumulation area), and slowed in the next period of 1992–2001 when lower displacements are observed (0.32 and 0.20 m/year, respectively), until the landslide seemed to stop in the period of 2001–2008, when the measured displacements are not significant. After that, in the last period of 2008–2010 a reactivation of the landslide occurred with displacement rates of about 0.21 and 0.12 m/year. The volumetric calculations confirm this evolution, with volume rates ranging between 10 to 20×10^3 m^3/year in the active periods and values near to 0 in the inactive periods.

At this point, we must mention the complementary results found in this landslide by the combined use of Terrestrial Laser Scanner (TLS) and Global Navigation Satellite Systems (GNSS) between 2008 and 2010 [60], which showed a reactivation in the period March 2009 to June 2010—coinciding with a year of intense rainfall—after an inactive period between September 2008 and March 2009. In the reactivation period a downward movement of 1.20 m at the top was measured, which corresponds to a descent rate of 0.96 m/year; meanwhile, below the middle part of the mass, 1.30 m of advance was established with a displacement rate of 1.04 m/year. These values of the annual rates about 1 m/year express periods of greater activity with a velocity near to that catalogued as "slow" (higher than 1 m/year) between periods of much lower activity in which the velocity is "very slow" to "extremely slow". These data suggest that other vertical displacements and volumes measured between the flights used in this study could have taken place in shorter intervals of time than the periods considered in this work.

Thus, in a more speculative way, if the date of 1989 were considered as the starting point of the landslide—triggered by a period of heavy rainfall [75]—the estimated rates of displacements calculated for the period of 1984–1992 (decimeters by year) would increase to 1.7 m/year in the depleted area and almost 1 m/year in the accumulation area (velocity catalogued as "slow"), on the same order of magnitude as those calculated with TLS measurements [60]. If the period in which the displacement occurred were even shorter, the displacement rates could be even higher, which expresses a large initial impulse for the landslide. The same situation could have occurred in other active periods such as 1992–2001, in relation to the heavy rainfall in the years 1995 to 1998, particularly in the winter of 1996–1997. Again, the displacement rates could increase to values near 1 m/year or higher.

For these landslides of discontinuous activity, an interesting concept is the activity duration or diachroneity [63]. According to the scale established in these works, the degree of diachroneity assigned to the studied landslide is VIII (10 to 100 years). Nevertheless, if only the periods of activity are considered, the degree of diachroneity would be VI (1 month to 1 year).

The triggering mechanism of the landslide was the rainfall, as in other regions near to it [55] and in general in Mediterranean countries [82]. In these areas the movements are triggered occasionally by "rambla" flash floods cutting down through the confined planar slide mass and dragging out the mass accumulation [75,83]. These flood events occur at the beginning of autumn with heavy rain (cold drop phenomenon) or at the beginning of spring after a rainy winter, in which heavy rains coincide with snow melting from the Sierra Nevada range.

Thus, in an approximate way, a relationship between the landslide activity and the rainfalls can be established. The landslide probably was triggered in relation to a rainy event of about 800 mm between November 1989 and January 1990 (more than 400% of the average rainfall for this period). Other equivalent periods occurred between November 1996 and January 1997 with more than 600 mm (more than 300% of the average) [53,55,59,75,83], and in the early months of 2010 with 500 mm between December 2009 and March 2010 (more than 200% of the average rainfall). These events and the general rainfalls of the corresponding years (1989–1990, 1995–1998, and 2009–2011) are related to the higher activity of the periods 1984–1992, 1992–2001, and 2008–2010, respectively. Meanwhile, the lesser activity in the period 2001–2008, when minimum displacements and volumetric changes were recorded, is associated with the absence of relevant precipitation.

This irregular pattern of landslides and rainfall coincides with other evidence such as field observations and landslide inventories along the Betic Cordillera since 1979 [54–57], particularly in the vicinity of this landslide [53,58,59,75,83]. The analysis of the available regional climatic data, reliable in the region since 1940, shows some intense rainy periods similar to those recorded in 1989, 1997, and 2010 [54], triggering slope failures in the study zone. These rainfall patterns, spanning periods of two or three years, register intense peaks in some months. First, we can see five periods exceeding 400 mm in three months, when the mean annual rainfall is 620 mm/y, and also another four periods of accumulated rainfall of more than 300 mm. Thus, the return period for rainfall triggering landslides in the region is around 15 years for the more intense events and eight years if less intense events are taken into account. Other analyses, based on average annual rainfall, show return periods of 18 years for annual precipitation higher than 950 mm, and five years for precipitation higher than 750 mm, these reaching the threshold of new landslides in the region [84]. Finally, recent studies based on the analysis of the anomalies in antecedent cumulative rainfall establish return periods of 12.4 years for these anomalies [85], a value similar to the previous studies.

Comparing these data with climatic studies on the origin of the precipitation in southern Spain, the intense rainfall in the region appears to be related to a negative value of the North Atlantic Oscillation index (NAO index, [86]). Positive values, which are more frequent, are related to more intense precipitation in Northern Europe and drier periods with rainfall in the Mediterranean area. As the periodicity of winter values of this NAO index is eight years [87], this can be seen as coincident with the return periods identified.

6. Conclusions

In conclusion, the real usefulness of LiDAR and digital photogrammetry techniques, as well as the use of differential DEMs, is demonstrated in landslide studies, particularly concerning the determination of the temporal evolution of landslides. The accuracy and consistency of these techniques enables the detection of very detailed slope features and their changes. In this work a new methodology has been developed, based on the use of a stereo photogrammetric workstation for outdating DTMs of more recent flights by editing them only in unstable zones (with clear changes between DTMs), over the stereoscopic models of previous flights. This permits us to reduce the time spent editing DTMs and especially to avoid noise in the differential model calculations.

Concerning the results, the methodology has enabled the detection and quantification of the extent of depletion and accumulation areas of the landslide, with average vertical displacement of about 8.5 m in the depletion area and 4 m in the accumulation area, attaining maximum displacements of almost 30 m in both areas. These displacements can be considered significant because they are an order of magnitude higher than the estimated uncertainty. Also, significant volumetric changes of about 300×10^3 m^3 in depletion area and 120×10^3 m^3 in accumulation areas were quantified, with some 175×10^3 m^3 of materials evacuated afterwards by the river flow.

The displacement and volumetric change rates are not constant over time, attaining maximum values in the initial period 1984–1992 (displacements of about 40 cm/year) and more moderate in the periods 1992–2001 and 2008–2010 (displacements of 15–30 cm/year), during seasons of heavy rainfalls, particularly in 1997 and 2010. The lowest rates were recorded in the period 2001–2008, when the displacement hardly reached 4 cm/year in a long dry period and they can be considered non-significant regarding the accuracy of the technique used. Other observations and data from TLS lead to estimates of displacement rates of higher than 1 m/year in some recent reactivation (2009–2010), similar to that found during the landslide-triggering phase. This difference in slide activity features diachronic slope movements alternating periods of variable activity depending on the influence of triggering factors. From these data, an "extremely to very slow and slow" slide velocity is calculated.

Nevertheless, current research could provide more detailed results on the slide displacements and evolution by using more flights, mainly available in the last decade, and incorporating TLS and UAV photogrammetric data, as well as displacement wireless sensor networks (WSN) and local rainfall data, for easier forecasts of future slide displacements. On the other hand, other techniques such as the determination of displacement vectors in significant slope points or the calculation of absolute distances provide more precise information on the slide kinematics. Finally, regional quantitative studies of other landslides are also needed to draw better quality landslide-hazard maps.

Acknowledgments: This research was funded by the projects P06-RNM-02125 and RNM-06862 (ISTEGEO) funded by the Andalusian Research Plan, projects CGL2008-04854 and TIN2009-09939 funded by the Ministry of Science and Innovation of Spain and Research Groups TEP-213, and RNM 221 of the Andalusian Research Plan.

Author Contributions: Tomás Fernández, José Luis Pérez, Javier Cardenal and Clemente Irigaray conceived and organized the research activity. José Luis Pérez, Javier Cardenal, and Carlos Colomo processed the photogrammetric data, generated the DTMs, and made the calculations (differential models and volumes). Tomás Fernández, José Antonio Palenzuela and Clemente Irigaray made the GIS analysis and Tomás Fernández interpreted the resulting data. Jorge Delgado and José Chacón supervised the research activity. All authors contributed to the writing of this manuscript.

Conflicts of Interest: The authors declare no conflicts of interest.

References

1. Fernández, T.; Pérez, J.L.; Colomo, C.; Mata, E.; Delgado, J.; Cardenal, J.; Irigaray, C.; Chacón, J. Digital photogrammetry and LiDAR techniques to study the evolution of a landslide. In Proceedings of the 8th International Conference on Geo-information for Disaster Management: Best Practices, Enschede, The Netherlands, 13–14 December 2012; Zlatanova, S., Peters, R., Fendel, E.M., Eds.; Universiteit Twente: Enschede, The Netherlands, 2012; pp. 95–104.
2. Mantovani, F.; Soeters, R.; van Westen, C.J. Remote Sensing Techniques for Landslide Studies and Hazard Zonation in Europe. *Geomorphology* **1996**, *15*, 213–225. [CrossRef]
3. Chacón, J.; Irigaray, C.; Fernández, T.; El Hamdouni, R. Engineering geology maps: Landslides and Geographical Information Systems (GIS). *Bull. Eng. Geol. Environ.* **2006**, *65*, 341–411. [CrossRef]
4. Metternicht, G.; Hurni, L.; Gogu, R. Remote sensing of landslides: An analysis of the potential contribution to geo-spatial systems for hazard assessment in mountainous environments. *Remote Sens. Environ.* **2005**, *98*, 284–303. [CrossRef]
5. Tofani, V.; Hong, Y.; Singhroy, V. Introduction: Remote Sensing Techniques for Landslide Mapping and Monitoring. In *Landslide Science for a Safer Geoenvironment*; Sassa, K., Canuti, P., Yin, Y., Eds.; Springer: Cham, Switzerland, 2014; Part III; pp. 301–303.
6. Hapke, C.J. Estimation of Regional Material Yield from Coastal Landslides Based on Historical Digital Terrain Modelling. *Earth Surf. Process. Landf.* **2005**, *30*, 679–697. [CrossRef]
7. Prokešová, R.; Kardoš, M.; Medveďová, A. Landslide dynamics from high-resolution aerial photographs: A case study from the Western Carpathians, Slovakia. *Geomorphology* **2010**, *115*, 90–101. [CrossRef]
8. Chandler, J.H.; Brunsden, D. Steady state behaviour of the Black Ven Mudslide: The application of archival analytical photogrammetry to studies of landform change. *Earth Surf. Process. Landf.* **1995**, *20*, 255–275. [CrossRef]
9. Gentili, G.; Giusti, E.; Pizzaferri, G. Photogrammetric Techniques for the Investigation of the Corniglio Landslide. In *Applied Geomorphology*; Allison, R.J., Ed.; John Wiley & Sons: Chichester, UK, 2002; pp. 39–48.
10. Casson, B.; Delacourt, C.; Allemand, P. Contribution of multi-temporal remote sensing images to characterize landslide slip surface: Application to the La Clapière landslide (France). *Nat. Hazards Earth Syst. Sci.* **2005**, *5*, 425–437. [CrossRef]
11. Casson, B.; Delacourt, C.; Baratoux, D.; Allemand, P. Seventeen Years of the "La Clapière" Landslide Evolution Analysed from Ortho-Rectified Aerial Photographs. *Eng. Geol.* **2003**, *68*, 123–139. [CrossRef]
12. Van Westen, C.J.; Getahun, F.L. Analyzing the evolution of the Tessina landslide using aerial photographs and digital elevation models. *Geomorphology* **2003**, *54*, 77–89. [CrossRef]
13. Walstra, J.; Chandler, J.H.; Dixon, N.; Dijkstra, T.A. Time for Change—Quantifying Landslide Evolution Using Historical Aerial Photographs and Modern Photogrammetric Methods. In *The International Archives of the Photogrammetry, Remote Sensing and Spatial Information Sciences, Proceedings of the XXth ISPRS Congress, Istanbul, Turkey, 12–23 July 2004*; Altan, O., Ed.; Copernicus Publications: Gottingen, Germany, 2004; Volume XXXV, Part B4; pp. 475–480.
14. Walstra, J.; Chandler, J.H.; Dixon, N.; Dijkstra, T.A. Extracting Landslide Movements from Historical Aerial Photographs. In *Landslides: Evaluation and Stabilization*; Lacerda, W., Erlich, M., Fontoura, S.A.B., Sayao, A.S.F., Eds.; Taylor & Francis: London, UK, 2004; pp. 843–850.
15. Walstra, J.; Chandler, J.H.; Dixon, N.; Dijkstra, T.A. Aerial Photography and Digital Photogrammetry for Landslide Monitoring. *Geol. Soc. (Lond.) Spec. Publ.* **2007**, *283*, 53–63. [CrossRef]
16. Walstra, J.; Dixon, N.; Chandler, J.H. Historical aerial photographs for landslide assessment: Two case histories. *Q. J. Eng. Geol. Hydrogeol.* **2007**, *40*, 315–332. [CrossRef]
17. Baldi, P.; Fabris, M.; Monticelli, R. Monitoring the morphological evolution of the Sciara Del Fuoco during the 2002–2003 Stromboli eruption using multi-temporal photogrammetry. *ISPRS J. Photogramm. Remote Sens.* **2005**, *59*, 199–211. [CrossRef]

18. Baldi, P.; Cenni, N.; Fabris, M.; Zanutta, A. Kinematics of a landslide derived from archival photogrammetry and GPS data. *Geomorphology* **2008**, *102*, 435–444. [CrossRef]
19. Cardenal, J.; Delgado, J.; Mata, E.; González-Díez, A.; Remondo, J.; Díaz de Terán, J.R.; Francés, E.; Salas, L.; Bonachea, J.; Olague, I.; et al. The use of digital photogrammetry techniques in landslide instability. In *Geodetic Deformation Monitoring: From Geophysical to Geodetic Roles*; Gil-Cruz, J., Sanso, F., Eds.; IAG Springer Series: New York, NY, USA; Heidelberg, Germany, 2006; pp. 259–264.
20. Kasperski, J.; Delacourt, C.; Allemand, P.; Potherat, P. Evolution of the Sedrun landslide (Graubünden, Switzerland) with ortho-rectified air images. *Bull. Eng. Geol. Environ.* **2010**, *69*, 421–430. [CrossRef]
21. Fabris, M.; Menin, A.; Achilli, V. Landslide displacement estimation by archival digital photogrammetry. *Ital. J. Remote Sens.* **2011**, *43*, 23–30. [CrossRef]
22. González-Díez, A.; Fernández-Maroto, G.; Doughty, M.W.; Díaz de Terán, J.R.; Bruschi, V.; Cardenal, J.; Pérez, J.L.; Mata, E.; Delgado, J. Development of a methodological approach for the accurate measurement of slope changes due to landslides, using digital photogrammetry. *Landslides* **2014**, *11*, 615–628. [CrossRef]
23. Dewitte, O.; Jasselette, J.C.; Cornet, Y.; Van Den Eeckhaut, M.; Collignon, A.; Poesen, J.; Demoulin, A. Decadal-scale analysis of ground movements in old landslides in western Belgium. *Eng. Geol.* **2008**, *99*, 11–22. [CrossRef]
24. Corsini, A.; Borgatti, L.; Cervi, F.; Dahne, A.; Ronchetti, F.; Sterzai, P. Estimating mass-wasting processes in active earth slides—Earth flows with time-series of High-Resolution DEMs from photogrammetry and airborne LiDAR. *Nat. Hazards Earth Syst. Sci.* **2009**, *9*, 433–439. [CrossRef]
25. Mora, P.; Baldi, P.; Casula, G.; Fabris, M.; Ghirotti, M.; Mazzini, E.; Pesci, A. Global Positioning Systems and Digital Photogrammetry for the Monitoring of Mass Movements: Application to the Ca' Di Malta Landslide (Nothern Apennines, Italy). *Eng. Geol.* **2003**, *68*, 103–121. [CrossRef]
26. Brückl, E.; Brunner, F.K.; Kraus, K. Kinematics of a deep-seated landslide derived from photogrammetric, GPS and geophysical data. *Eng. Geol.* **2006**, *88*, 149–159. [CrossRef]
27. De Bari, C.; Lapenna, V.; Perrone, A.; Puglisi, C.; Sdao, F. Digital photogrammetric analysis and electrical resistivity tomography for investigating the Picerno landslide (Basilicata region, southern Italy). *Geomorphology* **2011**, *133*, 34–46. [CrossRef]
28. Yeh, M.L.; Hsiao, Y.C.; Chen, Y.H.; Chung, J.C. A study on unmanned aerial vehicle applied to acquire terrain information of landslide. In Proceedings of the 32 Asian Conference Remote Sensing, Taipei, Taiwan, 3–7 October 2011; Volume 3, pp. 2210–2215.
29. Niethammer, U.; James, M.R.; Rothmund, S.; Travelletti, J.; Joswig, M. UAV-based remote sensing of the Super-Sauze landslide: Evaluation and results. *Eng. Geol.* **2012**, *128*, 2–11. [CrossRef]
30. Stumpf, A.; Malet, J.P.; Kerle, N.; Niethammer, U.; Rothmund, S. Image-based mapping of surface fissures for the investigation of landslide dynamics. *Geomorphology* **2013**, *186*, 12–27. [CrossRef]
31. Turner, D.; Lucieer, A.; de Jong, S.M. Time Series Analysis of Landslide Dynamics using an Unmanned Aerial Vehicle (UAV). *Remote Sens.* **2015**, *7*, 1736–1757. [CrossRef]
32. Fernández, T.; Pérez, J.L.; Cardenal, F.J.; Gómez, J.M.; Colomo, C.; Delgado, J. Analysis of landslide evolution affecting olive groves using UAV and photogrammetric techniques. *Remote Sens.* **2016**, *8*, 837.
33. Peterman, V. Landslide activity monitoring with the help of unmanned aerial vehicle. In *The International Archives of the Photogrammetry Remote Sensing and Spatial Information Sciences, Proceedings of the International Conference on Unmanned Aerial Vehicles in Geomatics, Toronto, Canada, 30 August–2 September 2015*; Armenakis, E., Ed.; Copernicus Publications: Gottingen, Germany, 2015; Volume XL-1/W4, pp. 215–218.
34. Vrublova, D.; Kapica, R.; Jirankova, E.; Strus, A. Documentation of landslides and inaccessible parts of a mine using an unmanned UAV system and methods of digital terrestrial photogrammetry. *GeoSci. Eng.* **2015**, *LXI*, 8–19.
35. Yang, Z.; Lan, H.; Liu, H.; Li, L.; Wu, Y.; Meng, Y.; Xu, L. Post-earthquake rainfall-triggered slope stability analysis in the Lushan area. *J. Mt. Sci.* **2015**, *12*, 232–242. [CrossRef]
36. Shi, B.; Liu, C. UAV for Landslide Mapping and Deformation Analysis. In Proceedings of the International Conference on Intelligent Earth Observing and Applications, Guilin, China, 23–24 October 2015; Zhou, G., Kang, C., Eds.; Volume 9808.
37. Lindner, G.; Schraml, K.; Mansberger, R.; Hübl, J. UAV monitoring and documentation of a large landslide. *Appl. Geomat.* **2016**, *8*, 1–11. [CrossRef]

38. Hsieh, Y.C.; Chan, Y.; Hu, J. Digital Elevation Model Differencing and Error Estimation from Multiple Sources: A Case Study from the Meiyuan Shan Landslide in Taiwan. *Remote Sens.* **2016**, *8*, 199. [CrossRef]
39. Al-Rawabdeh, A.; He, F.; Moussa, A.; El-Sheimy, N.; Habib, A. Using an Unmanned Aerial Vehicle-Based Digital Imaging System to Derive a 3D Point Cloud for Landslide Scarp Recognition. *Remote Sens.* **2016**, *8*, 95. [CrossRef]
40. Walstra, J.; Chandler, J.H.; Dixon, N.; Wackrow, R. Evaluation of the controls affecting the quality of spatial data derived from historical aerial photographs. *Earth Surf. Process. Landf.* **2010**, *36*, 853–863. [CrossRef]
41. Adams, J.C.; Chandler, J.H. Evaluation of LiDAR and Medium Scale Photogrammetry for Detecting Soft-Cliff Coastal Change. *Photogramm. Rec.* **2002**, *17*, 405–418. [CrossRef]
42. Glenn, N.F.; Streutker, D.R.; Chadwick, D.J.; Thackray, G.D.; Dorsch, S.J. Analysis of LiDAR-derived topographic information for characterizing and differentiating landslide morphology and activity. *Geomorphology* **2006**, *73*, 131–148. [CrossRef]
43. Derron, M.H.; Jaboyedoff, M. LIDAR and DEM techniques for landslides monitoring and characterization. *Nat. Hazards Earth Syst. Sci.* **2010**, *10*, 1877–1879. [CrossRef]
44. Palenzuela, J.A.; Marsella, M.; Nardinocchi, C.; Pérez, J.L.; Fernández, T.; Chacón, J.; Irigaray, C. Landslide detection and inventory by integrating LiDAR data in a GIS environment. *Landslides* **2015**, *12*, 1035–1050. [CrossRef]
45. McKean, J.; Roering, J. Objective landslide detection and surface morphology mapping using high resolution airborne laser altimetry. *Geomorphology* **2004**, *57*, 331–351. [CrossRef]
46. Kasai, M.; Ikeda, M.; Asahina, T.; Fujisawa, K. LiDAR-derived DEM evaluation of deep-seated landslides in a steep and rocky region of Japan. *Geomorphology* **2009**, *113*, 57–69. [CrossRef]
47. Burns, W.J.; Coe, J.A.; Kaya, B.S.; Ma, L. Analysis of Elevation Changes Detected from Multi-Temporal LiDAR Surveys in Forested Landslide Terrain in Western Oregon. *Environ. Eng. Geosci.* **2010**, *16*, 315–341. [CrossRef]
48. DeLong, S.B.; Prentice, C.S.; Hilley, G.E.; Ebert, Y. Multitemporal ALSM change detection, sediment delivery, and process mapping at an active earthflow. *Earth Surf. Proc. Land.* **2012**, *37*, 262–272. [CrossRef]
49. Roering, J.J.; Mackey, B.H.; Marshall, J.A.; Sweeney, K.E.; Deligne, N.I.; Booth, A.M.; Handwerger, A.L.; Cerovski-Darriau, C. "You are HERE": Connecting the dots with airborne lidar for geomorphic fieldwork. *Geomorphology* **2013**, *200*, 172–183. [CrossRef]
50. Cavalli, M.; Trevisani, S.; Comiti, F.; Marchi, L. Geomorphometric assessment of spatial sediment connectivity in small Alpine catchments. *Geomorphology* **2013**, *188*, 31–41. [CrossRef]
51. Tarolli, P. High-resolution topography for understanding Earth surface processes: Opportunities and challenges. *Geomorphology* **2014**, *216*, 295–312. [CrossRef]
52. Bossi, G.; Cavalli, M.; Crema, S.; Frigerio, S.; Quan Luna, B.; Mantovani, M.; Marcato, G.; Schenato, L.; Pasuto, A. Multi-temporal LiDAR-DTMs as a tool for modelling a complex landslide: A case study in the Rotolon catchment (eastern Italian Alps). *Nat. Hazards Earth Syst. Sci.* **2015**, *15*, 715–722. [CrossRef]
53. Fernández, T.; Irigaray, C.; El Hamdouni, R.; Chacón, J. Methodology for landslide susceptibility mapping by means of a GIS. Application to the Contraviesa area (Granada, Spain). *Nat. Hazards* **2003**, *30*, 297–308. [CrossRef]
54. Fernández, T.; Ureña, M.A.; Delgado, J.; Cardenal, J.; Irigaray, C.; Chacón, J. Examples of natural risk analysis from SDI. In Proceedings of the International Cartographic Conference (CO-022), Paris, France, 21–27 June 2011.
55. Irigaray, C.; Lamas, F.; El Hamdouni, R.; Fernández, T.; Chacón, J. The importance of the precipitation and the susceptibility of the slopes for the triggering of landslides along the roads. *Nat. Hazards* **2000**, *21*, 65–81. [CrossRef]
56. Irigaray, C.; Fernández, T.; El Hamdouni, R.; Chacón, J. Evaluation and validation of landslide susceptibility maps obtained by a GIS matrix method: Examples from the Betic Cordillera (southern Spain). *Nat. Hazards* **2007**, *41*, 61–79. [CrossRef]
57. Chacón, J. Landslide susceptibility, hazard & risk GIS mapping in the Betic Cordillera (Spain): Areas with limited information about triggering factors. In *Guidelines for Mapping Areas at Risk of Landslides in Europe, Proc. Experts Meeting Joint Research Center (JRC), Ispra, Italy, 23–24 October 2007*; Hervás, J., Ed.; Office for Official Publications of the European Communities: Luxembourg, Luxembourg, 2007; pp. 23–26.

58. El Hamdouni, R.; Irigaray, C.; Fernández, T.; Chacón, J.; Keller, E.A. Assessment of relative active tectonics, southwest border of the Sierra Nevada (Southern Spain). *Geomorphology* **2008**, *96*, 150–173. [CrossRef]
59. Jiménez-Perálvarez, J.D.; Irigaray, C.; El Hamdouni, R.; Chacón, J. Landslide-susceptibility mapping in a semi-arid mountain environment: An example from the southern slopes of Sierra Nevada (Granada, Spain). *Bull. Eng. Geol. Environ.* **2011**, *70*, 265–277. [CrossRef]
60. Palenzuela, J.A.; Irigaray, C.; Jiménez-Perálvarez, J.D.; Chacón, J. Application of Terrestrial Laser Scanner to the Assessment of the Evolution of Diachronic Landslides. In *Landslide Science and Practice*; Margottini, C., Canuti, P., Sassa, K., Eds.; Springer: Berlin/Heidelberg, Germany, 2013; Section 68; pp. 517–523.
61. Agliardi, F.; Crosta, G.; Zanchi, A. Structural constraints on deep-seated slope deformation kinematics. *Eng. Geol.* **2001**, *59*, 83–102. [CrossRef]
62. Aldaya, F.; Díaz de Federico, A.; García-Dueñas, V.; Martínez García, E.; Navarro-Vila, F.; Puga, E. *Mapa Geológico de España 1:50.000: Hoja 1042 Lanjarón*; 2ª serie; IGME: Madrid, Spain, 1981; p. 32.
63. Chacón, J.; Irigaray, C.; El Hamdouni, R.; Jiménez-Perálvarez, J.D. Diachroneity of landslides. In *Geologically Active*; Williams, A.L., Pinches, G.M., Chin, C.Y., McMorran, T.J., Massey, C.I., Eds.; CRC Press/Balkema, Taylor & Francis Group: Leiden, The Netherlands, 2010; pp. 999–1006.
64. Colomo-Jiménez, C.; Pérez-García, J.L.; Fernández-del-Castillo, T.; Gómez-López, J.M.; Mozas-Calvache, A.M. Methodology for orientation and fusion of photogrammetric and LiDAR data for multitemporal studies. In *The International Archives of the Photogrammetry, Remote Sensing and Spatial Information Sciences, Proceedings of the XXIII ISPRS Congress, Prague, Czech Republic, 12–19 July 2016*; Copernicus Publications: Gottingen, Germany, 2016; Volume XLI-B7, pp. 639–645.
65. *Socet Set 5.2*; Bae Systems Plc.: London, UK, 2009.
66. *Terramatch*; TerraSolid Ltd.: Helsinki, Finland, 2012.
67. Oshel, E.; Liddle, D. Photogrammetric applications: Space Application of Photogrammetry. In *Manual of Photogrammetry*, 6th ed.; McGlone, C., Mikhail, E., Bethel, J., Eds.; American Society for Photogrammetry and Remote Sensing (ASPRS): Bethesda, MD, USA, 2013; p. 1183.
68. Cavalli, M.; Tarolli, P. Application of LiDAR technology for rivers analysis. *Ital. J. Eng. Geol. Environ.* **2011**, *1*, 33–44.
69. *Maptek I-Site Studio 6.0*; Maptek Pty. Ltd.: Adelaide, Australia, 2016.
70. *ArcGIS 10.3*; Esri: Redlands, CA, USA, 2013.
71. Brasington, J.; Rumsby, B.T.; McVey, R.A. Monitoring and modelling morphological change in a braided gravel-bed river using high resolution GPS-based survey. *Earth Surf. Proc. Land.* **2000**, *25*, 973–990. [CrossRef]
72. Wheaton, J.M.; Brasington, J.; Darby, S.E.; Sear, D.A. Accounting for uncertainty in DEMs from repeat topographic surveys: Improved sediment budgets. *Earth Surf. Proc. Land.* **2010**, *35*, 136–156. [CrossRef]
73. WP/WLI. A suggested method for describing the rate of movement of a landslide. *Bull. Eng. Geol. Environ.* **1995**, *52*, 75–78.
74. Hungr, O.; Leroueil, S.; Picarelli, L. The Varnes classification of landslide types, an update. *Landslides* **2014**, *11*, 167–194. [CrossRef]
75. Thornes, J.B.; Alcántara-Ayala, I. Modelling mass failure in a Mediterranean mountain environment: Climatic, geological, topographical and erosional controls. *Geomorphology* **1998**, *24*, 87–100. [CrossRef]
76. García, A.F.; Zhu, Z.; Ku, T.L.; Sanz de Galdeano, C.; Chadwick, O.A.; Chacón, J. Tectonically driven landscape development within the eastern Alpujarran Corridor, Betic Cordillera, SE Spain (Almería). *Geomorphology* **2003**, *50*, 83–110. [CrossRef]
77. Varnes, D.J. Slope movement, types and processes. In *Landslides: Analysis and Control*; Schuster, R.L., Krizek, R.J., Eds.; Transportation Research Board Special Report National Academy of Sciences: Washington, DC, USA, 1978; Volume 176, pp. 12–33.
78. Fell, R. Landslide risk assessment and acceptable risk. *Can. Geotech. J.* **1994**, *31*, 261–272. [CrossRef]
79. Cardinali, M.; Reichenbach, P.; Guzzetti, F.; Ardizzone, F.; Antonini, G.; Galli, M.; Cacciano, M.; Castellani, M.; Salvati, P. A geomorphological approach to the estimation of landslide hazards and risks in Umbria, Central Italy. *Nat. Hazards Earth Syst. Sci.* **2002**, *2*, 57–72. [CrossRef]
80. Rodríguez-Ortiz, J.M.; Hinojosa, J.A.; Prieto, C. Regional studies on mass movements in Spain. In Proceedings of the 3th IAEG Congress, Madrid, Spain, 4–8 September 1978; Volume 1, pp. 267–277.
81. WP/WLI. A suggested method for describing the activity of a landslide. *Bull. Eng. Geol. Environ.* **1993**, *47*, 53–57.

82. Guzzeti, F. Landslide hazard assessment and risk evaluation: Limits and prospectives. In Proceedings of the 4th EGS Plinius Conference, Mediterranean Storms, Mallorca, Spain, 2–4 October 2002.
83. Alcántara-Ayala, I.; Thornes, J.B. Structure and hydrology in controlling mass failure in space and time: The case of the Guadalfeo failures. In Proceedings of the 8th International Conference & Field Trip on Landslides, Granada, Spain, 27–28 September 1996; Chacón, J., Irigaray, C., Fernández, T., Eds.; Balkema: Rotterdam, The Netherlands, 1996; pp. 89–96.
84. Jimenez-Perálvarez, J.D. Movimientos de ladera en la vertiente meridional de Sierra Nevada (Granada, España): Identificación, análisis y cartografía de susceptibilidad y peligrosidad mediante SIG. Ph.D. Thesis, University of Granada, Granada, Spain, July 2012.
85. Palenzuela, J.A.; Jiménez-Perálvarez, J.D.; Chacón, J.; Irigaray, C. Assessing critical rainfall thresholds for landslide triggering by generating additional information from a reduced database: An approach with examples from the Betic Cordillera (Spain). *Nat. Hazards* **2016**, *84*, 185–212. [CrossRef]
86. Trigo, R.M.; Pozo, D.; Timothy, C.; Osborn, J.; Castro, Y.; Gámiz, S.; Esteban, M.J. NAO influence on precipitation, river flow and water resources in the Iberian Peninsula. *Int. J. Climatol.* **2004**, *24*, 925–944. [CrossRef]
87. Pozo, D.; Esteban, M.J.; Rodrigo, F.S.; Castro, Y. An analysis of the variability of the North Atlantic Oscillation in the time and frequency domains. *Int. J. Climatol.* **2000**, *20*, 1675–1692. [CrossRef]

5

Comparing Manual and Semi-Automated Landslide Mapping based on Optical Satellite Images from Different Sensors

Daniel Hölbling [1,*], Clemens Eisank [2], Florian Albrecht [1], Filippo Vecchiotti [3], Barbara Friedl [1], Elisabeth Weinke [1] and Arben Kociu [3]

1 Department of Geoinformatics–Z_GIS, University of Salzburg, Schillerstrasse 30, 5020 Salzburg, Austria; florian.albrecht@sbg.ac.at (F.A.); barbara.friedl@sbg.ac.at (B.F.); elisabeth.weinke@sbg.ac.at (E.W.)
2 GRID-IT—Gesellschaft für angewandte Geoinformatik mbH, Technikerstraße 21a, 6020 Innsbruck, Austria; eisank@grid-it.at
3 Geological Survey of Austria (GBA), Neulinggasse 38, 1030 Vienna, Austria; filippo.vecchiotti@geologie.ac.at (F.V.); arben.kociu@geologie.ac.at (A.K.)
* Correspondence: daniel.hoelbling@sbg.ac.at

Abstract: Object-based image analysis (OBIA) has been increasingly used to map geohazards such as landslides on optical satellite images. OBIA shows various advantages over traditional image analysis methods due to its potential for considering various properties of segmentation-derived image objects (spectral, spatial, contextual, and textural) for classification. For accurately identifying and mapping landslides, however, visual image interpretation is still the most widely used method. The major question therefore is if semi-automated methods such as OBIA can achieve results of comparable quality in contrast to visual image interpretation. In this paper we apply OBIA for detecting and delineating landslides in five selected study areas in Austria and Italy using optical Earth Observation (EO) data from different sensors (Landsat 7, SPOT-5, WorldView-2/3, and Sentinel-2) and compare the OBIA mapping results to outcomes from visual image interpretation. A detailed evaluation of the mapping results per study area and sensor is performed by a number of spatial accuracy metrics, and the advantages and disadvantages of the two approaches for landslide mapping on optical EO data are discussed. The analyses show that both methods produce similar results, whereby the achieved accuracy values vary between the study areas.

Keywords: landslides; remote sensing; semi-automated mapping; object-based image analysis (OBIA); manual mapping; visual interpretation; spatial accuracy metrics; Alps

---

1. Introduction

Optical Earth Observation (EO) data has proven beneficial for mapping the location and spatial extent of landslides at various spatial scales, ranging from local to regional [1]. The interpretation of optical EO data for landslide recognition may be conducted using manual, semi-automatic, and automatic approaches [2].

Traditionally, manual visual image interpretation for landslide recognition has been based on aerial photographs and recently extended to high resolution (HR) and very high resolution (VHR) optical satellite imagery (e.g., WorldView-2/3, QuickBird) [2] and is still the most widely used method for landslide detection and inventory preparation. Imagery with lower resolutions (e.g., SPOT, Landsat) has also been successfully employed for delineating single, larger landslides [3], yet VHR optical data has become the best choice for landslide recognition on the basis of landform analysis over larger areas [1]. The comparison between landslide maps derived from aerial images and satellite images reveals that both data sources yield results of equal value [4,5]. Manual visual image interpretation of optical imagery turned out to be practicable for landslide mapping, especially where landslides led to significant changes in the land cover, as is usually the case for shallow landslides. Shape, size, pattern, and texture with characteristic contrast in brightness (for black and white photography) or color variation (in case of a multispectral image) are used for manual landslide interpretation [6]. One major advantage related to visual image interpretation is the use of stereo-vision for landslide recognition, as some landslide features are detectable only employing the third dimension [2]. However, manual interpretation reveals several drawbacks; it is resource and time consuming and the quality of the resulting landslide maps strongly depends on the experience and skills of the investigator [7–9]. The variability in human visual interpretation tasks across time and individuals is a common but understated problem and has also been investigated for remote sensing image interpretation of features in urban areas [10], in rural areas [11], and in slums [12]. This subjectivity issue is difficult to measure, but influences the reliability of landslide mapping [13,14], which further depends on the persistence of the landslide morphology, the quality and scale of the imagery, the complexity of the geological and geomorphological setting, the presence of vegetation coverage, and the final map scale [9]. It has been recognized that, when more than one compilation of landslide inventories is closely evaluated, different operators produce unequal results in terms of the number and density of landslides, which also vary in geometry and extent [7,8]. Ardizzone et al. [14] compared three independently produced landslide inventory maps and conclude that the detail and precision of landslide maps primarily depend on the experience and skills of the investigator and only secondly on the quality of the remote sensing data and base maps used.

During the past decade, advances in computer science and machine intelligence and the ever increasing number of satellite missions acquiring VHR data led to the development of novel technologies, fostering the semi-automated interpretation of EO data. The current approaches can be split into two categories [1,15], pixel-based and object-based, both containing techniques applicable to single and multi-temporal images and frequently making use of additional data, e.g., digital elevation models (DEMs) [16]. Pixel-based methods take into account the spectral information associated with each pixel without considering its actual neighborhood [2]. Considering the resolution of HR/VHR data and the size and spatial distribution of landslides, pixel-based methods tend to be additionally sensitive to errors [17], frequently resulting in salt-and-pepper classifications, with single pixels being demarcated as landslides.

Due to these circumstances, there has been a trend towards object-based landslide recognition in recent years, as demonstrated by a range of studies [8,16–29]. Object-based image analysis (OBIA) provides a set of innovative tools for semi-automatically delineating and classifying landslides based on EO data. Object-based algorithms rely on the concepts of image segmentation and classification. Regarding landslide recognition, landslides or landslide features are composed of aggregations of homogeneous pixels rather than spatially uncorrelated cells and are then classified according to specific characteristics. By supporting the use of a plethora of properties during the classification process, e.g., spectral, spatial, textural, and contextual, OBIA is somewhat mimicking human perception and thereby proves its potential for producing more accurate and realistic landslide maps than pixel-based procedures [15,17,18]. OBIA therefore can be considered a specific type of model of the human perception process [30]. This model type performs well in producing an understandable encoding of explicit expert knowledge in rulesets for the production of repeatable image classification results, e.g., as demonstrated by Eisank et al. [31]. At the same time, OBIA is also a simplification of the human perception process and has certain disadvantages. In practice, OBIA workflows are a tradeoff between keeping a low complexity and covering all details and tend to ignore special cases in favor of generally applicable classification rules. Therefore, like manual visual image interpretation, OBIA-derived classification results are not free of error either. Consequently, the accuracy of semi-automated landslide classifications should be documented, but, at the same time, the reliability of the reference should be critically scrutinized. Attainable accuracy values vary considerably between studies, often due to the varying complexity and appearance of landslides. Accuracy values reach up to 95% for the detection of landslide scars and range between 73% and 93.7% for the detection of landslide-affected areas and between 44.8% and 90% for the identification of different landslide types such as debris slides, rock slides, or debris flows [15–22,25–27]. Object-based landslide mapping routines are often tailored to specific study areas and data and thus reveal a lack of transferability and robustness. Also the selection of suitable segmentation parameters for the creation of "meaningful" image objects, which in turn affects the accuracy of image classification as well as the application of user driven thresholds, remain critical issues [27].

In this study, we aim to compare the results of two independently performed landslide mapping approaches; semi-automated detection of landslides by OBIA and manual mapping of landslides. The analyses are based on optical satellite imagery obtained from different sensors, i.e., Landsat 7, SPOT-5, WorldView-2, WorldView-3, and Sentinel-2. Based on five selected case studies in the Alps, spatial accuracy values are calculated to evaluate if the results are comparable when the same data basis is used for analyses.

2. Materials and Methods

2.1. Study Areas

Five landslide prone areas in the Alps with different geomorphic characteristics were selected as test sites for the development and comparison of the mapping approaches. Two study sites are located in the federal state of Salzburg, Austria, two in the federal state of Vorarlberg, Austria, and one in Südtirol-Alto Adige Autonomous Province, Italy (Figure 1). The Italian test site is divided into two sub-areas located in the Gader Valley.

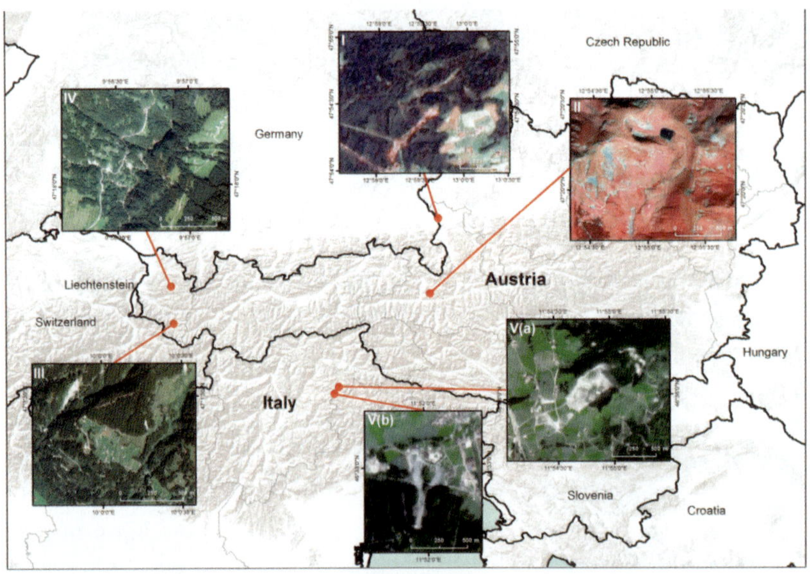

Figure 1. Study areas in Austria and Italy. Study areas I (Fürwag/Haunsberg) and II (Pinzgau) are located in the federal state of Salzburg, Austria; study areas III (Montafon) and IV (Bregenzerwald) are located in the federal state of Vorarlberg, Austria; and study area V (Gader Valley) is in Südtirol-Alto Adige Autonomous Province, Italy. The Italian study area in the Gader Valley is divided in two sub-areas (V(a) and V(b)).

2.1.1. Study Areas I and II in the Federal State of Salzburg

Study areas I and II (Salzburg; Figure 1) are located in a geologic domain that encompasses Quaternary until Permian units. They are respectively [32]:

- The quaternary alluvial and moraine deposits
- The tertiary Eocene-Miocene Molasse
- The Jura-mid Eocene continental margin Helvetic Zone
- The Lower Cretaceous-Eocene Penninic Rhenodanubian Flysch
- The Permo-Mesozoic Northern Calcareous Alps (NCA)

The Permo-Mesozoic Northern Calcareous Alps were deposited in a reef-submerged environment corresponding to the opening of the Penninic Sea. At the very end of the Cretaceous Period, on the active margin of the Austroalpine microplate, as a result of the subduction, an accretionary wedge was formed, and a new deep water flysch-like deposition facies, the Rhenodanubian Flysch, took place [32].

The erosional debris of the Alpine uplift occurring between the Alpine orogenic front and the European foreland formed the tertiary Molasse. During the Würmian, the extent of the Salzach glacier reached the Inn basin to the north. The Salzach valley, an over-deepened Alpine valley, was filled mainly with lacustrine and delta sediments. The melting outlet-glacier deposited big terminal moraines, and the melt water runoff caused strong lateral erosion along the foot of the slopes, which locally provoked deep-seated gravitational slope deformations [33].

The "Fürwag Landslide" (study area I) is located at the Haunsberg, approximately 10 km north of the city of Salzburg, and covers an area of approximately 170 ha. The landslide-prone area consists of various mudslides, which have been formed at different times. The sandstones, shales, and the Flysch, in particular the Zementmergelserie (marl cement series), of this pre-alpine zone are very prone to mass movements [34]. The latest sliding processes occurred between 1999 and 2003, and have almost reached the valley floor, representing a constant threat to major infrastructure facilities [35].

In the Pinzgau region (study area II) the greywacke zone is especially susceptible to sliding processes. The Central Alps are dominated by different metamorphic rock types, leading to a great variety of mass movement types [34]. In June 2013 parts of this region have been declared a disaster zone due to torrential rains, which caused severe flooding and numerous landslides. A mountainous area (the Hundstein Mountain) characterized by a high relief and sparse vegetation, where several geomorphological phenomena occur, has been selected as study area in the Pinzgau region.

2.1.2. Study Areas III and IV in the Federal State of Vorarlberg

The major tectonic units of study areas III and IV (Vorarlberg; Figure 1), can be subdivided from North to South into three groups [36]:

- Molasse
- Helvetic Nappes and Vorarlberg Flysch (Eastern Alps origin)
- Northern Calcareous Alps and Silvretta Crystalline (Western Alps origin)

The gneisses, schists, and amphibolites of the Silvretta Crystalline of the Variscan age are the oldest rocks in the area. The Helvetic limestones and marls (as well as the NCA) were formed in a shallow sea in Mesozoic times, whereas the deep turbiditic origin of the Vorarlberg Flysch dates back to the Late Cretaceous-Early Tertiary Period. The Molasse was deposited instead in the Alpine foreland basin as a residual product of the mountain building process and consists of conglomerate sandstones, claystone and marls of mixed marine and non-marine nature [36].

In the Montafon valley (study area III) tectonic deformation and weakening of rock strength during Alpine orogenesis favors the occurrence of landslides [37]. Additionally, glacial erosion led to the steepening of slopes, triggering tensional fissures and accelerating a number of geomorphological processes, e.g., rock slides and rock falls [38].

In Bregenzerwald (study area IV), the valley slopes were oversteepened by glacial erosion, leading to increased slope instability after deglaciation [39]. In May 2005, intense and prolonged rainfall induced inundations and numerous landslides, leading to severe infrastructure and property damages. Prominent examples of major landslides in the Bregenzerwald area are the Doren landslide and the landslide Rindberg/Sibratsgfäll [40,41].

2.1.3. Study Area V in Südtirol—Alto Adige Autonomous Province

The major lithological formations of study area V (Südtirol-Alto Adige Autonomous Province; Figure 1) are [42]:

- Schiliar dolomites (detritic massive dolomite of Anisian age)
- Wengen formation (marls and calciturbidites mixed with volcaniclastic conglomerates of Ladinian age)
- S. Cassian formation (grey mudstones and calciturbidites of Carnian age)

The geology of the Dolomite area is complex being the result of the dismantling of the Hercinian mountains and Permo-Trias rifting tectonics in Permian times, which gave way to the Athesian magmatism [42]. In Anisian times, the massive detritic Schiliar dolomite was formed as a consequence of a rifting event. After another upper Ladinian paroxistic volcanic phase, the Wengen formation was formed. The successive recovery of carbonate sedimentation allowed the right conditions for the sedimentation of the S. Cassiano Formation [43].

The Gader Valley (study areas V(a) and V(b)) is prone to gravitational processes such as landslides due to its geological and geomorphological conditions, particularly the upper parts of the valley [44]. In December 2012, after long and intensive rainfall, a large landslide destroyed several buildings and endangered critical infrastructure in Badia. Further large landslides are located near the village of Corvara. The Corvara landslide, which affects an area of more than 2.5 km^2 and has an overall volume of more than 300 million m^3 [45], repeatedly causes damages to infrastructure. The most recent landslide happened in April 2014 at the border between Corvara and Colfosco. Some areas on its western part were reactivated in August 2016.

2.2. Data

For each study area, selected subsets of optical satellite images with different properties from different sensors were used (Table 1). Digital elevation models (DEMs), and particularly the derived slope information, were used as supplementary data. The database for Montafon and Bregenzerwald was also complemented by a terrain roughness index (TRI) layer, computed as the sum of change in elevation between a grid cell and its eight neighboring grid cells [46]. For the study areas in Salzburg, a DEM derived from airborne laser scanning (ALS), which is freely available at a resampled resolution of 10 m from the geodata portal of Austria (www.geoland.at), was used. For the two study areas in Vorarlberg, an ALS-based DEM at 5 m spatial resolution provided by the federal government of Vorarlberg was available. The ALS campaigns over the Austrian Alps were conducted between 2006 and 2013 by the regional governments. In Südtirol-Alto Adige Autonomous Province, a DEM at 5 m resolution (acquired between 2004 and 2006) was provided by the Autonomous Province of Bolzano via their geodata portal (http://geoportal.buergernetz.bz.it/default.asp).

Table 1. Selected database for each study area.

Study Area	Size (km^2)	Optical Sensor	Acquisition Date	Spectral Resolution [1]	Spatial Resolution (m)	DEM Resolution (m)
I	4.7	Landsat 7	28 July 2002	1× pan 7× multispectral (blue, green, red, nir, swir-1, thermal, swir-2)	15 30	10
II	2.8	SPOT-5	10 September 2011	3× multispectral (green, red, nir)	2.5	10
III	1.4	WorldView-2	29 August 2015	1× pan 4× multispectral (blue, green, red, nir)	0.5 2	5
IV	1.8	WorldView-3	13 August 2015	1× pan 4× multispectral (blue, green, red, nir)	0.5 2	5
V(a) V(b)	3 1.2	Sentinel-2	27 August 2016	13× multispectral (coastal aerosol, blue, green, red, red edge 1-3, nir, red edge 4, water vapour, swir cirrus, swir 1-2)	10 (blue, green, red, nir) 20 (red edge 1-4, swir 1-2) 60 (other)	5

[1] refers to the available panchromatic (pan) and multispectral bands for each dataset.

The image selection process relied primarily on the availability of cloud-free data acquired in the summer season, possibly soon after a major landslide-triggering event or after the reactivation of landslide processes. Regarding DEM information, the supply of multi-temporal and post-event data was limited; the available data sets were therefore considered even if the acquisition date is not always consistent with the respective optical images. Pre-processing of the satellite images was implemented with ERDAS IMAGINE software (Hexagon Geospatial, Stockholm, Sweden) and included:

- Pansharpening if a panchromatic band was provided along with multispectral bands.
- Orthorectification supported by rational polynomial coefficients (RPCs) and/or ground control points (GCPs) to achieve sub-pixel geolocation accuracies.

2.3. Object-Based Landslide Mapping

For semi-automated landslide mapping, object-based approaches were developed and applied for each dataset and study area. Efforts were made to follow similar workflows; however, due to the different characteristics of the satellite images and study areas and because of the varying complexity and appearance of landslides, the transferability was limited. Thus, the classification rules were adapted for each test case. Moreover, two different but interchangeable software products, eCognition 9.2 (Trimble Geospatial, Sunnyvale, CA, USA) and InterIMAGE 1.43 (Pontifical Catholic University of Rio de Janeiro (PUC-Rio) and Brazilian National Institute for Space Research (INPE), Rio de Janeiro, Brazil), were used to develop the knowledge-based image analysis workflows. The two products are similar in their logic and functionality, meaning that a workflow implemented in eCognition can also be implemented in InterIMAGE and vice versa. Since the major objective of this study is the evaluation of OBIA mapping results in comparison to manual mapping results, rather than the development of sophisticated mapping routines, the implemented OBIA approaches were kept as simple as possible to produce acceptable results in a fast and efficient manner.

Table 2 shows the software that was used for mapping landslides in each study area, the segmentation method and the segmentation parameters that were used for creating image objects, and the main classification parameters for identifying landslide objects. Slightly different segmentation and classification parameters were used for mapping the two large landslides in study area V. Thus, the sub-areas (V(a), a landslide near the villages of Badia and La Villa, and V(b), a landslide near the village of Corvara) were treated separately during OBIA mapping. In general, the Normalized Difference Vegetation Index (NDVI) and the slope information were most useful for the detection of landslide bodies. For refining the classification, further parameters (e.g., brightness, length/width ratio, relation to neighboring objects, compactness, perimeter/area ratio) were used additionally or in combination with each other. Thresholds were selected based on (1) an exploratory analysis of image object properties (e.g., mean slope, mean NDVI) and (2) recommendations in the literature [17,31].

Table 2. Software, the segmentation method and parameters (SP—Scale Parameter, S—Shape criterion, C—Compactness criterion), and the main classification parameters used for object-based landslide mapping for each case study.

Study Area	Software	Segmentation Method and Parameters	Main Classification Parameters
I	eCognition 9.2 (Trimble Geospatial)	Multiresolution segmentation: SP: 25; S: 0.1; C: 0.5	Mean diff. to neighbors (NDVI) < 0 Mean slope > 10 Length/Width > 4
II	eCognition 9.2 (Trimble Geospatial)	Multiresolution segmentation: SP: 15; S: 0.3; C: 0.5	Mean NDVI < 0.04 Mean slope > 10 Mean brightness > 85
III	InterIMAGE 1.43	Threshold segmentations based on NDVI and brightness layer (Threshold = Mean of layer)	Mean slope > 35° Mean TRI > 1.5 Size > 100 m^2 Perimeter/Area Ratio (P/A) < 0.55 Mean NDVI < 0.55

Table 2. *Cont.*

Study Area	Software	Segmentation Method and Parameters	Main Classification Parameters
IV	InterIMAGE 1.43	1. Threshold segmentations based on NDVI and brightness layer (Threshold = Mean of layer) 2. Multiresolution segmentation based on slope; SP: 400; S: 0.5; C: 0.5	Same as III, except: Mean slope > 15° No use of Mean NDVI
V (a)	eCognition 9.2 (Trimble Geospatial)	Multiresolution segmentation: SP: 150; S: 0.5; C: 0.5	Mean NDVI < 0.4 or < 0.5 (depending on fine adjustment in combination with other features) Mean slope > 10
V (b)	eCognition 9.2 (Trimble Geospatial)	Multiresolution segmentation: SP: 150; S: 0.3; C: 0.8	Mean NDVI < 0.4 Mean slope > 8 Length/Width > 1.6 Shape index < 3.6

2.4. *Manual Landslide Mapping*

Three different types of features were manually mapped. First, points at the central area of landslide scars were located. Second, the same DEMs as for semi-automated mapping were used to recognize and digitize the crown scars (if possible) and at last, for most of the mass movements detected, polygons were drawn. Some basic information was recorded and stored in a GIS database assigned with several attribute fields covering the following aspects:

- Technical aspects connected to the intrinsic characteristics of the sensors adopted (Table 3)
- Geomorphologic and physical aspects of the phenomena mapped (Table 4)
- Statistical information showing, for example, the number of mapped landslides and their size (Table 5)

Table 3. Database fields that describe how the remote sensing image quality influences the manual landslide mapping.

Band Combination	Digitalization Level of Detail	Geometric Quality	Completeness Quality	Landslide Classification Quality
R-G-B	Mapping Scale (varies between 1:1.000–1:20,000) depends on spatial resolution	accurate (1)	complete (1)	certain (1)
		low degree of inaccuracy (2)	high degree of completeness (2)	low degree of uncertainty (2)
		high degree of inaccuracy (3)	low degree of completeness (3)	high degree of uncertainty (3)
		inaccurate (4)	incomplete (4)	uncertain (4)

The technical aspects concerning the scale used for the manual mapping can be distinguished between objective and subjective ones. The first group contains aspects such as the band combination (most of the time false color composite infrared was used) and the spatial resolution, which influenced the digitalization scale. For the second group, a broader distinction was proposed; for subjective factors, which are synthesized in three quality fields (Table 3), scores from 1 (best) to 4 (worst) were assigned to each polygon created.

The first quality field is related to the geometry that relies directly on the image resolution; for example, landslides mapped on Landsat 7 result in a lower geometric accuracy than landslides mapped on SPOT-5 and WorldView-2/3 images. The completeness quality field represents a purely subjective index, being related to the ability of the operator to map a phenomenon as completely as possible. In this case, the landscape context and local characteristics such as forest presence, which obliterated parts of the phenomena, or the re-growing of vegetation, which concealed previous landslide scars and deposits, played a very important role. Concerning the landslide classification quality, the best score was reached when a VHR image was available for the manual mapping.

Visual interpretation partly allowed distinguishing between different subtypes and parts of landslides (Table 4), depending on the specific geographical and geological settings of the five study areas. If a landslide scar could be identified, then it was represented as polyline; when the scar could not be recognized due to data constraints, the accumulation and the transport areas were mapped instead.

The presence of a DEM in order to delineate the exact boundaries of the landslides was useful; even if it was mainly used for the confirmation of the scar extension highlighted in the VHR images (by means of the hillshade and the contour lines). Bulges identification on the deposition area were also supported by the digital elevation data, whereas, since the trail left at the transport section of a landslide is enhanced by a difference in local re-growing of vegetation, only the optical images could be used for identifying these landslide sub-parts.

The landslide activity field was derived from additional alternative data sources (e.g., a public image Web Map Service (WMS) showing older aerial photography for the study areas I, II, and V), published landslide data such as the multi-temporal landslide inventory of Vorarlberg [47] for study areas III and IV, and from unpublished material such as the Austrian mass movement cadastre GEORIOS [48] for study area II, since it was not possible to obtain this information from satellite imagery from one point in time. Objects stored in the GEORIOS database were digitized from geological maps published at different scales (ranging between 1:50,000 and 1:5,000). Although this information represents historical evidence of mass movements, these kind of products only represent active (at the time of the map release) phenomena that are larger than the surrounding geological feature [49] and care should be taken when using this information in order to produce detailed (scale 1:10,000) landslide inventories.

Table 4. Database fields that describe the geomorphological aspects of the manually digitized landslides.

Landslide Type	Landslide Subtype	Landslide Accumulation	Landslide Transport	Landslide Activity [1]
Landslide process type classification [50]	Landslide process subtype classification [50,51]	accumulation area complete (1) incomplete (2) absent (3)	transport area complete (1) incomplete (2) absent (3)	active (1) probably still active (2) dormant (3) not active (4)

[1] based on additional archived data and literature.

2.5. Comparison Methods

For comparing the semi-automated mapping results from OBIA to the mapping results of the visual landslide interpretation, the traditional thematic accuracy assessment approach for land use/land cover image classifications has been employed and adapted [52–54]. It has been applied for landslide classifications by Hölbling et al. [23]. Accordingly, we calculated the landslide area that was mapped by both methods (i.e., the overlapping area). Based on this overlapping area, the respective producer's (overlap expressed as share of the manual mapping area) and user's (as share of the OBIA mapping area) accuracies were computed. Furthermore, the difference in the landslide area mapped by both methods was computed and expressed as a ratio relative to the manual mapping area.

Beyond the area-based comparison, we performed a comparison of the number of polygons mapped as landslides. The difference in the count of landslide polygons between the OBIA result and the manual mapping result was expressed as a ratio relative to the manual mapping count.

Additionally, a set of object-by-object comparison metrics was calculated, which has proven to be useful in the geomorphological context of object-based landform delimitation [55]. Spatial accuracy metrics were used to quantify the spatial match between the manually digitized landslide polygons (considered reference areas) and the landslide polygons extracted with OBIA workflows (considered test areas). In total, five established spatial accuracy metrics were selected [55–57]: (1) Quality Rate (QR), (2) Area-Fit-Index (AFI), (3) Over-Segmentation Rate (OR), (4) Under-Segmentation Rate (UR), and (5) a compound metric D, which combines OR and UR. All these metrics rely on area proportions, i.e., the area of manual polygons, the area of OBIA polygons, and the spatial overlap between manual and OBIA polygons. Except for AFI, the possible range of values is from 0 to 1. A value of 0 indicates a perfect spatial match between the manual and the OBIA area, i.e., the areas have exactly the same spatial extent. In contrast, if the manual and OBIA polygons do not produce any spatial overlap, a value of 1 is obtained. A value around 0.5 can be interpreted such that the majority of manual polygons are overlapped by OBIA polygons but that the spatial extent of the overlapping polygons is different and that some non-overlapping polygons exist. Only the values of AFI can be negative. For the presented landslide mapping tests, it is therefore desirable to obtain metric values as close as possible to 0. The computation of the metrics was automated in the open source software QGIS (Open Source Geospatial Foundation, Beaverton, OR, USA), based on a tool that has been developed previously for use in object-based software [55]. Metrics were computed (1) globally, i.e., for all landslides per study site, and (2) locally, i.e., separately for each landslide. The formulas for and mathematical notations of the spatial accuracy metrics can be found in Clinton et al. [56]. It is also important to mention that there exists a mathematical relation between UA and UR, as well as PA and OR; the total (not percentage) values of each pair (UA-UR, PA-OR) sum up to 1.

Next to the calculation of spatial accuracy metrics, a non-spatial accuracy metric, the so-called Miss Rate (MR), was used. MR quantifies the completeness of the semi-automated landslide extraction process. It is computed as the ratio between the number of undetected manual polygons and the total number of manual polygons per study site [55]. The range of values is between 0 and 1. The closer the values are to 0, the more complete is the OBIA landslide extraction.

3. Results

3.1. Landslide Mapping Results

3.1.1. OBIA Landslide Mapping Results

Figures 2–6 show the OBIA landslide mapping results in comparison to the manual mapping results for each study area. There is good coincidence between most of the results; however, several disparities can be observed that are often related to the ability of a human interpreter to delineate a landslide body as whole, even if parts of the landslide are covered by vegetation. This is the case, for example, for the large landslide near the villages of Badia and La Villa in Südtirol-Alto Adige Autonomous Province (study area V(a)). Similarly, landslides that are still visible on the images but are already overgrown by grass are difficult to detect semi-automatically due to the missing spectral difference to the surrounding areas. This can be recognized in the southwestern part of study area IV, where not all landslides, and particularly landslide tails, could be detected. The detection of landslides by OBIA was most challenging in study area II, since it was hardly possible to differentiate landslides from other processes using the SPOT-5 image and the DEM only. Statistics and detailed information about the accuracy of the semi-automated mapping in comparison to the manual mapping are given in Section 3.2.

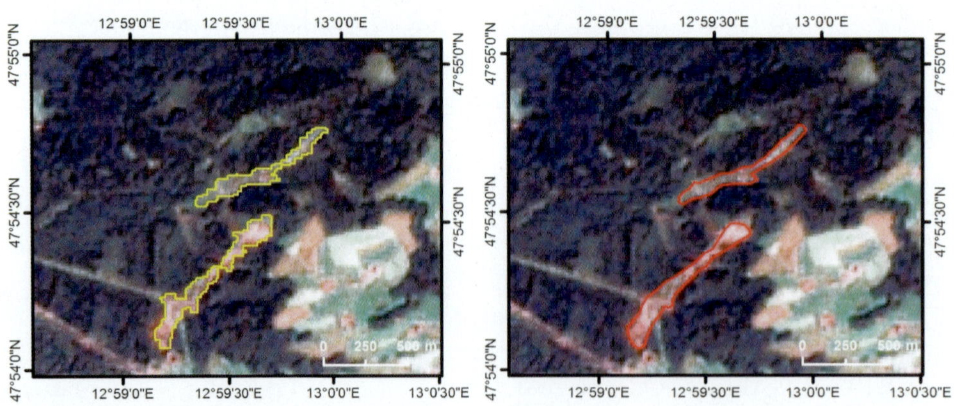

Figure 2. Semi-automatically mapped landslides using the object-based image analysis (OBIA) approach (in yellow; **left**) and manually mapped landslides (in red; **right**) for study area I based on the Landsat 7 image.

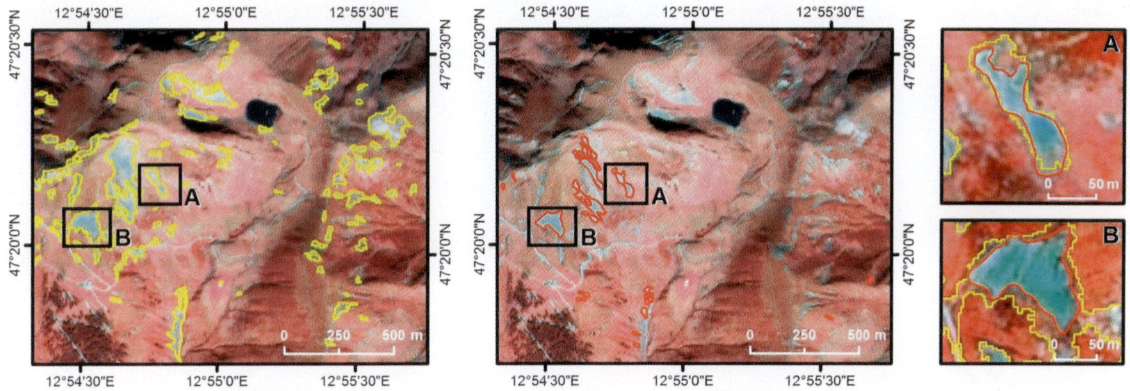

Figure 3. Semi-automatically mapped landslides using the object-based image analysis (OBIA) approach (in yellow; **left**) and manually mapped landslides (in red; **middle**) for study area II based on the SPOT-5 image. Two subsets demonstrating the variation of the mapping results for selected landslides are shown on the **right**. The black rectangles (A, B) indicate the location of the subsets.

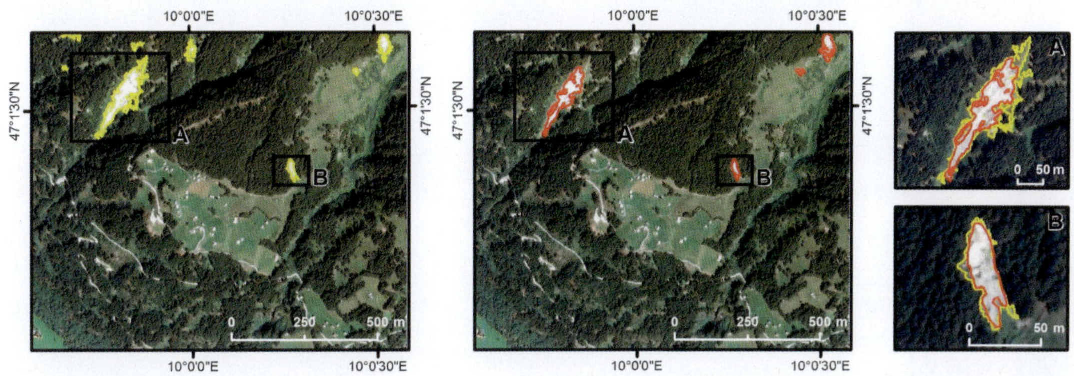

Figure 4. Semi-automatically mapped landslides using the object-based image analysis (OBIA) approach (in yellow; **left**) and manually mapped landslides (in red; **middle**) for study area III based on the WorldView-2 image. Two subsets demonstrating the variation of the mapping results for selected landslides are shown on the **right**. The black rectangles (A, B) indicate the location of the subsets.

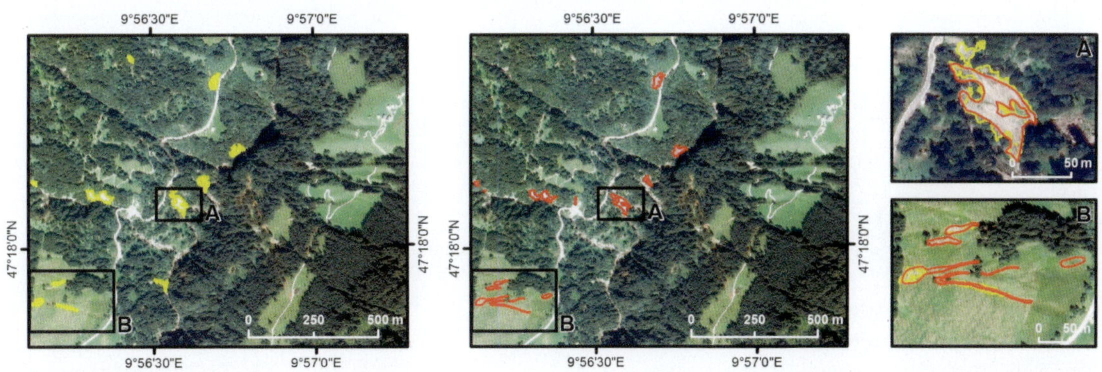

Figure 5. Semi-automatically mapped landslides using the object-based image analysis (OBIA) approach (in yellow; **left**) and manually mapped landslides (in red; **middle**) for study area IV based on the WorldView-3 image. Two subsets demonstrating the variation of the mapping results for selected landslides are shown on the **right**. The black rectangles (A, B) indicate the location of the subsets.

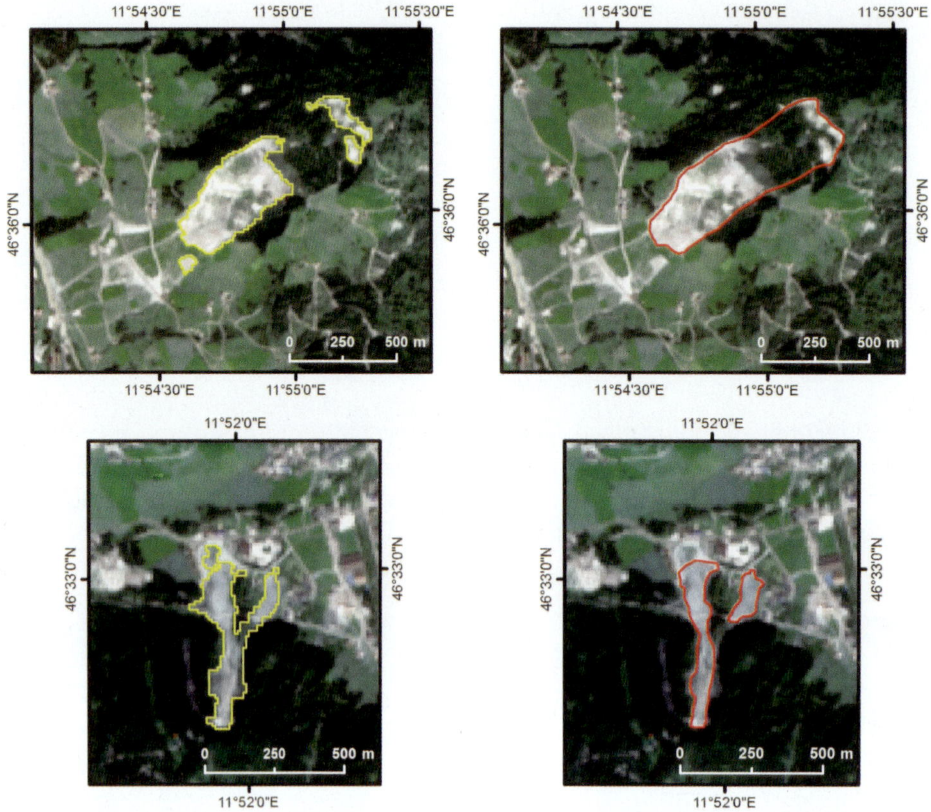

Figure 6. Semi-automatically mapped landslides using the object-based image analysis (OBIA) approach (in yellow; **left**) and manually mapped landslides (in red; **right**) for study areas V(a) (**top**) and V(b) (**below**) based on the Sentinel-2 image.

Table 5 presents some basic statistics for the OBIA mapping results for each study area.

Table 5. Statistics of the OBIA mapping results for each of the five study areas.

Statistical Property	Study Area I	Study Area II	Study Area III	Study Area IV	Study Area V
Number of landslides	2	85	9	11	5
Area affected by landslides (km^2)	0.148	0.203	0.019	0.011	0.250
Smallest mapped landslide (m^2)	59700	75	135	244	4200
Largest mapped landslide (m^2)	87900	38741	12577	2400	148800
Average size of mapped landslides (m^2)	73800	2393	2110	992	49960

3.1.2. Manual Landslide Mapping Results

Table 6 shows the statistics of the manual landslide mapping results for each study area. Probably the most notable case is the Pinzgau study area (II). This is a special area of interest, being a mountainous area characterized by a high relief and sparse vegetation, where several geomorphological phenomena such as snow scars induced by avalanches, gully erosions connected to torrent activity, and rock glacier deposits could be perceived and mapped as mass movements. Surprisingly, a high landslide density can be recognized. Since the differentiation of landslides from other mass movement types only based on the SPOT-5 image and the 10 m DEM was hardly possible in this case, even by visual interpretation, a 5 m DEM and aerial photography from 2011 (accessible through a free WMS service) were additionally used for manual landslide mapping. Furthermore, some GEORIOS polygons were used as guidance for the identification of landslide subtypes.

Table 6. Statistics of the manual mapping results for each of the five study areas.

Statistical Property	Study Area I	Study Area II	Study Area III	Study Area IV	Study Area V
Number of landslides	2	18	5	14	3
Area affected by landslides (km^2)	0.140	0.031	0.011	0.010	0.320
Smallest mapped landslide (m^2)	54855	247	66	30	10781
Largest mapped landslide (m^2)	85339	8374	7200	1882	270966
Average size of mapped landslides (m^2)	70097	1830	2157	709	106533

The three quality fields were introduced in order to trace back the logic used for the task of manual landslide mapping and to communicate the perceived difficulties encountered during the mapping process. The three quality indices are directly correlated to the spatial resolution of the used satellite images. For each of the analyzed images, the achievable score varies. The best score reached among all the quality categories is:

- Score of "1" for study area III and IV (1 m resolution)
- Score of "2" for area II (2.5 m resolution) and area V (10 m resolution)
- Score of "3" for study area I (15 m resolution)

In general, when considering all quality measures, the best results were achieved on the VHR images (including area II with 2.5 m resolution), always bearing in mind the subjectivity introduced by the interpreter.

3.2. *Accuracy Assessment*

A variety of different metrics was used to quantitatively capture the agreement between the OBIA mapping results and the manual delineation. Comparisons were performed in all study areas and for all datasets.

3.2.1. Comparison of Overlapping Areas

The metrics according to the general accuracy assessment approach for image classifications, as presented in Hölbling et al. [23], are summarized in Table 7. Concerning the number of landslides, the best agreement occurred in study area I (Fürwag), where both approaches detected two large landslides. The highest discrepancy occurred in study area II (Pinzgau), where a large number of objects were considered landslides only by the OBIA result but not by the manual delineation. In the other study areas, the OBIA and manual results deviated from each other by two to four landslides. Since the numbers of landslides were relatively low, with a count between three and 14, the difference by percentage still showed high values between −20% and 80%. The comparison of the landslide affected area reveals similar tendencies: again, study area I showed the best agreement with a difference of only 5.28%, and study area II showed the highest discrepancy, with OBIA detecting an affected area that is almost five times as large as the manual mapping. In study area III (Montafon), the difference in the landslide-affected area was 76% and reflected the agreement already shown by the difference in landslide numbers. For study area IV (Bregenzerwald) and study area V (Gader Valley), the OBIA deviation from the manually mapped area is relatively low, with 9.91% and −21.8%., but here the difference shows a contrasting direction related to the number of landslides. The producer's and user's accuracy values in study area I and IV were very similar. For study area II, the values show that most of the landslide affected areas identified with manual mapping have also been captured by OBIA (producer's accuracy 90.6%) but that many more areas were classified that were left out by manual mapping (user's accuracy 15.6%). Study areas III and V(b) show a similar situation to study area II but not as pronounced. Study area V(a) showed the opposite behavior because a significant portion of a landslide has been included in the manual mapping that was not covered by the OBIA result.

Table 7. Comparison of OBIA mapping results and manual mapping results per study area.

Comparison Metric	Study Area I	Study Area II	Study Area III	Study Area IV	Study Area V
Number of landslides (OBIA)	2	85	9	11	5
Number of landslides (manual)	2	18	5	14	3
Difference in numbers OBIA—manual (count; %)	0 0%	+67 372%	+4 80%	−3 −21%	+2 +67%
Landslide affected area (OBIA, km^2)	0.148	0.203	0.0190	0.0109	0.250
Landslide affected area (manual, km^2)	0.140	0.0351	0.0108	0.00993	0.320
Area difference OBIA—manual (%)	5.28%	480%	76 %	9.91%	−21.8%
Overlap area (km^2)	0.118	0.0318	0.0104	0.00691	0.208
Producer's Accuracy (%)	84.3%	90.6%	95.8%	69.6%	V(a): 60.6% V(b): 90.0%
User's Accuracy (%)	80.1%	15.6%	54.4%	63.3%	V(a): 91.2% V(b): 62.7%

3.2.2. Completeness of the Semi-Automated Extraction

Relatively low MR values in the range from 0 to 0.35 were obtained (Figure 7), demonstrating that at least two thirds of the manual polygons were also extracted by the OBIA method. For the Gader Valley and Fürwag, even all reference polygons were detected with the semi-automated workflows. This is indicated by a MR value of 0. However, for these areas, the total number of reference polygons was small (n = 1–2) and the spatial extent relatively large. Usually, the combination of low reference number and large reference areas increases the success of the automated landslide detection, which is also supported by the presented results.

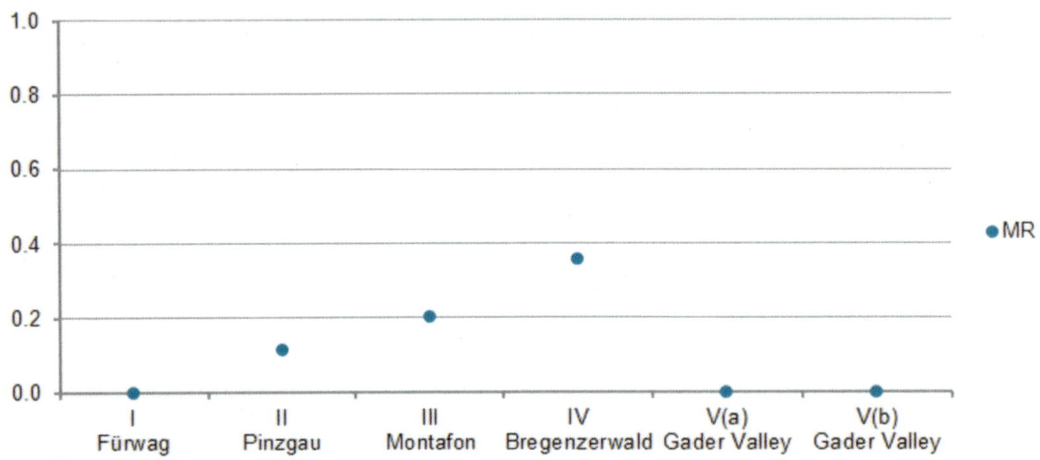

Figure 7. Completeness of the automated landslide extraction, quantified by the values of Miss Rate (MR).

3.2.3. Spatial Accuracy of the Semi-Automated Extraction

Generally, global spatial accuracy values are below 0.5, indicating good to high spatial agreements between the reference and automated landslide polygons for the selected study sites (Figure 8). Only for the Pinzgau area higher values, and hence, lower spatial accuracies were achieved. For Pinzgau, the high values of QR, D, and UR show that there is a high difference between the size of reference and test polygons. As a result, small spatial overlaps are produced. Since OR is close to 0 and AFI is negative (Figures 8 and 9), it can be concluded that the extracted landslide polygons are far larger than the reference polygons and/or that landslides were extracted at the wrong locations, i.e., the automated extraction over-estimates the spatial extent of the reference. To a lesser extent, the afore-mentioned conclusion also holds true for the Montafon study site (III). From Figures 8 and 9, it is also obvious that study area V(a) is the only area where the value of UR is lower than OR and where AFI is positive. This means that the extracted landslide polygon is smaller than the reference polygon (n = 1), i.e., the extraction under-estimates the reference. The best spatial agreement between reference and test polygons was obtained for the Fürwag area ($n = 2$). The spatial accuracy values are relatively low, and the deviation between them is minimal. These characteristics are achieved if the over- and under-estimated areas are relatively small and the area proportions of the overlap, test, and reference polygons are similar.

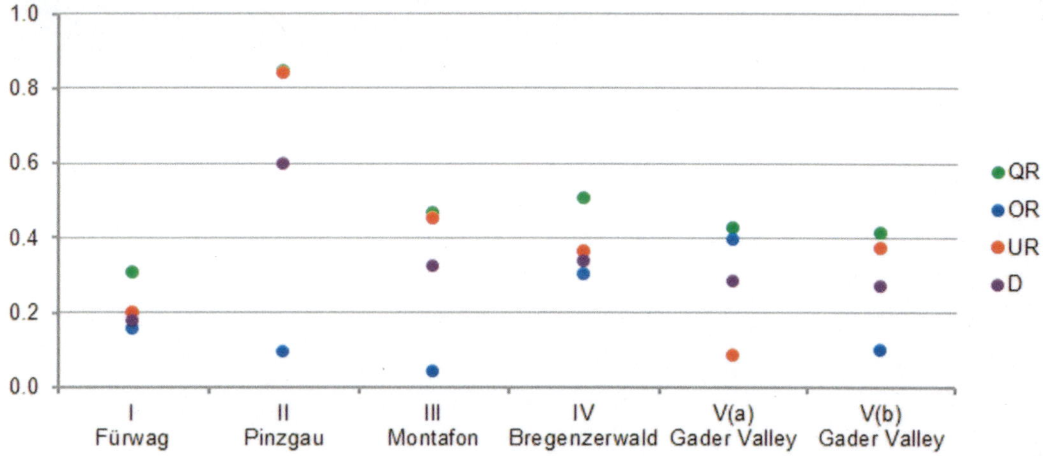

Figure 8. Global spatial accuracies for the selected study sites. Results presented for four metrics; Quality Rate (QR), Under-Segmentation Rate (UR), Over-Segmentation Rate (OR), and the UR-OR compound metric D.

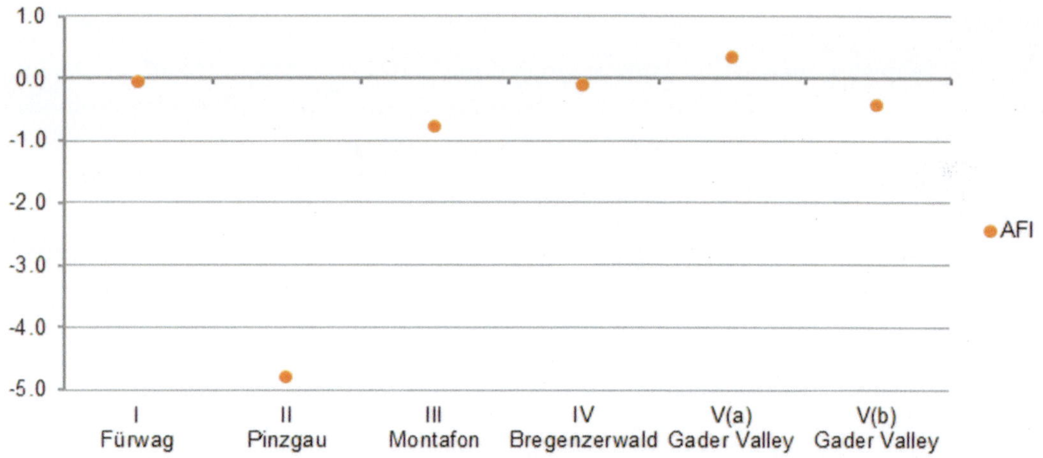

Figure 9. Global values of Area-Fit-Index (AFI) for the selected study sites.

Figure 10 depicts the spatial accuracy signatures of individual landslides per study site. The signatures are obtained by plotting the four metrics, QR, UR, OR, and D; each on one side of the X- and Y-axis. In case of high spatial agreement, the shape of the graph is diamond-like and the diamond is positioned as close as possible around the zero point. The afore-mentioned characteristics are best met for the two Fürwag landslides. When looking at the deviation of spatial accuracy signatures for the landslides per study area, it can be concluded that the lowest values are obtained for Fürwag and Montafon, whereas for Pinzgau and Bregenzerwald higher deviations are depicted in Figure 10. Higher deviations indicate a higher variation of results for a study area, i.e., some reference landslides are under-estimated by the test polygons, while others are over-estimated.

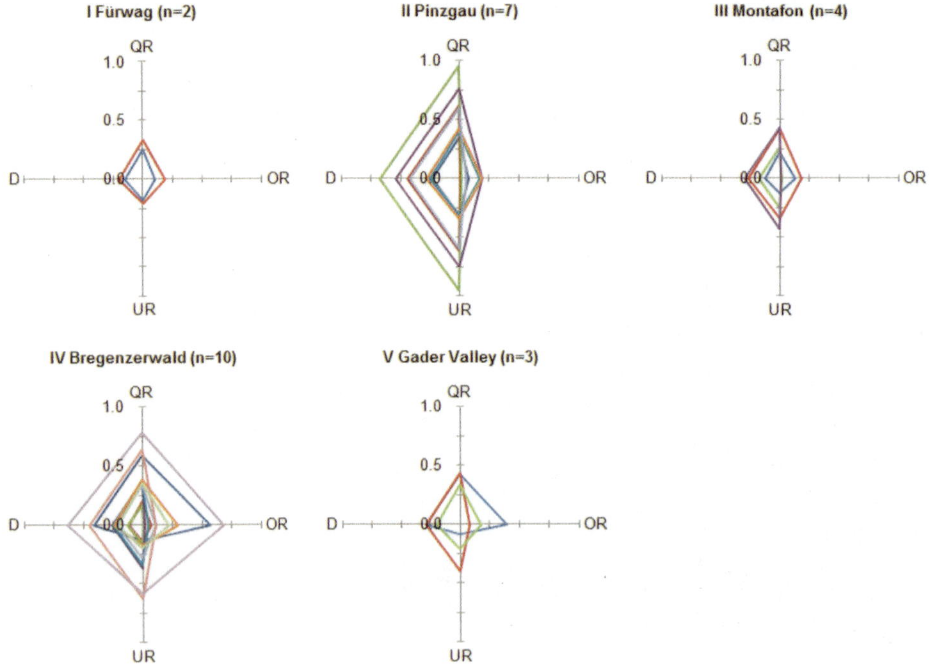

Figure 10. Landslide-specific spatial accuracy values per study site I–V. Each graph in I–V depicts the spatial accuracy signature of a detected landslide. Results presented for four metrics; Quality Rate (QR), Under-Segmentation Rate (UR), Over-Segmentation Rate (OR), and the UR-OR compound metric D.

4. Discussion

Landslide maps produced with semi-automated methods are commonly compared to manually mapped landslides in order to quantitatively assess the accuracy of results. Visual interpretation, however, while still the most common method for mapping landslides on optical remote sensing data, has some disadvantages: it is a time-consuming process, depends on various factors, and cannot constitute a completely "true" reference. However, often it is the only available reference data [8,23]. A "true" reference, i.e., ground truth reference data, would require the reference dataset to be of a significantly higher accuracy and to be collected independently from the tested dataset [58]. The manual delineation and the OBIA mapping are two independently performed processes, but they employ the same satellite images, and thus their results cannot be considered fully independent. For the same reason, the manual interpretation did not result in landslide delineation with a significantly higher accuracy. Consequently, manual mapping implies some degree of uncertainty that cannot be avoided, even if created by an expert. Therefore, if some of the disagreement between the two compared datasets can be attributed to the uncertainty in the manual delineation, the accuracy of the OBIA results may be larger than the calculated accuracy values suggest. At least they would fall in the same range of acceptable uncertainty that commonly used manual delineations imply. Of course, this could only be confirmed if a valid ground truth reference dataset was available that also allows the accuracy of the manual delineation to be assessed. In the current practice of assessing the accuracy of OBIA results with manual delineations, only strong disagreements can clearly be attributed to mapping errors, e.g., in the case of study area II where a user's accuracy value of only 15.6% occurred. The conceptual problem remains that the manual mapping does not completely fulfill the standards as a reference for a classical accuracy assessment. Due to the uncertainty inherent in the manual mapping, we suggest to use the term "agreement" instead of "accuracy" when comparing the two classifications. The concept of agreement for evaluating the quality of OBIA landslide mapping results also applies to other fields of applications where a fully reliable reference is missing. The approach of assessing the agreement may not yield as sharp a quality indicator as the classical approach of an accuracy assessment, but it can still be highly valuable. The selection of appropriate quality indicators is a subject for a necessary discussion within the landslide community. It enables the design of a validation approach for EO-based landslide mapping that conforms to the Quality Assurance Framework for Earth Observation (QA4EO; http://QA4EO.org/).

The accuracy assessment achieved a number of different metrics that were used for assessing the accuracy of the OBIA landslide maps. In addition to classical PA and UA, which were computed separately for each site, this study proposes to integrate spatial accuracy metrics for quantifying the spatial agreement between manual and automated landslide polygons on an object-by-object basis. The use of landslide-specific accuracies allows a more detailed interpretation of the results. For instance, the deviation of landslide accuracy values within a study area may give clues about the stability/robustness of the automated extraction workflow. The best results where all measures show similar values were achieved for study area I, with PA and UA of approximately 80% and QR, UR, OR, and D of approximately 0.2. However, for area I only two reference polygons were extracted. The results of this study confirm that lower accuracies are achieved for areas where a larger number of landslides have to be mapped by OBIA methods. We also compared delineations of single landslides that performed equally well or even better. However, what makes it difficult to interpret the calculated accuracy values is the lack of information about the accuracy thresholds accepted within the landslide community. Such information would be highly valuable for estimating the quality of landslide maps derived from EO data with semi-automated methods.

The results in this study obviously show that the very high spatial resolution of the optical satellite image does not necessarily guarantee a high quality of the automated landslide mapping. The relative quality of semi-automated landslide mapping can be better for Landsat 7 data than for WorldView-2/3 images, as exemplified by the Fürwag test case. However, when drawing such conclusions, one has to keep in mind that there is a direct relationship between the spatial resolution of the optical satellite image and the reliability of the reference polygons digitized therefrom. It is not surprising that VHR images allow the mapping of smaller landslides and to some degree also the differentiation of subtypes, particularly during manual mapping. On the other hand, Landsat 7 and Sentinel-2 proved their applicability for identifying and delineating larger landslides, whereby the agreement between both results was high (apart from the special case of study area V(a), as described below). Especially Sentinel-2 can be considered a relevant data source for semi-automated landslide mapping and landslide change detection, also because of its high revisiting frequency.

Both methods produced acceptable and comparable results. However, the differentiation of landslide types and sub-parts with OBIA was not possible with the available data. Very high resolution DEM data from different dates would be helpful for that, but data availability was limited. Also, optical data acquired directly after a landslide triggering event and another image taken several weeks or months later could be helpful for the semi-automated differentiation of source, transport, and accumulation area, since landslide tails tend to revegetate faster than landslide scars [23]. These difficulties become obvious when looking at the results from the Pinzgau study area, where different mass movement processes are present. While the OBIA approach classified too many areas as landslides due to similar spectral properties, only a few landslides were identified by manual mapping. In such a case, ground truthing would be necessary to create a reliable landslide inventory. An example that requires expert interpretation can be found in the southwestern part of the Bregenzerwald study site where some small shallow landslides were manually mapped, which could not be identified by OBIA despite the use of WorldView-3 since they were already grassed over. Study area V(a) is an interesting case because a significant portion of a landslide has been included in the manual mapping that was not covered by the OBIA mapping. This part of the landslide is covered by vegetation. In the manual interpretation, the inclusion of this part was logically inferred by expert knowledge, an element that is difficult to implement in automated mapping workflows. Eisank et al. [31] proposed a knowledge-based landslide mapping system, but further research is needed in this direction to compete with a human interpreter.

Two OBIA software products with a different range of functionalities were used for semi-automated landslide mapping, both delivering acceptable results for the presented case studies. OBIA workflows were developed in such a way that they can potentially be set up in any of the two software products, but each image analysis tool consists of different implemented approaches for image segmentation and classification. As a consequence, the workflows may produce different results, particularly related to the object delineations. Further studies could focus on a systematic comparison of the applicability of different software products to identify the most adequate tool for each sensor and landscape characteristic. However, this also depends on the data used and on the preferences and skills of the software user.

By combining both approaches, i.e., starting with the semi-automated detection of landslide candidates and then refining this classification manually, the whole mapping process could be improved while achieving adequate mapping results. Particularly large-scale mapping tasks would probably benefit from a degree of automation [59]. By implementing such a combined approach in an interactive and easy-to-use web service intended for practitioners with at least a certain knowledge in landslide recognition, landslide mapping could become faster and more efficient [60,61]. Such a web service can be relevant, especially for users and practitioners who might not be familiar with sophisticated image analysis software.

5. Conclusions

In this study we compared two landslide mapping approaches, OBIA and manual mapping, in five selected test areas with various geological and geomorphological settings in the Alps. Satellite images with different resolutions from different sensors were used. The analyses show that both methods produce similar results, whereby the achieved accuracy values vary between the study areas. It is important to note that using manual mapping for reference is critical, and the associated uncertainty should be considered when interpreting the accuracy values. While visual expert interpretation has advantages for differentiating landslides subtypes and for delineating single landslides as single objects and thus to identify the number of mapped landslides, it is a subjective and time-consuming task. This subjectivity issue is difficult to measure but influences the reliability of the landslide mapping. We proposed a scheme for including different quality measures in manual mapping. This might be useful for evaluating the reliability of manual mapping as reference data, even if some of these measures are subjective and rely on the interpreter. On the other hand, the OBIA approach is more transparent due to the use of replicable classification rules, but the expert knowledge required for the mapping of complex landslides or landslides that are partly covered by vegetation is difficult to implement in automated mapping workflows. Semi-automated landslide mapping approaches show high potential and would find even more acceptance if a set of generally agreed quality indicators and accuracy thresholds for the validation of EO-based landslide maps were commonly used.

Acknowledgments: This research has been supported by the Austrian Research Promotion Agency FFG in the Austrian Space Applications Program (ASAP 11) through the project "Land@Slide" (contract no: 847970).

Author Contributions: Daniel Hölbling had the idea for the paper and performed the object-based landslide mapping for study areas I, II, and V. Clemens Eisank did the OBIA mapping for study areas IV and V and calculated the spatial accuracy values. Florian Albrecht contributed to the accuracy assessment and the interpretation of results. Filippo Vecchiotti performed the visual landslide interpretation and provided the geological background information. Barbara Friedl wrote major parts of the introduction and contributed to the discussion. Elisabeth Weinke revised the manuscript and contributed to the discussion and conclusion. Arben Kociu provided valuable background information and revised the manuscript. All authors contributed to writing the paper.

Conflicts of Interest: The authors declare no conflict of interest. The founding sponsors had no role in the design of the study; in the collection, analyses, or interpretation of data; in the writing of the manuscript; or in the decision to publish the results.

References

1. Van Westen, C.J.; Castellanos, E.; Kuriakose, S.L. Spatial data for landslide susceptibility, hazard, and vulnerability assessment: An overview. *Eng. Geol.* **2008**, *102*, 112–131. [CrossRef]
2. Scaioni, M.; Longoni, L.; Melillo, V.; Papini, M. Remote Sensing for Landslide Investigations: An Overview of Recent Achievements and Perspectives. *Remote Sens.* **2014**, *6*, 9600–9652. [CrossRef]
3. Singhroy, V. Remote sensing of landslides. In *Landslide Hazard and Risk*; Glade, T., Anderson, M.G., Crozier, M.J., Eds.; Wiley & Sons: West Sussex, UK, 2005; pp. 469–492. ISBN 9780471486633.
4. Fiorucci, F.; Cardinali, M.; Carlà, R.; Rossi, M.; Mondini, A.C.; Santurri, L.; Ardizzone, F.; Guzzetti, F. Seasonal landslide mapping and estimation of landslide mobilization rates using aerial and satellite images. *Geomorphology* **2011**, *129*, 59–70. [CrossRef]
5. Gao, J.; Maro, J. Topographic controls on evolution of shallow landslides in pastoral Wairarapa, New Zealand, 1979–2003. *Geomorphology* **2010**, *114*, 373–381. [CrossRef]
6. Morgan, J.L.; Gergel, S.H.; Coops, N.C. Aerial Photography: A Rapidly Evolving Tool for Ecological Management. *BioScience* **2010**, *60*, 47–59. [CrossRef]
7. Galli, M.; Ardizzone, F.; Cardinali, M.; Guzzetti, F.; Reichenbach, P. Comparing landslide inventory maps. *Geomorphology* **2008**, *94*, 268–289. [CrossRef]
8. Hölbling, D.; Friedl, B.; Eisank, C. An object-based approach for semi-automated landslide change detection and attribution of changes to landslide classes in northern Taiwan. *Earth Sci. Inform.* **2015**, *8*, 327–335. [CrossRef]
9. Guzzetti, F.; Mondini, A.C.; Cardinali, M.; Fiorucci, F.; Santangelo, M.; Chang, K.-T. Landslide inventory maps: New tools for an old problem. *Earth-Sci. Rev.* **2012**, *112*, 42–66. [CrossRef]
10. Van Coillie, F.M.B.; Gardin, S.; Anseel, F.; Duyck, W.; Verbeke, L.P.C.; De Wulf, R.R. Variability of operator performance in remote-sensing image interpretation: The importance of human and external factors. *Int. J. Remote Sens.* **2014**, *35*, 754–778. [CrossRef]
11. Albrecht, F.; Lang, S.; Hölbling, D. Spatial accuracy assessment of object boundaries for object-based image analysis. In Proceedings of the GEOBIA 2010—Geographic Object-Based Image Analysis, Ghent, Belgium, 29 June–2 July 2010; Addink, E., Van Coillie, F.M.B., Eds.; ISPRS Vol. No. XXXVIII-4/C7, Archives ISSN No 1682-1777.
12. Kohli, D.; Stein, A.; Sliuzas, R. Uncertainty analysis for image interpretations of urban slums. *Comput. Environ. Urban Syst.* **2016**, *60*, 37–49. [CrossRef]
13. Guzzetti, F.; Cardinali, M.; Reichenbach, P.; Carrara, A. Comparing Landslide Maps: A Case Study in the Upper Tiber River Basin, Central Italy. *Environ. Manag.* **2000**, *25*, 247–263. [CrossRef]
14. Ardizzone, F.; Cardinali, M.; Carrara, A.; Guzzetti, F.; Reichenbach, P. Impact of mapping errors on the reliability of landslide hazard maps. *Nat. Hazards Earth Syst. Sci.* **2002**, *2*, 3–14. [CrossRef]
15. Moosavi, V.; Talebi, A.; Shirmohammadi, B. Producing a landslide inventory map using pixel-based and object-oriented approaches optimized by Taguchi method. *Geomorphology* **2014**, *204*, 646–656. [CrossRef]
16. Stumpf, A.; Kerle, N. Object-oriented mapping of landslides using Random Forests. *Remote Sens. Environ.* **2011**, *115*, 2564–2577. [CrossRef]
17. Martha, T.R.; Kerle, N.; Jetten, V.; van Westen, C.J.; Kumar, K.V. Characterising spectral, spatial and morphometric properties of landslides for semi-automatic detection using object-oriented methods. *Geomorphology* **2010**, *116*, 24–36. [CrossRef]
18. Barlow, J.; Franklin, S.; Martin, Y. High spatial resolution satellite imagery, DEM derivatives, and image segmentation for the detection of mass wasting processes. *Photogramm. Eng. Rem. S.* **2006**, *72*, 687–692. [CrossRef]
19. Behling, R.; Roessner, S.; Kaufmann, H.; Kleinschmit, B. Automated Spatiotemporal Landslide Mapping over Large Areas Using RapidEye Time Series Data. *Remote Sens.* **2014**, *6*, 8026–8055. [CrossRef]
20. Blaschke, T.; Feizizadeh, B.; Hölbling, D. Object-Based Image Analysis and Digital Terrain Analysis for Locating Landslides in the Urmia Lake Basin, Iran. *IEEE J. Sel. Top. Appl.* **2014**, *7*, 4806–4817. [CrossRef]
21. Heleno, S.; Matias, M.; Pina, P.; Sousa, A.J. Semiautomated object-based classification of rain-induced landslides with VHR multispectral images on Madeira Island. *Nat. Hazards Earth Syst. Sci.* **2016**, *16*, 1035–1048. [CrossRef]

22. Hölbling, D.; Füreder, P.; Antolini, F.; Cigna, F.; Casagli, N.; Lang, S. A Semi-Automated Object-Based Approach for Landslide Detection Validated by Persistent Scatterer Interferometry Measures and Landslide Inventories. *Remote Sens.* **2012**, *4*, 1310–1336. [CrossRef]
23. Hölbling, D.; Betts, H.; Spiekermann, R.; Phillips, C. Identifying Spatio-Temporal Landslide Hotspots on North Island, New Zealand, by Analyzing Historical and Recent Aerial Photography. *Geosciences* **2016**, *6*, 48. [CrossRef]
24. Kurtz, C.; Stumpf, A.; Malet, J.-P.; Gançarski, P.; Puissant, A.; Passat, N. Hierarchical extraction of landslides from multiresolution remotely sensed optical images. *ISPRS J. Photogramm.* **2014**, *87*, 122–136. [CrossRef]
25. Lahousse, T.; Chang, K.-T.; Lin, Y. Landslide mapping with multiscale object-based image analysis—A case study in the Baichi watershed, Taiwan. *Nat. Hazards Earth Syst. Sci.* **2011**, *11*, 2715–2726. [CrossRef]
26. Lu, P.; Stumpf, A.; Kerle, N.; Casagli, N. Object-oriented change detection for landslide rapid mapping. *IEEE Geosci. Remote Sens.* **2011**, *8*, 701–705. [CrossRef]
27. Martha, T.R.; Kerle, N.; van Westen, C.J.; Jetten, V.; Kumar, K.V. Segment Optimization and Data-Driven Thresholding for Knowledge-Based Landslide Detection by Object-Based Image Analysis. *IEEE Geosci. Remote Sens.* **2011**, *49*, 4928–4943. [CrossRef]
28. Martha, T.R.; Kerle, N.; van Westen, C.J.; Jetten, V.; Kumar, K.V. Object-oriented analysis of multi-temporal panchromatic images for creation of historical landslide inventories. *ISPRS J. Photogramm.* **2012**, *67*, 105–119. [CrossRef]
29. Rau, J.-Y.; Jhan, J.-P.; Rau, R.-J. Semiautomatic object-oriented landslide recognition scheme from multisensor optical imagery and DEM. *IEEE Trans. Geosci. Remote Sens.* **2014**, *52*, 1336–1349. [CrossRef]
30. Lang, S. Object-based image analysis for remote sensing applications: modeling reality—Dealing with complexity. In *Object-Based Image Analysis*; Blaschke, T., Lang, S., Hay, G.J., Eds.; Springer: Heidelberg/Berlin, Germany; New York, NY, USA, 2008; pp. 1–25. ISBN 978-3-540-77058-9.
31. Eisank, C.; Hölbling, D.; Friedl, B.; Chen, Y.C.; Chang, K.T. Expert knowledge for object-based landslide mapping in Taiwan. *South-East. Eur. J. Earth Obs. Geomat.* **2014**, *3*, 347–350.
32. Wagreich, M.; Lukeneder, A.; Egger, H. Cretaceous History of Austria. In Proceedings of the International Meeting on Correlation of Cretaceous Micro—And Macrofossils—Berichte der Geologischen Bundesanstalt, Vienna, Austria, 16–18 April 2008; Volume 74, pp. 12–30.
33. Vecchiotti, F.; Kociu, A. Geohazard Description for Salzburg, Public Pangeo FP7 Final Report, 2013. Available online: http://www.pangeoproject.eu/eng/coverage_map (accessed on 30 March 2017).
34. Embleton-Hamann, C. Geomorphological hazards in Austria. In *Geomorphology for the Future*; Kellerer-Pirklbauer, A., Keiler, M., Embleton-Hamann, C., Stötter, J., Eds.; Innsbruck University Press: Innsbruck, Austria, 2007; pp. 33–56; ISBN 978-3-902571-18-2.
35. Fiebiger, G. Die Rutschung Fürwag im Norden von Salzburg/Österreich. Prozess und Maßnahmen. In Proceedings of the International Congress Interpraevent, Matsumoto, Japan, 14–18 October 2002; pp. 629–639.
36. Friebe, J.G. *Geologie der Österreichischen Bundesländer. Vorarlberg. Erläuterungen der Geologischen Karte Vorarlberg 1:100,000*; Geologische Bundesanstalt: Vienna, Austria, 2007.
37. Seijmonsbergen, A.C.; Anders, N.; Bouten, W. Geomorphological change detection using object-based feature extraction from multi-temporal LiDAR data. In Proceedings of the 4th GEOBIA, Rio de Janeiro, Brazil, 7–9 May 2012; Feitosa, R., Costa, G., Almeida, C., Eds.; pp. 484–489.
38. Dorren, L.K.; Maier, B.; Seijmonsbergen, A.C. Improved Landsat-based forest mapping in steep mountainous terrain using object-based classification. *Forest Ecol. Manag.* **2003**, *183*, 31–46. [CrossRef]
39. De Graaff, L.W.S.; Seijmonsbergen, A.C. Postglacial landslides and their impact on Pleistocene lake floor deposits in the Balderschwang Valley as witnessed by geomorphological, sedimentological and geophysical evidence (Vorarlberg, Austria). *Vorarlb. Naturschau* **2001**, *9*, 237–251.
40. Ghuffar, S.; Szekely, B.; Roncat, A.; Pfeifer, N. Landslide Displacement Monitoring Using 3D Range Flow on Airborne and Terrestrial LiDAR Data. *Remote Sens.* **2013**, *5*, 2720–2745. [CrossRef]
41. Jaritz, W.; Supper, R.; Wöhrer-Alge, M. Beurteilung geogener Gefahren in Hinblick auf eine Risikominderung in der Gemeinde Sibratsgfäll (Österreich). In Proceedings of the 11th Interpraevent Congress, Dornbirn, Austria, 26–30 May 2008; pp. 171–182.

42. Brandner, R.; Gruber, A.; Morelli, C.; Mair, V. Field trip 1—Pulses of Neotethys-Rifting in the Permomesozoic of the Dolomites. In *Geo.Alp*; Institute of Geology—University of Innsbruck: Innsbruck, Austria, 2016; Volume 13, pp. 7–70.
43. Gianolla, P.; Andreetta, R.; Furin, S.; Furlanis, S.; Riva, A. Geology of the Dolomites. In *Nomination of the Dolomites for Inscription on the World Natural Heritage List Unesco*; Artimedia: Trento, Italy, 2008; pp. 3–77.
44. Larcher, V.; Notarnicola, C.; Piacentini, D.; Pinter, T.; Schneiderbauer, S.; Soldati, M.; Strada, C. Analyse flachgründiger Massenbewegungen mittels Verwendung zweier statistischer Methoden im Gadertal (Südtirol). In Proceedings of the COGeo, Salzburg, Austria, 11 June 2010; Marschallinger, R., Wanker, W., Zobl, F., Eds.; pp. 1–11.
45. Corsini, A.; Pasuto, A.; Soldati, M.; Zannoni, A. Field monitoring of the Corvara landslide (Dolomites, Italy) and its relevance for hazard assessment. *Geomorphology* **2005**, *66*, 149–165. [CrossRef]
46. Riley, S.J.; De Gloria, S.D.; Elliot, R. A terrain ruggedness that quantifies topographic heterogeneity. *Interm. J. Sci.* **1999**, *5*, 23–27.
47. Tilch, N. Identifizierung gravitativer Massenbewegungen mittels multitemporaler Luftbildauswertung in Vorarlberg und angrenzender Gebiete. *Jahrb. Geol. Bundesanstalt* **2014**, *154*, 21–39.
48. Kociu, A.; Kautz, H.; Tilch, N.; Grösel, K.; Horst, H.; Reischer, J. Massenbewegungen in Österreich. *Jahrb. Geol. Bundesanstalt* **2007**, *147*, 215–220.
49. Wills, C.J.; McCrink, T.P. Comparing Landslide Inventories: The Map Depends on the Method. *Environ. Eng. Geosci.* **2002**, *8*, 279–293. [CrossRef]
50. Hungr, O.; Leroueil, S.; Picarelli, L. The Varnes classification of landslide types, an update. *Landslides* **2014**, *11*, 167–194. [CrossRef]
51. Cruden, D.M.; Varnes, D.J. Landslide Types and Processes. In *Landslides: Investigation and Mitigation*; Turner, A.K., Schuster, R.L., Eds.; Sp. Rep. 247; Transportation Research Board, National Research Council, National Academy Press: Washington, DC, USA, 1996; pp. 36–75. ISBN 978-0309062084.
52. Congalton, R.G.; Green, K. *Assessing the Accuracy of Remotely Sensed Data—Principles and Practices*, 2nd ed.; CRC Press: Boca Raton, FL, USA, 2008; p. 183. ISBN 978-1420055122.
53. Foody, G.M. Status of land cover classification accuracy assessment. *Remote Sens. Environ.* **2002**, *80*, 185–201. [CrossRef]
54. Pontius, R.G.; Millones, M. Death to Kappa: Birth of quantity disagreement and allocation disagreement for accuracy assessment. *Int. J. Remote Sens.* **2011**, *32*, 4407–4429. [CrossRef]
55. Eisank, C.; Smith, M.; Hillier, J. Assessment of multiresolution segmentation for delimiting drumlins in digital elevation models. *Geomorphology* **2014**, *214*, 452–464. [CrossRef] [PubMed]
56. Clinton, N.; Holt, A.; Scarborough, J.; Yan, L.; Gong, P. Accuracy Assessment Measures for Object-based Image Segmentation Goodness. *Photogramm. Eng. Rem. S.* **2010**, *76*, 289–299. [CrossRef]
57. Belgiu, M.; Draguţ, L. Comparing supervised and unsupervised multiresolution segmentation approaches for extracting buildings from very high resolution imagery. *ISPRS J. Photogramm.* **2014**, *96*, 67–75. [CrossRef] [PubMed]
58. Lang, S.; Albrecht, F.; Kienberger, S.; Tiede, D. Object validity for operational tasks in a policy context. *J. Spat. Sci.* **2010**, *55*, 9–22. [CrossRef]
59. Hölbling, D.; Betts, H.; Spiekermann, R.; Phillips, C. Semi-automated landslide mapping from historical and recent aerial photography. In Proceedings of the 19th AGILE Conference on Geographic Information Science, Helsinki, Finland, 14–17 June 2016; p. 5.
60. Hölbling, D.; Eisank, C.; Friedl, B.; Weinke, E.; Kleindienst, H.; Kociu, A.; Vecchiotti, F.; Albrecht, F. EO-based landslide mapping: From methodological developments to automated web-based information delivery. In Proceedings of the 13th Congress Interpraevent—Extended Abstracts, Lucerne, Switzerland, 30 May–2 June 2016; pp. 102–103.
61. Weinke, E.; Albrecht, F.; Hölbling, D.; Eisank, C.; Vecchiotti, F. Verfahren zur Implementierung eines Kartierungsdienstes für Rutschungen auf Basis von Fernerkundungsdaten und Nutzereinbindung. *AGIT J. Angew. Geoinform.* **2016**, 46–55. [CrossRef]

6

Integrating Expert Knowledge with Statistical Analysis for Landslide Susceptibility Assessment at Regional Scale

Christos Chalkias [1,*], Christos Polykretis [1,†], Maria Ferentinou [2,†] and Efthimios Karymbalis [1,†]

1 Department of Geography, Harokopio University, El. Venizelou 70, Athens 17671, Greece; chpolycretis@hua.gr (C.P.); karymba@hua.gr (E.K.)

2 Department of Geological Sciences, University of KwaZulu-Natal, Westville, Private Bag X54001, South Africa; Ferentinou@ukzn.ac.za

* Correspondence: xalkias@hua.gr

† These authors contributed equally to this work.

Academic Editors: Ruiliang Pu and Jesus Martinez-Frias

Abstract: In this paper, an integration landslide susceptibility model by combining expert-based and bivariate statistical analysis (Landslide Susceptibility Index—LSI) approaches is presented. Factors related with the occurrence of landslides—such as elevation, slope angle, slope aspect, lithology, land cover, Mean Annual Precipitation (MAP) and Peak Ground Acceleration (PGA)—were analyzed within a GIS environment. This integrated model produced a landslide susceptibility map which categorized the study area according to the probability level of landslide occurrence. The accuracy of the final map was evaluated by Receiver Operating Characteristics (ROC) analysis depending on an independent (validation) dataset of landslide events. The prediction ability was found to be 76% revealing that the integration of statistical analysis with human expertise can provide an acceptable landslide susceptibility assessment at regional scale.

Keywords: integration modeling; landslide susceptibility; expert-based fuzzy weighting; bivariate statistics; GIS; Greece

1. Introduction

Natural hazards have been studied continuously, and systematically, and during the last three decades, their frequency seems to be on the rise [1]. The need for reliable hazard maps is uncontested, in order to mitigate their effects in both natural and manmade environments. From 1900 to 2014, nearly 6% of global natural hazards were associated with landslides [2]. Mountainous or rural environments are subject to landslides as a major natural hazard which causes substantial economic, human, and environmental losses throughout the world each year [3].

Landslide Susceptibility (LS) is defined as the likelihood of a landslide occurring in an area on the basis of local terrain conditions [4]. LS zonation refers to the division of the land in homogeneous areas or domains according to the degree of actual or potential hazard [5]. In the last years, LS mapping has become a valuable tool for assessing disaster phenomena. The cartographic products of LS modeling can be used for developing mitigation plans and selecting the more suitable locations for construction.

Given the powerful abilities of Geographic Information Systems (GIS) technology to deal with a large amount of spatial data, LS is analyzed through effective spatial analysis tools. Many GIS-based approaches have been developed for LS assessment. In general, they are divided into two groups: qualitative and quantitative methods. The most important difference between these two types of methods is their degree of objectivity. The qualitative methods, used extensively during the 1970s and

1980s, depend on the knowledge and previous experience of the experts, resulting in a high degree of subjectivity. They include the geomorphological analysis [6,7] and the use of index or parameter maps. The use of index or parameter maps is subdivided into two methods: the combination or overlay of index maps [8,9] and the logical analytical models [10,11]. Some qualitative approaches, however, incorporate the idea of ranking and weighting the parameters involved, and may be semi-quantitative in nature [12]. Examples of this kind of approach are the Analytical Hierarchy Process (AHP) [13,14], the Weighted Linear Combination (WLC) [15], and their combination [16].

The quantitative methods are based on numerical expressions of the relationship between controlling factors and landslide occurrence. They include statistical analysis, geotechnical engineering approaches [17,18], and soft computing techniques such as Artificial Neural Network (ANN) [19–22], neuro-fuzzy [23,24], support vector machine [16,23] and decision tree [23,25] methods. In statistical analysis methods, the combination of parameters that have led to landslides in the past are determined statistically, and quantitative predictions are made for areas not affected by landslides but where similar background conditions exist [26]. Several researchers have applied these methods, which involve either bivariate [27–30] or multivariate analysis [16,31–33]. In bivariate analysis, each individual landslide-influencing factor is combined with a landslide inventory map, and weight values based on landslide densities are calculated for its corresponding categories. A well-known and widely used bivariate analysis method for LS zonation is the Landslide Susceptibility Index (LSI) [14,22,34,35]. It is increasingly recognized that in both qualitative and quantitative models, the results are prone to the inherent uncertainties related to various analysis parameters such as the errors and variability in model choice, weighting factors, data used, and human judgment [15].

In recent years, another soft computing technique, known as fuzzy logic, has been applied to map LS. Considering the determination of weightings (fuzzy membership values), this technique is either expert-based (semi-quantitative) approach or data-driven (quantitative) approach or a combination of both [36]. In expert-based approaches [37,38], the weightings are assigned based on the field knowledge of the experts, whereas in data-driven approaches [39,40] the weightings are determined according to correlations between landslide locations and landslide-conditioning factors. Such an expert-based approach, referred as Trapezoidal Fuzzy Number Weighting (TFNW) [12,41,42], was used in this study.

In Greece, previous studies concerning LS assessment have been implemented on moderate scale (1:25,000–1:500,000) using semi-quantitative methods such as AHP, WLC and TFNW [12,43–45], statistical analysis [22,35,46,47], ANN [22,48] and integration methods [49], and on national scale (<1:1,000,000) using statistical analysis [50,51], ANNs [50] and fuzzy logic [52].

The main aim of this paper was to improve our previous research works [12,35], integrating two different and already tested LS assessment methods: LSI and TFNW. Thus, a GIS-based integration method was developed and presented at regional scale (1:500,000) for Peloponnese peninsula, Greece. Furthermore, a validation analysis was implemented to estimate the prediction ability of the proposed method.

2. Study Area

Peloponnese was selected as the study area for the proposed method. It is one of the nine geographical departments of Greece and is located in its southern part (Figure 1). Peloponnese constitutes the largest peninsula of the country with an extent of 21,439 km^2. It is divided into seven administrative prefectures: Achaia, Argolida, Arkadia, Ilia, Korinthia, Lakonia and Messinia (Figure 1), with a population of 1,086,935 inhabitants [53]. Peloponnese has a typical Mediterranean climate with a hot and relatively dry summer, and a wet season during autumn, winter and spring [54]. The climate as well as the geotectonic status, the lithology and the seismicity have mainly formed the geomorphology of the study area. In general, its geomorphology can be characterized as complex, with a mountainous interior, many coastal cliffs in the south, and basins, coastal beaches, lakes and inland basins on the west and southeast coasts.

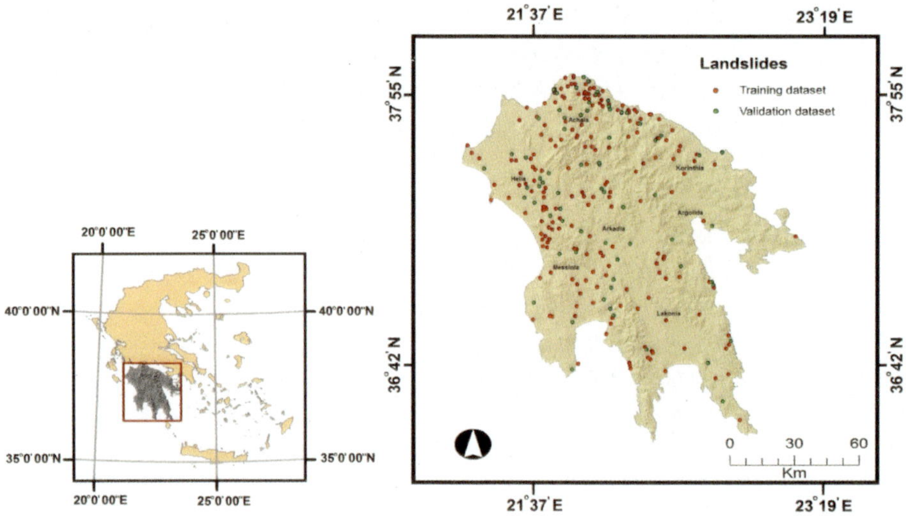

Figure 1. The location of the study area (Peloponnese peninsula) and the landslide inventory.

Peloponnese is a region highly vulnerable to the occurrence of severe natural disasters such as earthquakes, floods and forest fires. However, due to its complex geomorphology, all causal and triggering factors of landslide occurrence present a high spatial variability. Considering the morphological and geological conditions of Peloponnese, it is evident that slope stability is one of the most severe and frequent hazards for public safety as well as land-use planning and management of the area.

3. Materials and Methods

In order to accomplish the LS analyses in the study area, a spatial database was designed and developed, and spatial analysis tools were implemented within GIS environment with the use of an ArcGIS (version 9.3) software package. This database comprises two main parts: (a) the landslide inventory dataset and (b) the datasets for the landslide-influencing factors.

3.1. Landslide Inventory

In the present study, two main landslide inventories were used: (a) an inventory maintained by Institute of Geology and Mineral Exploration (IGME) covering a time period from 1910 to 1995 [55] and (b) an inventory, developed on the basis of fieldwork and aerial photography [56] interpretation, which contains landslide events occurring from 1995 to 2003. The final landslide dataset consists of 282 landslides (217 and 65 landslides from the first and second inventory, respectively) presented as point features in the centroid of each pixel (Figure 1). This dataset was randomly split into two separate groups: a training dataset and a validation dataset. The training dataset (75% of the landslide inventory) was used for the implementation of the model, whereas the validation dataset (25% of the landslide inventory) was used during the verification of the results produced from the model (see among others [57–59]). Our intention was to keep the same sampling strategy (*i.e.*, the same percentages for training and validation datasets) for both the integration model and the single models [12,35] combined for its creation in order to increase their comparability.

The landslides under investigation have more or less the same characteristics: lateral based and downslope movement of soils or rocks. In line with [60], in this study, the term landslide is used for translational and rotational earth slides, which were recorded in landslide inventories. These events vary consistently in volume, from some thousands of m^3 to several million m^3 [51], and depicting small to extremely large magnitude according to [61] classification. Subaqueous and liquefaction events are not examined.

3.2. Landslide-Influencing Factors

Although there are no standard guidelines for selecting the landslide-influencing factors, the nature of the study area, the scale of the analysis, the data availability and the general literature guidelines were taken into account [26,62–65]. In this study, land cover, lithology, elevation, slope angle, slope aspect, MAP and PGA have been selected as the landslide-influencing factors.

Slope stability is highly influenced by land cover. The land cover layer was classified by following the level-1 classification scheme of the CORINE Coordinate of Information on the Environment) program data [66]. Agricultural, forest and semi-natural areas cover the main part of Peloponnese, whereas urban is the dominant land cover in the coastal zone of the peninsula.

Lithology plays a key role in landslide activity since different lithologic units have different slope stability performances [67]. The lithological layer of the study area was created from the Geological Map of Greece [68]. The main lithological formations are carbonate rocks (44%): limestones, dolomites and marbles, and Neogene sediments (22%): usually marls, sandstones and mudstones. The central and western part of Peloponese belongs to the External Hellenides, which is characterized by Mesozoic–early Cenozoic sedimentary units (mainly carbonate rocks, schists-cherts and flysch) [69]. The most critical landslide-prone formations are flysch and neogene sediments, while schists and cherts significantly contribute in landslide phenomena [51].

Among all parameters for LS zonation, elevation, slope angle and slope aspect have been recognized as the most important [70–72]. The elevation dataset is useful to classify the local relief and locate points of maximum and minimum heights within terrains. Furthermore, it is well supported in the literature [43,46] that slope gradients have a large impact on landsliding in Peloponnese. In the study area, the slopes vary from gentle to very steep. Most of the landslides occur on gentle natural slopes where the translational type predominates. The aspect parameter is also related to differential weathering, exposure to sunlight and drying winds and soil moisture [62]. The Digital Elevation Model (DEM) was the key to generating the topographic factors related to the landslide activity of the area under investigation. We used the DEM from SRTM (Shuttle Radar Topography Mission) database. Elevation (0–2367 m), slope angle (0°–54°) and slope aspect were extracted from this DEM.

Previous studies emphasized the need of incorporating "dynamic" factors (MAP, PGA) with other "static" factors for LS zonation mapping in areas, whereas these factors are playing an important role not only in the reactivation of old landslides but in the development of new ones [73,74]. It is, therefore, imperative to incorporate these parameters, while carrying out LS analysis in seismically active and frequently rainfall-influenced regions. The precipitation data (339–1655 mm) used in this study refers to the mean annual precipitation during the period from 1950 to 1974 (source: Public Power Company, Athens, Greece). MAP is the average of the available long-term records [75]. The study area belongs to a seismically active zone with tectonism expressed through faults, thrust zones and folds. The seismic factor (0.05–3.95 m/s^2) was produced from the map of expected peak ground acceleration with a 475-year return period (10% probability of exceedance in 50 years). This map, from which the corresponding layer was derived, was launched in 1992 from the Technical Chamber of Greece. PGA is the absolute maximum amplitude of recorded ground acceleration [76]. Heavy rainfall and earthquakes have triggered several landslide events, mainly in northern and western areas of the peninsula [43,77,78]. Rainfall triggered landslides usually are rapid short moving events, while slow short-moving type also occur including extensive instability zones [79]. On the other hand, seismic triggered landslides occur in the vicinity of active faults, and usually related to other secondary seismic events, like soil liquefaction, subsidence of the coastal strip, and rock falls [80,81].

Most of the final layers were in raster format (grid), while others were converted from vector (point, line or polygon) to raster format (with cell size 250 m × 250 m for land cover, lithology, MAP and PGA factors, and 90 m × 90 m for elevation, slope angle and slope aspect factors).

3.3. Methodology

A GIS-based integration approach was developed to create a LS map using two different methods: LSI and TFNW. Such integration can adjust the parameters included in this study by means of the precise LS assessment and effectively reduce the subjectivity and uncertainty resulting from the use of single method. The LSI method was proposed by [82] and is based on the relationship between the spatial distribution of landslides and each conditioning factor [62]. Here, in order to quantify the impact of various categories for the standardization of each factor, we used the TFNW. The details of this approach have been described by [41]. The main steps of this methodology are presented below:

(a) Categorization of all landslide-influencing factors. In this step, the "Natural Breaks (Jenks)" categorization (five categories) was implemented for the "dynamic" factors (MAP and PGA), whereas for the other two factors with continuous values (elevation and slope angle) the categorization was executed in a manual way based on their presented values. "Natural Breaks (Jenks)" method is a data classification method that minimizes variance within groups of data and maximizes variance between groups of data [83]. For land cover, lithology, and aspect, all the categories of the nominal scale were preserved.

(b) Calculation of LSI for all factor categories (Table 1) based on Equation (1) [26]:

$$LSI_{i,j} = ln\left(\frac{N_{i,j}}{A_{i,j}} / \frac{N_T}{A_T}\right) \quad (1)$$

with $N_{i,j}$ the number of landslides in category j of the factor i, $A_{i,j}$ the area of this category, N_T the total number of landslides and A_T the total area under investigation. Thus, the calculation of the landslide density in each category and of the area for each category was implemented using GIS-based overlay functions.

Table 1. Categories, landslide densities, susceptibilities (LSI), fuzzy values and weights for landslide related factors.

Layers (Factors)	Categories (Classes)	Landslide Density	LSI	(Experts), STAT Fuzzy Value	LSI/TFNW Weight
Land cover	Artificial surfaces	0.09	2.04	(H, M, M), VH	0.45
	Agricultural areas	0.66	0.42	(M, H, H), H	0.42
	Forest and semi-natural land	0.25	−0.79	(L, L, M), VL	0.13
Lithology	Phyllites/Gneiss (metamorphic)	0.07	0.16	(L, M, L), H	0.14
	Limestones—Marbles	0.20	−0.77	(L, L, L), VL	0.05
	Schists (metamorphic)	0.01	−0.51	(M, M, M), VL	0.08
	Neogene	0.44	0.69	(H, VH, M), VH	0.22
	Quaternary	0.12	−0.06	(H, VH, M), M	0.16
	Flysch	0.09	0.06	(VH, VH, VH), M	0.19
	Cherts—Schists	0.06	0.36	(M, L, M), H	0.15
MAP	<663 mm	0.03	−1.69	(L, VL, M), VL	0.07
	663–884 mm	0.30	0.14	(M, L, H), M	0.19
	885–1079 mm	0.43	0.29	(M, M, H), H	0.25
	1080–1295 mm	0.19	−0.03	(H, H, VH), M	0.25
	>1295 mm	0.05	−0.22	(VH, VH, VH), L	0.23
PGA	<1.16 m/s^2	0.02	−0.14	(L, VL, L), M	0.16
	1.16–1.88 m/s^2	0.01	−0.98	(L, L, L), VL	0.08
	1.89–2.53 m/s^2	0.17	−0.81	(M, M, M), VL	0.13
	2.54–3.11 m/s^2	0.67	0.53	(H, H, H), VH	0.37
	>3.11 m/s^2	0.13	−0.28	(VH, VH, VH), L	0.27

Table 1. *Cont.*

Layers (Factors)	Categories (Classes)	Landslide Density	LSI	(Experts), STAT Fuzzy Value	LSI/TFNW Weight
Elevation	<234 m	0.36	0.17	(L, VL, L), H	0.25
	234–524 m	0.27	0.11	(L, L, M), M	0.22
	525–851 m	0.22	–0.03	(H, M, M), M	0.28
	>851 m	0.15	–0.43	(M, H, H), L	0.24
Slope angle	<5°	0.31	–0.16	(VL, VL, L), L	0.09
	5°–10°	0.30	0.19	(L, L, L), H	0.21
	11°–15°	0.22	0.23	(M, M, M), H	0.25
	16°–20°	0.12	0.08	(H, H, H), M	0.25
	>20°	0.05	–0.77	(VH, VH, VH), VL	0.21
Slope aspect	North	0.31	0.31	(H, H, M), H	0.31
	East	0.21	–0.15	(M, M, M), M	0.23
	South	0.16	–0.47	(M, M, M), L	0.18
	West	0.32	0.16	(H, H, H), M	0.28

LSI: Landslide Susceptibility Index, LSI/TFNW: The integration method, VL: Very Low, L: Low, M: Moderate, H: High, VH: Very High.

(c) The maximum value, the range and the standard deviation of LSI values for all factors were calculated in order to interpret the importance of each factor. The range is the difference between the maximum and minimum LSI values. The standard deviation measures the spread of the LSI values about the mean value.

(d) Definition of linguistic variables and fuzzy numbers for LS categories in order to incorporate uncertainty in the analysis. All fuzzy numbers were expressed as (a_k, b_k, c_k, d_k). The definition of these fuzzy numbers is presented in [12].

(e) Three independent experts, with scientific background in the fields of geosciences and engineering geology, and with experience and scientific knowledge of the study area, were invited to assign a linguistic importance weighting for every category of each factor. From these linguistic judgments we obtained the corresponding fuzzy numbers. Such judgments are inevitably subjective, but, by proposing several possible scenarios, followed by the systematic testing and elimination of options, as a result of additional investigation and discussion, it is possible to develop reliable estimates. Experimental evidence suggests that group judgments appear to be more accurate than judgments of a typical group member [84]. In this case, we considered a homogeneous group of experts with equal degree of importance for each one. Accordingly, the overall expert-based judgment is the mean value of the three judgment values. The sum of these numbers is still a fuzzy number. Thus, we proceeded to the computation of the aggregated fuzzy weights of individual categories (Table 1).

(f) A fourth "statistical expert" judgment was added in the above list on the basis of the statistical analysis (Table 1). This statistical interpretation was based on the LSI values for each category according to the following rules:

- In a category with LSI value between –0.15 and 0.15, a "Moderate" susceptibility was assigned.
- Categories with negative (<–0.15) LSI values were defined as zones of "Very Low" or "Low" susceptibility according to their ascending order.
- Categories with positive (>0.15) LSI values were defined as zones of "Very High" or "High" susceptibility according to their descending order.

Accordingly, the range of LSI values was interpreted for each factor according to the following rules:

- In a factor with range between 1.00 and 1.50, a "Moderate" susceptibility was assigned.
- Factors with range <1 were characterized with "Very Low" or "Low" susceptibility according to their ascending order.

- Factors with range >1.50 were characterized with "Very High" or "High" susceptibility according to their descending order.

It is worth mentioning that in our case, we assigned an equal degree of importance between the "statistical expert" and the group of experts.

(g) After the defuzzification of the fuzzy weights of individual LS categories, we proceed to the computation of the normalized weights and the construction of the weight vector (Table 1). More details of this step could be found in [12].

By implementing the same procedure described previously, we used expert-based and statistical-based linguistic variables to calculate the weight of importance for each conditioning factor (Table 2).

Table 2. Weight values for each factor according to the integration LSI/TFNW method.

Parameter	Parameter Importance (Expert Judgment), STAT	LSI/TFNW Weight
Land cover	(M, M, H), M	0.12
Lithology	(VH, VH, H), H	0.17
MAP	(VH, VH, VH), VH	0.20
PGA	(H, H, H), H	0.16
Elevation	(M, L, H), L	0.08
Slope angle	(VH, VH, VH), H	0.18
Slope aspect	(H, M, M), L	0.09

(h) The last step is the aggregation of relative values, and the generation of the final LS map (Figure 2). This step was implemented by using the WLC method [85]. We classified the final LS map into five discrete categories: "Very Low", "Low", "Moderate", "High" and "Very High" landslide susceptibility according to the standard deviation classification method. This method uses the mean value to generate class breaks by adding or subtracting one standard deviation at a time [70]. Moreover, in order to maintain five categories, we embedded low and high outliers into "Very Low" and "Very High" susceptibility categories, respectively.

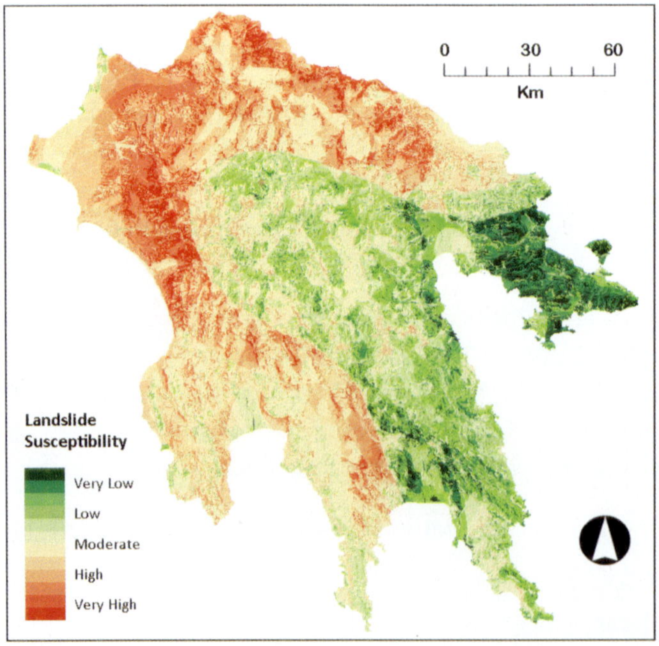

Figure 2. The landslide susceptibility map produced from the integration LSI/TFNW model.

4. Results—Validation

As mentioned previously, we developed an integration LS model according to the LSI/TFNW approach. This approach allowed for the assessment of landslide susceptibility by integrating expert-based and statistical modelling methods.

The results from the LSI/TFNW model are presented in Tables 1 and 2. The most important causal factors were found to be MAP, slope angle and lithology with weight values 0.20, 0.18 and 0.17, respectively. Elevation was the weakest factor in the scale of importance (weight value = 0.08).

The output LS map (Figure 2) from the integration model shows that 26% and 7% of the study area were classified as "High" and "Very High" susceptibility zones, respectively. The same map also shows that the zones of "High" and "Very High" susceptibility are mainly located around the northern (coastal) and western part of Peloponnese, with pockets of "High" susceptibility in the rest part of the peninsula.

The overlay of the final LS map with the landslide training dataset indicated that 28%, 42% and 24% (total: 94%) of the landslide events fall within "Very High", "High" and "Moderate" LS zones (in total 66% of the study area), respectively. It is notable that according to the used model only 6% of the dataset falls in "Low" susceptibility zone, with no landslides into the "Very Low" susceptibly zone.

We have also proceeded with a standard validation analysis using data from landslide events, not included in the initial spatial database (validation dataset), in order to estimate the overall performance of the model in the study area. Validation is essential to know the predictive value of the model [86]. For the validation of the output, the Receiver Operating Characteristics (ROC) curve was drawn, and the Area Under Curve (AUC) value was calculated for the proposed model. ROC analysis is considered as a powerful method for the validation of LS models [87,88]. In practice, the AUC performs very well and is often used when a general measure of predictiveness is desired [89]. The AUC values range from 0.5 to 1.0. The ideal model yields a value close to 1.0 (perfect fit), whereas a value close to 0.5 indicates an inaccurate model (random fit).

Figure 3 shows the ROC curve of LSI/TFNW model for the validation dataset. The AUC value of 0.76 indicates a reasonable prediction ability of the model.

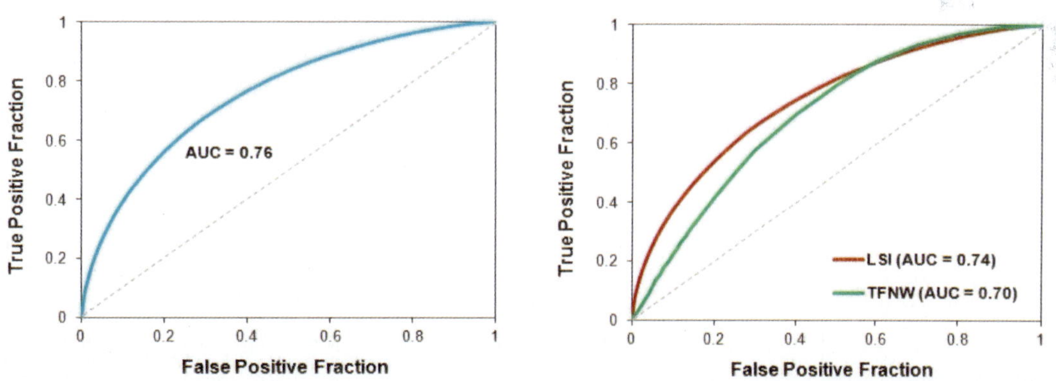

Figure 3. ROC curves for the integration LSI/TFNW model and the combined single models, LSI and TFNW.

Previous studies [12,35] in the same area with the use of the single methods combined in this integration model (TFNW and LSI, respectively), produced slightly worse results (Figure 3). The AUC values for these models were 0.70 and 0.74, respectively, based on the standard deviation classification method for their final LS maps.

Specifically, comparing the final map of the proposed integrated method with the landslide susceptibility map produced from TFNW model [12], it is shown that in both maps 14% (3039 km^2), 9% (2001 km^2) and 1% (301 km^2) of the study area are defined as zones of "Moderate", "High" and "Very High" susceptibility, respectively. On the other hand, the differences between the LS maps for these three susceptibility zones are presented to cover 32% (6773 km^2) of the study area. The differences for the remaining susceptibility zones ("Very Low" and "Low") cover a restricted part of the study area (7% or 1377 km^2).

In addition, comparing the final LS map of this study with this one created from LSI model [35], it is derived that in both maps 21% (4427 km^2), 18% (3829 km^2) and 4% (881 km^2) of the study area belong to "Moderate", "High" and "Very High", respectively, susceptibility zones. On the contrary, the coverage of differences between the final maps for these three susceptibility zones is found to be 16% (3291 km^2) of the study area. Finally, the coverage of differences for the remaining susceptibility zones is more or less similar with the relative coverage of the aforementioned comparison between integration and TFNW models.

5. Discussion

The idea of implementing an integration method for LS zonation is not new. Recent examples are the efforts of [90–92]. In the first study, AHP and fuzzy sets were combined to determine LS levels for areas of the Metro Vancouver region, British Columbia, Canada. The parameters were standardized to a common measurement scale using fuzzy set membership functions and the weight value for each parameter was calculated using AHP. In the second study, the authors produced a LS zonation map in the Chehelchay Basin (northeastern Iran). They applied Geographically-Weighted Principal Component Analysis (GWPCA) to calculate LS using a fuzzy gamma operator. Finally, the third study focused on the methodology of the urgent LS assessment right after the 2013 Lushan earthquake (southwest of Sichuan Province, China). An assessment approach was developed integrating bivariate statistical analysis (LSI) and AHP methods for the assignment of factor class weights and factor weights, respectively.

This study combined an expert-based semi-quantitative approach (TFNW) with a data-driven statistical method (LSI) to prepare a LS map at a regional scale on the Peloponnese peninsula, Greece. The main motivation is to create an integration model that preserves and integrates the advantages of the combined models at the specific scale of analysis. To achieve this objective, seven conditioning factors (elevation, slope angle, slope aspect, lithology, land cover, MAP and PGA) were taken into consideration.

The LSI model is a data-driven bivariate statistical approach. The bivariate statistical technique is one of the most preferable models at this scale [93]. In this LS modeling, all the problems of landslide inventory maps will automatically be inherited to the final susceptibility maps [94]. In general, a reliable, accurate and complete landslide inventory map will provide high-quality statistical models [95]. In many countries—including Greece—this kind of extended landslide inventory is not available.

On the other hand, for the expert-based fuzzy weighting model, the landslide inventory map is not needed. This kind of analysis is purely subjective. To some extent, opinions may change for every individual expert and thus may be subjected to cognitive limitations with uncertainty and subjectivity. However, methods that depend on expert opinions are often useful for regional assessments [96]. A significant part of the semi-quantitative LS research follows a similar strategy as the proposed one by inviting a limited number of experts (see among others [42,97]). The main issue is not to invite as many scientists as possible but to invite experts with detailed knowledge of the landslide events in the area under investigation and to ensure an overall consensus of their evaluation about the importance of the factors involved.

Recently, ROC analysis has been widely used in the international literature to evaluate the prediction capability of the LS mapping methods. In this study, the empiric ROC area for the integration model was estimated to be 0.76 for the validation dataset. Thus, there is 76% agreement between prepared LS map and landslide locations of the validation dataset, which is a reasonable result, taking into consideration the scale of analysis. Moreover, this model proved to be slightly more effective for the overall LS assessment of the study area as it recorded the highest percentage in our validation test in comparison with LSI and TFNW methods (74% and 70%, respectively). Therefore, the reliability of the proposed integration model is directly related to the reliability of both data-driven and expert-based methods combined. In other words, it seems that a combination of two quite satisfactory methods according to the proposed integration model generates even better results. However, this preliminary finding needs more research to be confirmed through the implementation of the proposed model in more studies areas with various geographical and geotectonic settings.

The implementation of the integration model in the study area revealed that there are different zones within Peloponnese which seem to configure various landslide susceptibility clusters. The "High" susceptibility areas are concentrated in the western Peloponnese with few scattered pockets in the remaining parts of the peninsula. Additionally, a linear concentration of "High" susceptibility values is located in the northern coastal zone of the study area. According to the final LS map, the "Very High" susceptibility zone covers a significant part of the study area (7% of the total area).

The comparison of the proposed method with the expert-based and the statistical methods showed that the integrated method illustrates more emphatically the zones of "High" or "Low" susceptibility in both of the individual methods. Accordingly, the main part the areas of "High" and "Very High" susceptibility in the hybrid map (classes "4" and "5"), is within the class "5" ("Very high" susceptibility) in the statistical and the expert-based LS maps (99% and 67%, respectively). Concerning the spatial distribution of the areas with high LS values in the integrated method, despite the preservation of the general spatial pattern (high LS values in the northern and the western part of Peloponnese) it has to be noticed that there are some pockets of high LS in areas with limited existence of previous landslides. This fact is due to the expert-based classification and it may be useful to identify high LS zones in areas with lack of landslide events.

The most important factors for the zonation of LS in the study area are MAP, slope angle and lithology. It seems that the incorporation of "dynamic" factors (MAP, PGA) in this regional analysis was more or less beneficial to the LS assessment. Among the categories of these two factors, the categories with high levels of annual precipitation (885–1079 mm) and seismic acceleration (2.54–3.11 m/s^2) have the highest LSI values combining the highest landslide density and coverage area. In contrast with aforementioned categories of MAP and PGA factors, the category with $11°–15°$ slope angle presents the highest LSI value combining a lower landslide density and coverage area than the categories with low levels of slope angle ($<10°$).

Some limitations and assumptions of the proposed method have to be pointed out. First of all, as the analysis is based on medium-scale datasets, the results are unsuitable for detailed site-oriented specific analysis. At large scales, more exhaustive datasets and detailed geotechnical information are required. A second limitation is related to the landslide inventory dataset. At regional scale, this dataset does not include the total amount of the landslide events within the area under investigation. In this paper, we considered equal importance between the statistical and expert-based models for the construction of the integration model. Although this is a reasonable starting point, alternative importance ratios should be further tested. Finally, another limitation of this study is that the proposed method does not take into account the mutual relationships between causal factors. Future work will focus on this issue by combining expert-based with multivariate statistical models. Furthermore, it should be mentioned that, according to our analysis, the output LS map presents only the predicted spatial distribution of landslides. It does not present the temporal probability of landsliding.

Therefore, the results from this paper should be used in the first stage of hazard mapping, which is crucial in an essential part of quantitative risk analysis [98,99]. Despite its limitations, the proposed method can produce reliable landslide susceptibility maps at regional scale. This is very useful information for local authorities and decision makers in order to choose suitable locations for future land-use planning and implementation of developments, and target their mitigation strategies.

6. Conclusions

Statistical analysis and expert-based methods are often used to evaluate LS zonation. Considering that these general approaches have their pros and cons when applied at regional scale, we investigate the performance of an integration model. The results obtained show that it is possible to produce satisfactory LS maps with the proposed hybrid model. It seems that the overall assessment provided by the experts efficiently incorporates the knowledge of the study area and of the factors involved in the LS zonation. Thus, to some extent, the expertise of specialists can contribute to the enhancement of the standard statistical method especially in moderate and small scale analyses in areas with incomplete landslide event archives.

Further research may include the validation of the proposed model in other areas characterized by different geomorphological characteristics, and its comparison with other advanced LS models (such as logistic regression and ANN models) at regional scale.

Acknowledgments: Authors thank the three experts for their contribution in this paper. Also, authors are grateful to the anonymous reviewers for their useful comments and constructive manuscript reviewing.

Author Contributions: Christos Chalkias conceived of the study, designed the analysis, and participated in its implementation. Christos Polykretis participated in the spatial database creation and the implementation of statistical analysis, and drafted the manuscript. Maria Ferentinou and Efthimios Karymbalis participated in the design and implementation of the study and in the preparation of the validation dataset. Christos Chalkias, Christos Polykretis, Maria Ferentinou and Efthimios Karymbalis participated in the finalization of the manuscript. All authors read and approved the final manuscript.

Conflicts of Interest: The authors declare no conflict of interest.

References

1. Smith, K.; Petley, D.N. *Environmental Hazards: Assessing risk and reducing disaster*, 5th ed.; Routledge: Abingdon, UK, 2009.
2. Centre for Research on the Epidemiology of Disasters (CRED). *EM-DAT: International Disaster Database*; Université Catholique de Louvain: Brussels, Belgium, 2014.
3. Aleotti, P.; Chowdhury, R. Landslide hazard assessment: Summary review and new perspectives. *Bull. Eng. Geol. Environ.* **1999**, *58*, 21–44. [CrossRef]
4. Brabb, E.E. Innovative approaches to landslide hazard mapping. In Proceedings of the 4th International Symposium on Landslides, Toronto, ON, Canada, 16–21 September 1984; Volume 1, pp. 307–324.
5. Varnes, D.J. *Landslide Hazard Zonation: A Review of Principles and Practice*; UNESCO: Paris, France, 1984; Volume 3.
6. Santangelo, M.; Cardinali, M.; Rossi, M.; Mondini, A.C.; Guzzetti, F. Remote landslide mapping using a laser rangefinder binocular and GPS. *Nat. Hazards Earth Syst. Sci.* **2010**, *10*, 2539–2546. [CrossRef]
7. Listo, F.L.R.; Vieira, B.C. Mapping of risk and susceptibility of shallow-landslide in the city of Sao Paulo, Brazil. *Geomorphology* **2012**, *169–170*, 30–44. [CrossRef]
8. Castellanos Abella, E.A.; van Westen, C.J. Qualitative landslide susceptibility assessment bymulticriteria analysis: A case study from San Antonio del Sur, Guantánamo, Cuba. *Geomorphology* **2008**, *94*, 453–466. [CrossRef]
9. Avtar, R.; Singh, C.K.; Singh, G.; Verma, R.L.; Mukherjee, S.; Sawada, H. Landslide susceptibility zonation study using remote sensing and GIS technology in the Ken-Betwa River Link area, India. *Bull. Eng. Geol. Environ.* **2011**, *70*, 595–606. [CrossRef]
10. Intarawichian, N; Dasananda, S. Analytical hierarchy process for landslide susceptibility mapping in lower Mae Chaem watershed, northern Thailand. *Suranaree J. Sci. Technol.* **2010**, *17*, 277–292.
11. Demir, G.; Aytekin, M.; Akgün, A.; Ikizler, S.B.; Tatar, O. A comparison of landslide susceptibility mapping of the eastern part of the North Anatolian Fault Zone (Turkey) by likelihood-frequency ratio and analytic hierarchy process methods. *Nat. Hazards* **2013**, *65*, 1481–1506. [CrossRef]
12. Chalkias, C.; Ferentinou, M.; Polykretis, C. GIS supported landslide susceptibility modeling at regional scale: An expert-based fuzzy weighting method. *ISPRS Int. J. Geo-Inf.* **2014**, *3*, 523–539. [CrossRef]

13. Kayastha, P.; Dhital, M.R.; De Smedt, F. Application of the Analytical Hierarchy Process (AHP) for landslide susceptibility mapping: A case study from the Tinau watershed, west Nepal. *Comput. Geosci.* **2013**, *52*, 398–408. [CrossRef]
14. Pourghasemi, H.R.; Moradi, H.R.; Fatemi Aghda, S.M. Landslide susceptibility mapping by binary logistic regression, analytical hierarchy process, and statistical index models and assessment of their performances. *Nat. Hazards* **2013**, *69*, 749–779. [CrossRef]
15. Feizizadeh, B.; Blaschke, T. An uncertainty and sensitivity analysis approach for GIS-based multicriteria landslide susceptibility mapping. *Int. J. Geogr. Inf. Sci.* **2014**, *28*, 610–638. [CrossRef]
16. Kavzoglu, T.; Sahin, E.K.; Colkesen, I. Landslide susceptibility mapping using GIS-based multi-criteria decision analysis, support vector machines, and logistic regression. *Landslides* **2014**, *11*, 425–439. [CrossRef]
17. Michel, G.P.; Kobiyama, M.; Goerl, R.F. Comparative analysis of SHALSTAB and SINMAP for landslide susceptibility mapping in the Cunha River basin, southern Brazil. *J. Soils Sediments* **2014**, *14*, 1266–1277. [CrossRef]
18. Dahal, R.K.; Bhandary, N.P.; Hasegawa, S.; Yatabe, R. Topo-stress based probabilistic model for shallow landslide susceptibility zonation in the Nepal Himalaya. *Environ. Earth Sci.* **2014**, *71*, 3879–3892. [CrossRef]
19. Wu, X.; Niu, R.; Ren, F.; Peng, L. Landslide susceptibility mapping using rough sets and backpropagation neural networks in the Three Gorges, China. *Environ. Earth Sci.* **2013**, *70*, 1307–1318. [CrossRef]
20. Bi, R.; Schleier, M.; Rohn, J.; Ehret, D.; Xiang, W. Landslide susceptibility analysis based on ArcGIS and Artificial Neural Network for a large catchment in Three Gorges region, China. *Environ. Earth Sci.* **2014**, *72*, 1925–1938. [CrossRef]
21. Conforti, M.; Pascale, S.; Robustelli, G.; Sdao, F. Evaluation of prediction capability of the artificial neural networks for mapping landslide susceptibility in the Turbolo River catchment (northern Calabria, Italy). *Catena* **2014**, *113*, 236–250. [CrossRef]
22. Polykretis, C.; Ferentinou, M.; Chalkias, C. A comparative study of landslide susceptibility mapping using landslide susceptibility index and artificial neural networks in the Krios River and Krathis River catchments (northern Peloponnesus, Greece). *Bull. Eng. Geol. Environ.* **2015**, *74*, 27–45. [CrossRef]
23. Pradhan, B. A comparative study on the predictive ability of the decision tree, support vector machine and neuro-fuzzy models in landslide susceptibility mapping using GIS. *Comput. Geosci.* **2013**, *51*, 350–365. [CrossRef]
24. Sdao, F.; Lioi, D.S.; Pascale, S.; Caniani, D.; Mancini, I.M. Landslide susceptibility assessment by using a neuro-fuzzy model: A case study in the Rupestrian heritage rich area of Matera. *Nat. Hazards Earth Syst. Sci.* **2013**, *13*, 395–407. [CrossRef]
25. Park, I.; Lee, S. Spatial prediction of landslide susceptibility using a decision tree approach: A case study of the Pyeongchang area, Korea. *Int. J. Remote Sens.* **2014**, *35*, 6089–6112. [CrossRef]
26. He, Y.; Beighley, R.E. GIS-based regional landslide susceptibility mapping: A case study in southern California. *Earth Surf. Process. Landf.* **2008**, *33*, 380–393. [CrossRef]
27. Pourghasemi, H.R.; Pradhan, B.; Gokceoglu, C.; Mohammadi, M.; Moradi, H.R. Application of weights-of-evidence and certainty factor models and their comparison in landslide susceptibility mapping at Haraz watershed. *Iran. Arab. J. Geosci.* **2013**, *6*, 2351–2365. [CrossRef]
28. Epifânio, B.; Zêzere, J.L.; Neves, M. Susceptibility assessment to different types of landslides in the coastal cliffs of Lourinhã (Central Portugal). *J. Sea Res.* **2014**, *93*, 150–159. [CrossRef]
29. Sujatha, E.R.; Kumaravel, P.; Rajamanickam, G.V. Assessing landslide susceptibility using Bayesian probability-based weight of evidence model. *Bull. Eng. Geol. Environ.* **2014**, *73*, 147–161. [CrossRef]
30. Youssef, A.M.; Al-Kathery, M.; Pradhan, B. Landslide susceptibility mapping at Al-Hasher Area, Jizan (Saudi Arabia) using GIS-based frequency ratio and index of entropy models. *Geosci. J.* **2015**, *19*, 113–134. [CrossRef]
31. Bathrellos, G.D.; Kalivas, D.P.; Skilodimou, H.D. GIS-based landslide susceptibility mapping models applied to natural and urban planning in Trikala, Central Greece. *Estud. Geol.* **2009**, *65*, 49–65. [CrossRef]
32. Bai, S.B.; Wang, J.; Thiebes, B.; Cheng, C.; Chang, Z.Y. Susceptibility assessments of the Wenchuan earthquake-triggered landslides in Longnan using logistic regression. *Environ. Earth Sci.* **2014**, *71*, 731–743. [CrossRef]
33. Reichenbach, P.; Busca, C.; Mondini, A.C.; Rossi, M. The influence of land use change on landslide susceptibility zonation: The Briga catchment test site (Messina, Italy). *Environ. Manag.* **2014**, *54*, 1372–1384. [CrossRef] [PubMed]

34. Bourenane, H.; Bouhadad, Y.; Guettouche, M.S.; Braham, M. GIS-based landslide susceptibility zonation using bivariate statistical and expert approaches in the city of Constantine (Northeast Algeria). *Bull. Eng. Geol. Environ.* **2015**, *74*, 337–355. [CrossRef]
35. Chalkias, C.; Ferentinou, M.; Polykretis, C. GIS-based landslide susceptibility mapping on the Peloponnese Peninsula, Greece. *Geosciences* **2014**, *4*, 176–190. [CrossRef]
36. Tien Bui, D.; Pradhan, B.; Lofman, O.; Revhaug, I.; Dick, O.B. Spatial prediction of landslide hazards in Hoa Binh province (Vietnam): A comparative assessment of the efficacy of evidential belief functions and fuzzy logic models. *Catena* **2012**, *96*, 28–40. [CrossRef]
37. Sujatha, E.R.; Rajamanickam, V. Landslide susceptibility mapping of Tevankarai Ar sub-watershed, Kodaikkanal taluk, India, using weighted similar choice fuzzy model. *Nat. Hazards* **2011**, *59*, 401–425. [CrossRef]
38. Zhu, A.X.; Wang, R.; Qiao, J.; Qin, C.Z.; Chen, Y.; Liu, J.; Du, F.; Lin, Y.; Zhu, T. An expert knowledge-based approach to landslide susceptibility mapping using GIS and fuzzy logic. *Geomorphology* **2014**, *214*, 128–138. [CrossRef]
39. Pourghasemi, H.R.; Pradhan, B.; Gokceoglu, C. Application of fuzzy logic and Analytical Hierarchy Process (AHP) to landslide susceptibility mapping at Haraz watershed, Iran. *Nat. Hazards* **2012**, *63*, 965–996. [CrossRef]
40. Kayastha, P.; Bijukchhen, S.M.; Dhital, M.R.; De Smedt, F. GIS based landslide susceptibility mapping using a fuzzy logic approach: A case study from Ghurmi-Dhad Khola area, eastern Nepal. *J. Geol. Soc. India* **2013**, *82*, 249–261. [CrossRef]
41. Wang, W.D.; Xie, C.M.; Du, X.G. Landslides susceptibility mapping based on geographical information system, GuiZhou, south-west China. *Environ. Geol.* **2009**, *58*, 33–43. [CrossRef]
42. Wang, W.D.; Guo, J.; Fang, L.G.; Chang, X.S. A subjective and objective integrated weighting method for landslides susceptibility mapping based on GIS. *Environ. Earth Sci.* **2012**, *65*, 1705–1714. [CrossRef]
43. Rozos, D.; Bathrellos, G.D.; Skillodimou, H.D. Comparison of the implementation of rock engineering system and analytic hierarchy process methods, upon landslide susceptibility mapping, using GIS: A case study from the Eastern Achaia County of Peloponnesus, Greece. *Environ. Earth Sci.* **2011**, *63*, 49–63. [CrossRef]
44. Kritikos, T.; Davies, T.R.H. GIS-based multi-criteria decision analysis for landslide susceptibility mapping at northern Evia, Greece. *Z. Deutsch. Ges. Geowiss.* **2011**, *162*, 421–434.
45. Kouli, M.; Loupasakis, C.; Soupios, P.; Rozos, D.; Vallianatos, F. Landslide susceptibility mapping by comparing the WLC and WofE multi-criteria methods in the West Crete Island, Greece. *Environ. Earth Sci.* **2014**, *72*, 5197–5219. [CrossRef]
46. Papathanassiou, G.; Valkaniotis, S.; Ganas, A.; Pavlides, S. GIS-based statistical analysis of the spatial distribution of earthquake-induced landslides in the island of Lefkada, Ionian Islands, Greece. *Landslides* **2013**, *10*, 771–783. [CrossRef]
47. Chalkias, C.; Kalogirou, S.; Ferentinou, M. Landslide susceptibility, Peloponnese Peninsula in South Greece. *J. Maps* **2014**, *10*, 211–222. [CrossRef]
48. Tsangaratos, P.; Benardos, A. Estimating landslide susceptibility through an artificial neural network classifier. *Nat. Hazards* **2014**, *74*, 1489–1516. [CrossRef]
49. Gemitzi, A.; Falakakis, G.; Eskioglou, P.; Petalas, C. Evaluating landslide susceptibility using environmental factors, fuzzy membership functions and GIS. *Glob. NEST J.* **2011**, *13*, 28–40.
50. Ferentinou, M.; Chalkias, C. Mapping mass movement susceptibility across Greece with GIS, ANN and statistical methods. *Landslide Sci. Pract.* **2013**, 321–327.
51. Sabatakakis, N.; Koukis, G.; Vassiliades, E.; Lainas, S. Landslide susceptibility zonation in Greece. *Nat. Hazards* **2013**, *65*, 523–543. [CrossRef]
52. Gournellos, T.; Chalkias, C.; Tsagas, D. Landslide susceptibility zonation of Greece using fuzzy logic and GIS. *Geographies* **2006**, *12*, 114–126.
53. Hellenic Statistical Authority (ELSTAT). *Population and Housing Census: Resident Population*; ELSTAT: Piraeus, Greece, 2011.
54. Pnevmatikos, J.D.; Katsoulis, B.D. The changing rainfall regime in Greece and its impact on climatological means. *Meteorol. Appl.* **2006**, *13*, 331–345. [CrossRef]

55. Koukis, G.; Tsiambaos, G.; Sabatakakis, N. Landslides movements in Greece: Engineering geologicalcharacteristics and environmental consequences. In Proceedings of the 7th International Symposium of Engineering Geology and Environment, Athens, Greece, 23–27 June 1997; Volume 1, pp. 789–792.
56. Hellenic Military Geographical Service (HMGS). *Aerial Photography of Various Areas of Greece, Scale 1:40,000*; HMGS: Athens, Greece, 1969.
57. Che, V.B.; Kervyn, M.; Suh, C.E.; Fontijn, K.; Ernst, G.G.J.; del Marmol, M.A.; Trefois, P.; Jacobs, P. Landslide susceptibility assessment in Limbe (SW Cameroon): A field calibrated seed cell and information value method. *Catena* **2012**, *92*, 83–98. [CrossRef]
58. De Guidi, G.; Scudero, S. Landslide susceptibility assessment in the Peloritani Mts. (Sicily, Italy) and clues for tectonic control of relief processes. *Nat. Hazards Earth Syst. Sci.* **2013**, *13*, 949–963. [CrossRef]
59. Youssef, A.M.; Pradhan, B.; Pourghasemi, H.R.; Abdullahi, S. Landslide susceptibility assessment at Wadi Jawrah Basin, Jizan region, Saudi Arabia using two bivariate models in GIS. *Geosci. J.* **2015**, *19*, 449–469. [CrossRef]
60. Cruden, D.M.; Varnes, D.J. Landslide types and processes. In *Landslides: Investigation and Mitigation, Special Report*; Turner, A.K., Schuster, R.L., Eds.; National Academy Press: Washington, DC, USA, 1996.
61. Fell, R. Landslide risk assessment and acceptable risk. *Can. Geotech. J.* **1994**, *31*, 261–272. [CrossRef]
62. Magliulo, P.; Lisio, A.; Russo, F.; Zelano, A. Geomorphology and landslide susceptibility assessment using GIS and bivariate statistics: A case study in southern Italy. *Nat. Hazards* **2008**, *47*, 411–435. [CrossRef]
63. Das, O.; Sonmez, H.; Gokceoglu, C.; Nefeslioglu, H. A study on the effect of the ground acceleration on landslide susceptibility maps. *Geophys. Res. Abstr.* **2010**, *12*, EGU2010-5885-1.
64. Yalcin, A. GIS-based landslide susceptibility mapping using analytical process and bivariate statistics in Ardesen (Turkey): Comparisons of results and confirmations. *Catena* **2008**, *72*, 1–12. [CrossRef]
65. Hasekiogullari, G.D.; Ercanoglu, M. A new approach to use AHP in landslide susceptibility mapping: A case study at Yenice (Karabuk, NW Turkey). *Nat. Hazards* **2012**, *63*, 1157–1179. [CrossRef]
66. Heymann, Y.; Steenmans, C.; Croisille, G.; Bossard, M. *CORINE Land Cover: Technical Guide*; Office for Official Publications of the European Communities: Luxembourg City, Luxembourg, 1994.
67. Carrara, A.; Cardinali, M.; Detti, R.; Guzzetti, F.; Pasqui, V.; Reichenbach, P. GIS techniques and statistical models in evaluating landslide hazard. *Earth Surf. Process. Landf.* **1991**, *16*, 427–445. [CrossRef]
68. Institute of Geology and Mineral Exploration (IGME). *Geological Map of Greece, Scale 1:500,000*, 2nd ed.; IGME: Athens, Greece, 1983.
69. Underhill, J.R. Late Cenozoic deformation of the Hellenide foreland, western Greece. *Geol. Soc. Am. Bull.* **1989**, *101*, 613–634. [CrossRef]
70. Ayalew, L.; Yamagishi, H. The application of GIS-based logistic regression for landslide susceptibility mapping in the Kakuda-Yahiko Mountains Central Japan. *Geomorphology* **2005**, *65*, 15–31. [CrossRef]
71. Rahmani, M.; Hosieni, S.A.; Molla Aghajanzadeh, A.; Mashayekhan, A.; Mahalleh, S.M.M. Landslide hazard zonation using the MCE method (Case study: Agh Mashhad, Mazandaran Province, Iran). *Int. J. Agric. Sci.* **2013**, *3*, 240–244.
72. Ahmed, M.F.; Rogers, J.D.; Ismail, E.H. A regional level preliminary landslide susceptibility study of the upper Indus river basin. *Eur. J. Remote Sens.* **2014**, *47*, 343–373. [CrossRef]
73. Sato, H.P.; Hasegawa, H.; Fujiwara, S.; Tobita, M.; Koarai, M.; Une, H.; Iwahashi, J. Interpretation of landslide distribution triggered by the 2005 Northern Pakistan earthquake using SPOT 5 imagery. *Landslides* **2007**, *4*, 113–122. [CrossRef]
74. Owen, L.A.; Kamp, U.; Khattak, G.A.; Harp, E.L.; Keefer, D.K.; Bauer, M.A. Landslides triggered by the 8 October 2005 Kashmir earthquake. *Geomorphology* **2008**, *94*, 1–9. [CrossRef]
75. Huade, G.; Wilson, J.L.; Makhnin, O. Geostatistical mapping of mountain precipitation incorporating autosearched effects of terrain and climatic characteristics. *J. Hydrometeorol.* **2005**, *6*, 1018–1031.
76. Arjun, C.R.; Kumar, A. Artificial neural network-based estimation of peak ground acceleration. *ISET J. Earthq. Technol.* **2009**, *46*, 19–28.
77. Stournaras, G.; Tsimpidis, G.; Tsoumanis, P.; Yannatos, G.; Guillande, R. Instability Phenomena in the Neogenne Deposits of Northern Peloponnese. *Bull. Eng. Geol. Environ.* **1998**, *57*, 65–68. [CrossRef]
78. Papathanassiou, G. Estimating slope failure potential in an earthquake prone area: A case study at Skolis Mountain, NW Peloponnesus, Greece. *Bull. Eng. Geol. Environ.* **2012**, *71*, 187–194. [CrossRef]

79. Dounias, G.; Fikiris, I.F.; Kollios, A.; Kabbadas, M.; Marinos, P. The Nemea Landslide. Trigger Mechanism–Remedial Measures. In Proceedings of the 5th Panellenic Conference of Geotechnics and Geoenvironmental Engineering, Xanthi, Greece, 31 May–2 June 2006.
80. Papadopoulos, G.A.; Plessa, A. Magnitude-distance relations for earthquake-induced landslides in Greece. *Eng. Geol.* **2000**, *58*, 377–386. [CrossRef]
81. Chousianitis, K.; Del Gaudio, V.; Sabatakakis, N.; Kavoura, K.; Drakatos, G.; Bathrellos, G.D.; Skilodimou, H.D. Assessment of earthquake-induced landslide hazard in Greece: From Arias Intensity to spatial distribution of slope resistance demand. *Bull. Seismol. Soc. Am.* **2016**. [CrossRef]
82. Van Westen, C.J. Statistical landslide hazard analysis. In *ILWIS 2.1 for Windows Application Guide*; ITC Publication: Utrecht, The Netherlands, 1997; pp. 73–84.
83. Jenks, G.F. The data model concept in statistical mapping. *Int. Yearb. Cartogr.* **1967**, *7*, 186–190.
84. Lee, E.M.; Jones, D.K.C. *Landslide Risk Assessment*; Thomas Telford: London, UK, 2004; p. 161.
85. Malczewski, J. *GIS and Multicriteria Decision Analysis*; John Wiley & Sons: New York, NY, USA, 1999.
86. Remondo, J.; González-Díez, A.; Díaz de Terán, J.R.; Cendrero, A. Landslide susceptibility models utilizing spatial data analysis techniques. A case study from the Lower Deba Valley, Guipúzcoa (Spain). *Nat. Hazards* **2003**, *30*, 267–279. [CrossRef]
87. Nandi, A.; Shakoor, A. Preparation of a landslide susceptibility map of summit county, Ohio, USA, using numerical models. In Proceedings of the 10th IAEG Congress, Nottingham, UK, 6–10 September 2006.
88. Frattini, P.; Crosta, G.; Carrara, A. Techniques for evaluating the performance of landslide susceptibility models. *Eng. Geol.* **2010**, *111*, 62–72. [CrossRef]
89. Fawcett, T. An introduction to ROC analysis. *Pattern Recognit. Lett.* **2006**, *27*, 861–874. [CrossRef]
90. Dragićević, S.; Lai, T.; Balram, S. GIS-based multicriteria evaluation with multiscale analysis to characterize urban landslide susceptibility in data-scarce environments. *Habitat Int.* **2015**, *45*, 114–125. [CrossRef]
91. Faraji Sabokbar, H.; Shadman Roodposhti, M.; Tazik, E. Landslide susceptibility mapping using geographically-weighted principal component analysis. *Geomorphology* **2014**, *226*, 15–24. [CrossRef]
92. Yang, Z.H.; Lan, H.X.; Gao, X.; Li, L.P.; Meng, Y.S.; Wu, Y.M. Urgent landslide susceptibility assessment in the 2013 Lushan earthquake-impacted area, Sichuan Province, China. *Nat. Hazards* **2015**, *75*, 2467–2487. [CrossRef]
93. Van Westen, C.J.; Rengers, N.; Terlien, M.T.J.; Soeters, R. Prediction of the occurrence of slope instability phenomena through GIS-based hazard zonation. *Geol. Rundsch.* **1997**, *86*, 404–414. [CrossRef]
94. Ayalew, L.; Yamagishi, H.; Maruib, H.; Takami, K. Landslides in Sado Island of Japan: Part II. GIS-based susceptibility mapping with comparisons of results from two methods and verifications. *Eng. Geol.* **2005**, *81*, 432–445. [CrossRef]
95. Tien Bui, D.; Lofman, O.; Revhaug, I.; Dick, O. Landslide susceptibility analysis in the Hoa Binh province of Vietnam using statistical index and logistic regression. *Nat. Hazards* **2011**, *59*, 1413–1444.
96. Van Westen, C.J.; Rengers, N.; Soeters, R. Use of geomorphological information in indirect landslide susceptibility assessment. *Nat. Hazards* **2003**, *30*, 399–419. [CrossRef]
97. Saboya, F., Jr.; da Glória Alves, M.; Dias Pinto, W. Assessment of failure susceptibility of soil slopes using fuzzy logic. *Eng. Geol.* **2006**, *86*, 211–224. [CrossRef]
98. Bathrellos, G.D.; Gaki-Papanastassiou, K.; Skilodimou, H.D.; Papanastassiou, D.; Chousianitis, K.G. Potential suitability for urban planning and industry development by using natural hazard maps and geological—Geomorphological parameters. *Environ. Earth Sci.* **2012**, *66*, 537–548. [CrossRef]
99. Rozos, D.; Skilodimou, H.D.; Loupasakis, C.; Bathrellos, G.D. Application of the revised universal soil loss equation model on landslide prevention. An example from N. Euboea (Evia) Island, Greece. *Environ. Earth Sci.* **2013**, *70*, 3255–3266. [CrossRef]

7

A Web GIS Framework for Participatory Sensing Service: An Open Source-Based Implementation

Yu Nakayama [1,*], Kazuhiko Nakamura [2], Hitoshi Saito [3] and Rui Fukumoto [4]

1 Institute of Industrial Science, The University of Tokyo, 4-6-1 Komaba, Meguro-ku, Tokyo, Japan
2 Center for Spatial Information Science, The University of Tokyo, 5-1-5 Kashiwanoha, Kashiwa-shi, Chiba, Japan; k_nakamura@csis.u-tokyo.ac.jp
3 College of Economics, Kanto Gakuin University, 1-50-1 Mutsuura-higashi, Kanazawa-ku, Yokohama, Japan; hsaito@kanto-gakuin.ac.jp
4 Open Concierge, #307 3-21-5 Kandasakumacho, Chiyoda-ku, Tokyo, Japan; rui@openconcierge.org
* Correspondence: yu-nakayama@mcl.iis.u-tokyo.ac.jp

Abstract: Participatory sensing is the process in which individuals or communities collect and analyze systematic data using mobile phones and cloud services. To efficiently develop participatory sensing services, some server-side technologies have been proposed. Although they provide a good platform for participatory sensing, they are not optimized for spatial data management and processing. For the purpose of spatial data collection and management, many web GIS approaches have been studied. However, they still have not focused on the optimal framework for participatory sensing services. This paper presents a web GIS framework for participatory sensing service (FPSS). The proposed FPSS enables an integrated deployment of spatial data capture, storage, and data management functions. In various types of participatory sensing experiments, users can collect and manage spatial data in a unified manner. This feature is realized by the optimized system architecture and use case based on the general requirements for participatory sensing. We developed an open source GIS-based implementation of the proposed framework, which can overcome financial difficulties that are one of the major problems of deploying sensing experiments. We confirmed with the prototype that participatory sensing experiments can be performed efficiently with the proposed FPSS.

Keywords: participatory sensing; web GIS; crowdsensing; urban sensing

---

1. Introduction

There are millions of mobile phones in use around the world. Mobile phone sensing is an emerging area of research interest [1]. Today's mobile phones have a number of embedded sensors, including an accelerometer, digital compass, gyroscope, GPS, microphone, and camera. In addition, by affixing a sensory device, mobile sensing provides the opportunity to collect various types of information. Thus, mobile phone sensing holds great promise for a wide range of research fields, such as health monitoring, human behavior, social interaction, traffic monitoring, commerce, and environment monitoring [2]. There are two mobile phone sensing paradigms: participatory sensing and opportunistic sensing. They can be discussed in the literature of mobile crowdsensing [3,4]. In participatory sensing, mobile phone users actively participate in sensing activities. In opportunistic sensing, sensing activities are automated, without the users' involvement. This paper focuses on participatory sensing, which enables public and professional users to gather, analyze and share local knowledge [5].

The common architectural components for participatory sensing applications include mobile device data capture, personal data stream storage, and leveraged data processing [6]. To efficiently develop participatory sensing services, some server-side technologies have been proposed [7,8].

Although they provide a good platform for participatory sensing, they are not optimized for the spatial data management and processing. If the spatiality of collected data has important meaning, then the data need to be managed as GIS data. For the purpose of spatial data collection and management, many web GIS approaches have been studied [9–13]. However, they still have not focused on the optimal framework for participatory sensing services. Participatory sensing has not been sufficiently discussed in GIS literature, and vice versa. Moreover, the existing solutions have not given a general framework for providing participatory sensing services based on the comprehensive analysis of the system requirements.

Therefore, this paper defines the general web GIS framework for participatory sensing service (FPSS) based on the comprehensive analysis of the requirements. The proposed FPSS enables an integrated deployment of a platform for participatory sensing services that provides spatial data capture, storage, and data management functions. In various types of participatory sensing experiments, users can collect and manage spatial data in a unified manner. To show the validity of the proposed framework, we introduce an open source-based implementation of the FPSS. The proposed concept can be implemented in other ways, e.g., the spatial optimizations of Apache Spark Streaming can be employed to improve the scalability of real-time processing.

This paper is organized as follows. Section 2 describes the related work. In Section 3, we discuss and define the requirements for the data structure and workflow of the FPSS. Based on the requirements, we describe the architecture and the use case of the proposed FPSS in Section 4. Then, Section 5 introduces an open source-based implementation of the FPSS based on [13]. The effectiveness of the proposed FPSS is confirmed with the application study described in Section 6. We conclude this work in Section 7.

2. Related Work

2.1. Participatory Sensing Framework

To efficiently develop participatory sensing services, some frameworks have been developed. To provide services via a cloud computing system, Sensing-as-a-Service (S^2aaS) was introduced [7]. The requirements for an S^2aaS cloud are (1) supporting various applications on different smartphone platforms; (2) energy-efficiency; and (3) incentive mechanisms that attract mobile users to engage in sensing activities. A middleware framework was developed to build large-scale mobile phone sensing test beds [8]. This framework gives researchers a subset of available mobile devices on which to deploy experiments, and gives users fine-grained control over what sensor information they want to share to guard their privacy. An experimental result on top of the Android platform was reported. To help in the data collection process of participatory sensing, a framework for data quality and feedback was proposed [14]. In this work, they proposed a model of sensing behavior of individuals and appropriate feedback mechanisms. As an evolution of participatory sensing, mobile crowd sensing (MCS) is proposed as a new sensing paradigm, which leverages the power of citizens for large-scale sensing [15]. MCS has two features: (1) involving both implicit and explicit participation; and (2) collecting data from mobile social networks and mobile sensing. An integrated framework was proposed to enable cloud-based and people-centric applications on mobile devices [16]. The proposed approach exploited cloud computing to provide a delivery model targeted to mobile applications. It achieved significant energy saving in the mobile device, but it does not focus on participatory sensing.

In addition, privacy is one of the major problems in participatory sensing systems. In [17], they identified the unique challenges assuming an un-trusted central data server model, and proposed a privacy-aware framework. An anonymity-preserving reputation framework called IncogniSense was proposed to assess the user contributions without privacy threats [18]. It utilizes periodic pseudonyms generated using blind signature and relies on reputation transfer between them. The ParticipAct [19,20] is based on an Android framework [21] for simplifying the quick design of mobile sensing campaigns/projects and offering app developers a set of functions to quickly design their own mobile sensing services.

2.2. Crowdsensing Framework

Mobile crowdsensing is a strong paradigm for practically the same problem as participatory sensing [3,4,22,23]. The crowdsensing concepts to location tracking and collection of spatial data have been proposed in [24,25]. It was pointed out that the barriers for engaging people in crowdsensing are the impact on their battery lifetime and phone usage habits in [24]. To address these requirements, they proposed an energy-efficient application for measuring the location and signal strength data. For vehicular mobile crowd sensing systems, a heterogeneous sensing vehicle selection (HVS) method for the collection of comprehensive tempo-spatial sensing data was proposed in [25]. In this method, a utility function is determined to estimate the sensing capacity of the vehicles based on the spatial distribution and sensing interfaces of sensing vehicles and the coverage of collected data.

2.3. Web GIS Approach

For the purpose of spatial data collection and management, many web GIS approaches have been proposed. An open source GIS-based Java web application that automatically writes HTML and JavaScript code was developed [9]. WAMP-based and Apache Tomcat-based web GIS frameworks were developed and compared [10]. A web browser-based system for online fatal disease mapping was developed and it enables spatial visualization, detection and monitoring of incidence in real-time, without the need for any additional software [11]. An open source-based web GIS application was proposed to share the information and spatial data, allowing users with limited GIS knowledge to access the information customized for specific applications [12]. A field survey assisting system was developed in [13]. OWGIS version 2.0 is a Java and JavaScript application that builds easily configurable Web GIS sites [26]. This OWGIS generates mobile interfaces based on HTML5 technology, and can be used to create mobile applications.

However, these works have not discussed the optimal framework for participatory sensing services.

2.4. Contribution

The contribution of this paper is defining the general framework for providing participatory sensing services based on the comprehensive analysis of the system requirements. The proposed FPSS provides an efficient platform for participatory sensing services, which enables an integrated deployment of spatial data capture, storage, and management functions. The related existing solutions have not provided such a general framework. The goal of the FPSS is for mobile sensing organizers to easily launch sensing projects, not for app developers as [19,20]. Moreover, the FPSS can be used across multiple scales; an individual, a group, or a community [1], not only crowd sensing, and can coexist with privacy-aware frameworks.

3. Requirement

First, we explain the system requirements for the FPSS.

3.1. Definition

First, we define the participatory sensing experiments on which this work focuses. A sensing experiment is performed by an individual, a group, or a community [1]. The features of sensing experiments, including study area, survey items, or the purpose, are not limited to specific types. The spatial data type is point; participants record the measured data along with the location where they are obtained. The FPSS needs to provide any sensing experiment service that satisfies this definition.

3.2. Data Structure

To efficiently record data in a sensing experiment, we define that the recorded information consists of two factors: basic information about the experiment which is called metadata and spatial data obtained for each recording point in the field. This definition is depicted in Figure 1.

Metadata are uniquely recorded for an experiment and contain information about the experiment itself. They consist of title, date, study area, item, keywords, notes, and basemaps.

Spatial data are recorded along with the location and the values for the items linked; for example, the species, height, and an image file of a tree at each recording point. It is usual to obtain multiple spatial data for an experiment.

The survey items can be set as any data type, because the proposed framework should support various types of participatory sensing experiments. Thus, the table structure in the database should be optimized for the above data structure.

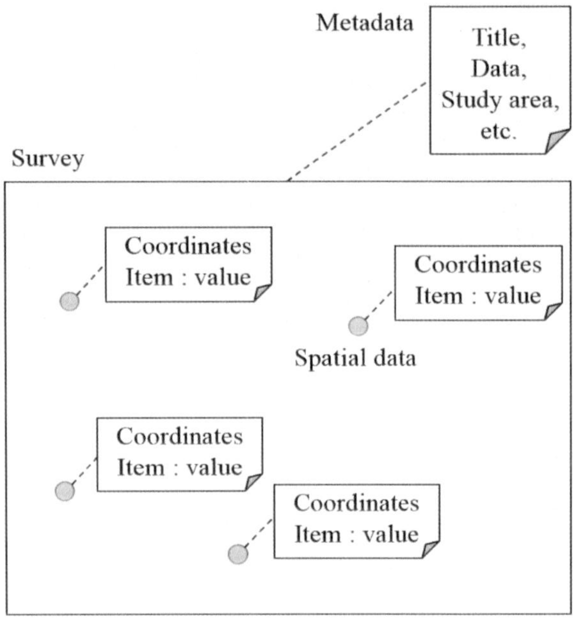

Figure 1. Data structure.

3.3. *Workflow*

Second, we analyze the system requirements based on the workflow of a sensing experiment. It is defined that a sensing experiment consists of three steps: (A) Preparation, (B) Data-acquisition, and (C) Application. This workflow is shown in Figure 2.

step (A) Preparation First, an organizer plans and prepares for the experiment. The organizer decides the operating procedure based on the purpose, including the study area, the survey items, and the schedule. The metadata of the experiment should be stored in the database in this step.

step (B) Data-acquisition Participants of the experiment obtain spatial data in the field, on the basis of the operating procedure decided in step (A). They can focus on the spatial data-acquisition process, and do not need to edit the metadata. As a sensing experiment is usually carried out by multiple fieldworkers at the same time, it is desirable that they can access the system and record/display the obtained data in real-time without an error or overlap.

step (C) Application After the data-acquisition, the obtained data are displayed, analyzed, exported, and used by users for their own purposes. The data users can include the organizer, participants, and others. It is important to select the data sharing policy based on the intention of the organizer. Some data are assumed to be private such as those for commercial use, and others are assumed to be public such as a CC-BY-SA license.

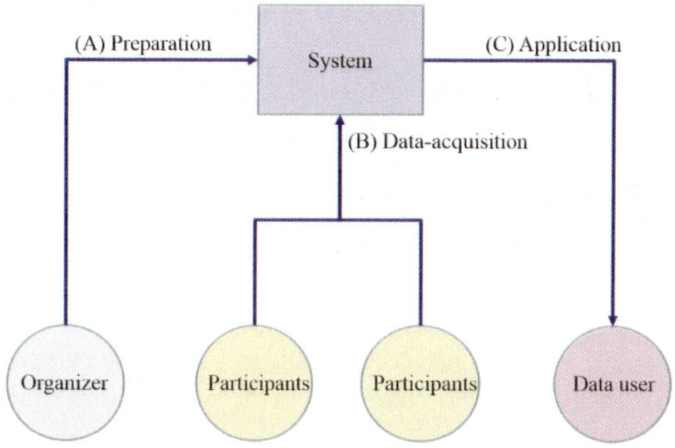

Figure 2. Workflow of the sensing experiment.

The use case of the FPSS should be optimized for this general workflow of a sensing experiment.

3.4. System Architecture

A client–server model is suitable for the system architecture, because multiple fieldworkers are expected to participate in the same sensing experiment at the same time. The client terminals are assumed to be smartphones and tablet PCs equipped with GNSS (Global Navigation Satellite System) including GPS, and other sensors. The fieldworkers travel to various fields with the client terminals, and access the server with wireless LANs or a mobile network such as 3G and LTE (Long Term Evolution). Although the client terminals need to access a remote server, network connectivity to a server is not always guaranteed in an outside field. Therefore, it is desirable for the system to compensate for the lack of network connectivity.

4. Framework for Participatory Sensing Service

The proposed FPSS is described in this section.

4.1. Architecture

We describe the architecture of the proposed framework, which is designed to meet the requirements above. Figure 3 shows the architecture of the proposed framework. It consists of the spatial database, the GIS server, the PSS module, and the client application.

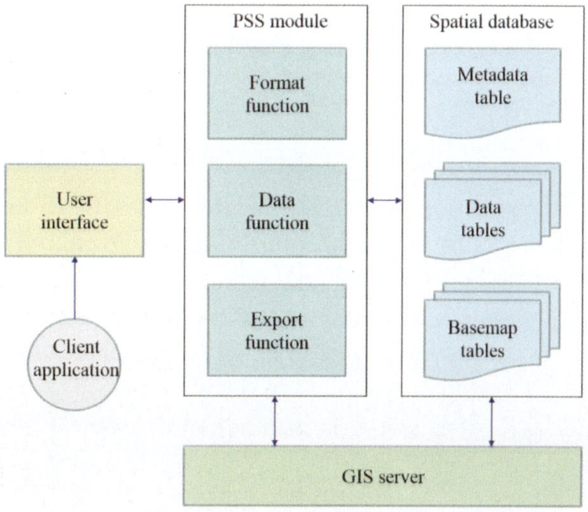

Figure 3. Architecture of the proposed framework.

4.1.1. Spatial Database

The spatial database stores metadata, spatial data, and basemaps. It has a metadata table, data tables, and basemap tables.

Metadata are stored in the metadata table, in which each record represents respective experiments.

Spatial data are stored in the data tables. The spatial data include the coordinates and values of survey items. The proposed framework creates a new data table with optimized fields for each experiment. The created table is linked to the corresponding metadata. With this table structure, the proposed framework supports various types of sensing experiments.

Basemaps are stored in the basemap tables. Basemaps include general open data which are registered in advance, and specific basemap layers of each study area uploaded by system users.

4.1.2. GIS Server

The GIS server provides a GIS engine and protocols for serving georeferenced map images, i.e., Web Map Service (WMS). This works in the back-end of the system and the interface to users is provided by the PSS module.

4.1.3. PSS Module

The PSS module is composed of three components: the format function, the data function, and the export function. They generate user interfaces, store data into the database, select data from the database, and render maps using the GIS server. Each component corresponds to the three steps in the workflow of a sensing experiment described in Section 3.3. The detailed functions are described along with the use case in Section 4.2.

4.1.4. Client Application

The client-side application provides a format-download function and a data-store function, to realize the off-line use of the client terminals.

4.2. Use Case

The sequence of the use case of the proposed framework is shown in Figure 4. The sequence consists of three steps: (1) Format creation; (2) Data input; (3) Export. These steps correspond to the workflow of a sensing experiment in Section 3.3; step (1) corresponds to step (A); step (2) corresponds to step (B); and step (3) corresponds to step (C).

In step (1) users upload and register basemaps to the database and create an experiment format.

In step (2) users download the created format and input the obtained data into the database using the format.

In step (3) users select the format to display as a map and download the selected data.

The novelty of the proposed framework is that it enables fieldworkers of various types of participatory sensing experiments to log data easily in an integrated way. With this architecture and use case, the users are able to create the optimal format for various types of sensing experiments. They can also input and export spatial data in a unified manner. The created format and stored data can be shared with other users. Accordingly, multiple fieldworkers are able to participate in the same experiment simultaneously. Each step is described in detail below.

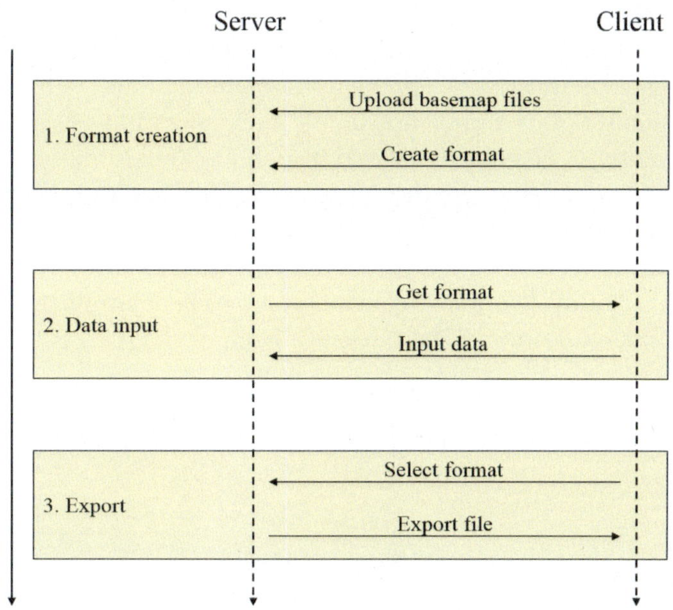

Figure 4. Sequence of the participatory sensing experiment.

4.2.1. Format Creation

The user accesses the server and creates a format for the experiment in the (A) Preparation step. Figure 5 shows the data flow between the system components. When the user starts this step, the format function creates a formatting web form as a user interface. The user fills out the formatting form with the experiment information. The metadata, i.e., the title, date, and notes, are stored in the metadata table. The basemaps are selected from the basemap table and the study area is set using map images. If the user wants to add a new basemap, a new basemap file can be uploaded and stored in the database in advance. The survey items are added designating their name and data types, and a corresponding table is created as a data table. The proposed framework can be employed for various types of sensing experiments because any survey item can be added in this step. Then, the sharing policy of this format is selected.

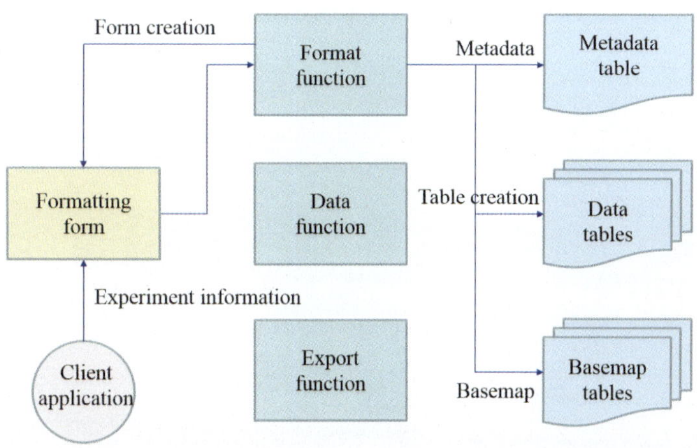

Figure 5. Use case (1): Format creation.

4.2.2. Data Input

In the (B) Data-acquisition step, the user stores the obtained data using the client terminals. Figure 6 shows the data flow in this step. When the user selects a survey from the list of editable surveys, an input form is created by the data function. The data function creates it by reading the metadata from the metadata table. The stored spatial data are read from the data table of this survey and displayed on the basemap images. When the user obtains a datum in the field, the user stores it using the input form. The user plots the location on the map image and fills out the required fields for this data table. An image file can be uploaded if a corresponding web form is created. If the user selects the displayed datum, the stored values are displayed in the form and they can be updated or deleted. Multiple fieldworkers can share the latest result in real-time, because the displayed spatial data are refreshed when the input form is reloaded. If the network connection to the server is unavailable, the obtained data are temporarily stored in the memory of the client terminals. They are automatically uploaded to the server when the network becomes available.

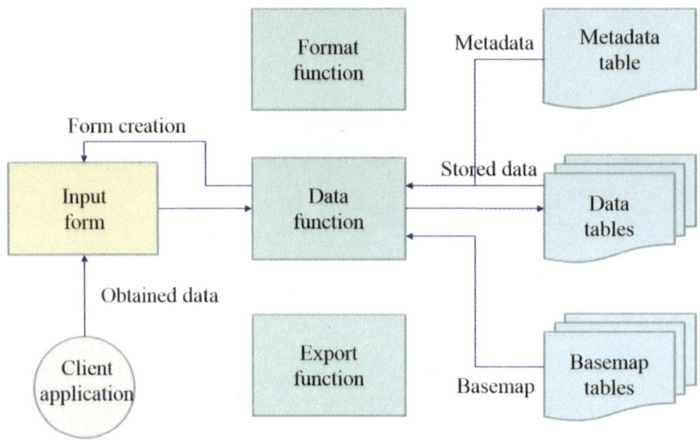

Figure 6. Use case (2): Data input.

4.2.3. Export

In the (C) Application step, the user displays and exports the stored data and uses them for the purpose of the experiment. Figure 7 shows the data flow in this step. When the user selects an experiment from the list of viewable experiments, an export form is created by the export function. The export function creates it by reading the metadata from the metadata table and the stored data from the data tables. The user can view the stored data on the basemap images. When the user requests to download the stored spatial data, the export module translates the spatial data to a file format.

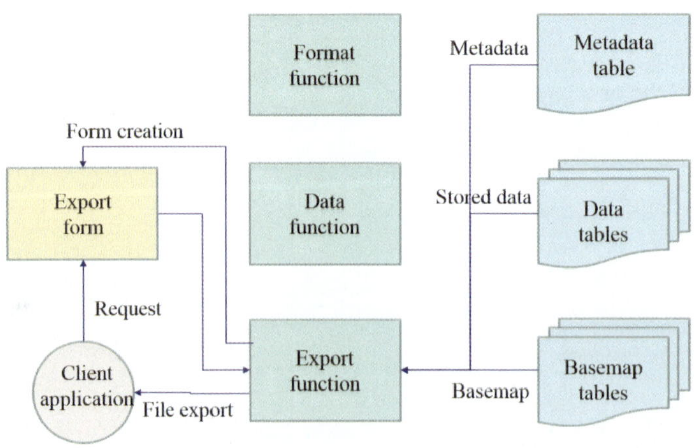

Figure 7. Use case (3): Export.

5. Prototype Implementation

In this section, we describe an open source-based prototype implementation of the proposed framework. As stated before, the purpose of the prototype is to confirm the validity of the proposed framework. Thus, the implementation of the proposed concept is not limited to this; different styles using different tools can be considered.

5.1. Overview

Here, we provide an overview of the implementation. As a prototype of the proposed framework, an open source-based approach is adopted to reduce the system installation cost. We implemented the prototype based on a filed survey assisting system in [13].

Here, we provide an outline of the filed survey assisting system. It is an open source-based web GIS package. It has two features that improve the usability for non-technical users [27]. One is web GIS architecture and the other is that open source GIS applications are wrapped with a content management system (CMS). Web GIS encourages users to employ the system because they can do so simply with web browsers, with which they are familiar [28]. In general, a CMS makes it easy for users to employ the system. The CMS manages web content in an integrated fashion and conceals complicated GIS functions from users. Users work with interfaces formed by the CMS, without having to struggle with or even be conscious of open source GIS applications.

5.2. Architecture

The architecture of the implemented system is presented in Figure 8.

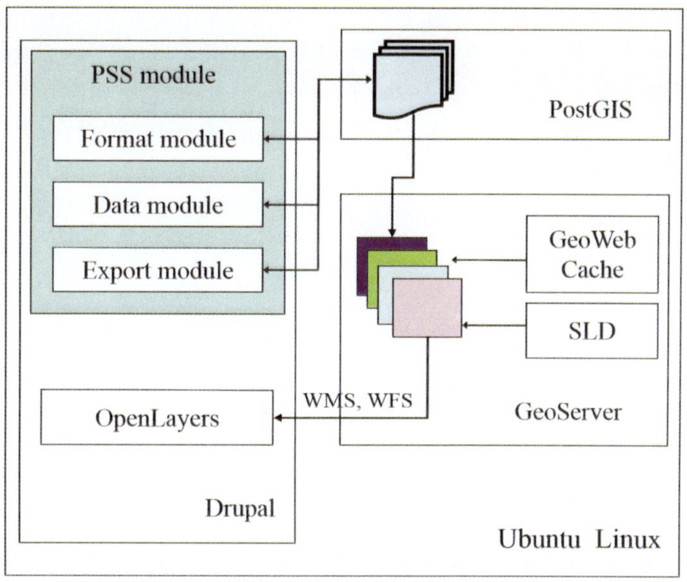

Figure 8. Architecture of the implemented system.

In the server, open source GIS applications and Drupal (CMS) run on Ubuntu Linux. Spatial data are stored using PostGIS. PostGIS is a spatial extension of PostgreSQL, which is one of the most popular open source database management systems. GeoServer works as a GIS server, which is a major web GIS application built on open standards and enables the interoperability of spatial data. It supports WMS, which delivers maps as images, and Web Feature Service (WFS), which delivers maps in Geography Markup Language (GML). These data formats are standardized by Open Geospatial Consortium (OGC) and used in many GIS applications. A Styled Layer Descriptor (SLD) describes the appearance of the map layers. GeoWebCache accelerates processing by caching requests. Drupal works in an integrated way as a user interface and manages GIS applications. We employed Drupal because it provides spatial modules including OpenLayers, which displays and renders maps on web pages using WMS and WFS.

The PSS module is developed as a module of Drupal and is written in PHP scripts. It creates user interfaces and controls database queries. The PSS module has three sub-modules: a format module, a data module, and an export module. The format module corresponds to the use case (1) Format creation; the data module corresponds to (2) Data input; and the export module corresponds to (3) Export.

The users access the server with their client terminals. They can view the downloaded formats using a client application, which provides the data-store function. The location of the user is displayed in the map images using GPS. The transmission protocol used between the server and clients is HTTP(S).

5.3. *Use Case*

Here, we describe the detail of the PSS module along with the use case. The PSS module is developed and installed as a Drupal module.

5.3.1. Format Creation

Figure 9 shows an example screenshot of the formatting form, which is created by the format module. The basic web functions such as account management and database queries are developed using the Drupal functions. The sharing policy of a created format is also easily controlled with the Drupal functions. The map images are provided with WMS using GeoServer and are displayed with OpenLayers. The basemaps are selected from Google Maps (Google Inc., Mountain View, CA, USA), VirtualEarth (Microsoft Corporation, Redmond, WA, USA), and OpenStreetMap (OSM, [29]). The users can also upload Shapefiles and GeoTIFFs as basemaps. The uploaded files are imported into the database using a shp2pgsql or a raster2pgsql tool.

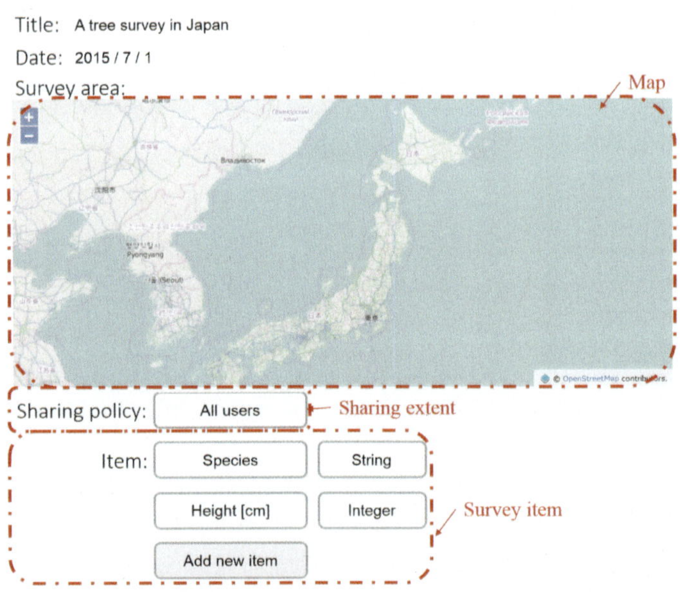

Figure 9. Example screenshot of format creation.

5.3.2. Data Input

When the user logs in to the system and requests the editable experiment list, the list of editable experiments is generated using the account and content management function of Drupal. Figure 10 shows an example screenshot of the input form, which is created by the data module. The data module calls queries to the corresponding tables in the database, and displays the map images with GeoServer and OpenLayers. When the user plots a new datum on the map, the system automatically obtains the coordinates and sends a query to the database to store it. The location is automatically plotted in the map, if the GPS function is available.

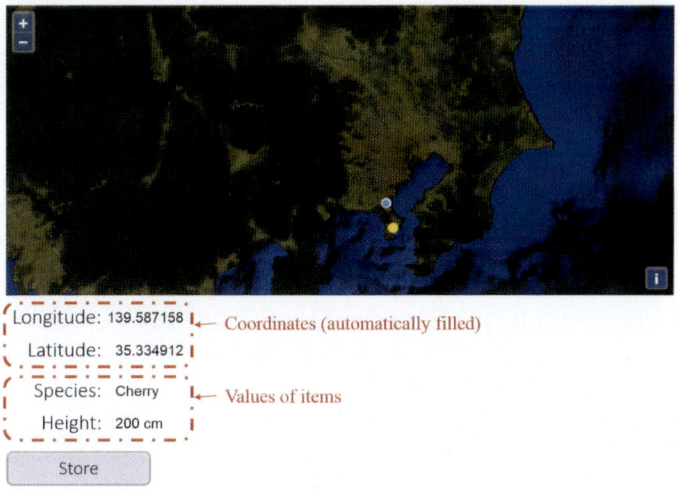

Figure 10. Example screenshot of data input.

If the network connection between the server and the client terminal becomes unavailable, the client application stops the request to the data module and continues to display the last information it downloaded from the server. Until the network connection becomes available, new data are stored in the client terminal's own memory. Then, the stored data are automatically uploaded to the server.

5.3.3. Export

When the user requests the viewable experiment list, it is generated with the CMS functions based on the user's account information. The export module calls queries to the database and displays the map images with GeoServer and OpenLayers. If the user wants to export the stored data, either CSV (Comma-Separated Values) or KML (Keyhole Markup Language) can be selected as the export file type. The export module translates the spatial data to the file using the PostGIS functions.

6. Evaluation

6.1. Outline

We evaluated the implemented framework through a test experiment. We confirmed that the participatory sensing experiment was performed smoothly by multiple fieldworkers with the proposed framework.

6.1.1. Study Area and Item

As an application study of the proposed framework, we undertook a growth survey of Kawazu-zakura tree (Prunus lannesiana Wils.) in Miura Peninsula, Kanagawa, Japan. The study area, shown in Figure 11, covers an area of about 500 m^2 from Keikyu-Miurakaigan Station to Komatsugaike Park. The location and height of the Kawazu-zakura trees were adopted as survey items.

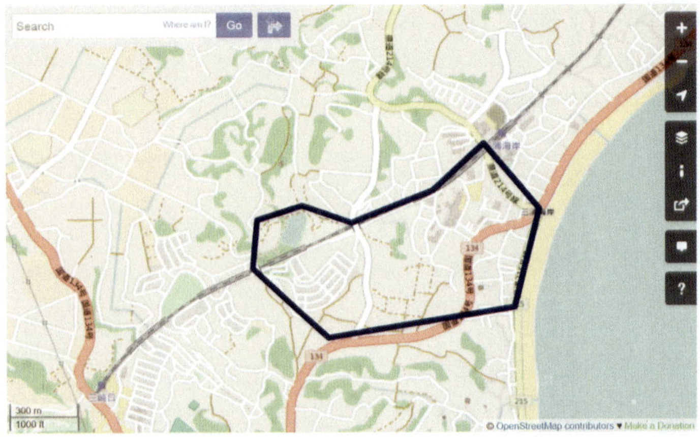

Figure 11. Study area (on OpenStreetMap).

6.1.2. Server

We located the server in a remote building. We stored the IP address in the server and connected it to the Internet. The client terminals accessed the server via the Internet. OSM tiles were installed in the server and used as the basemap.

6.1.3. Client

There were ten participants in the survey. They were divided into four groups. Each group member had an Android 4.2 based smartphone (LTE network), an Android 4.2 based tablet (3G network), an iPhone 5 (LTE network), and a laptop PC (WiMAX network). At first, the Android terminals were used without network services. For the latter half of the survey, they were used with the network services. Other client terminals were always used with Internet access. With this configuration, the data-store function of the client application was verified.

6.1.4. Procedure

One of the participants created an experiment format in the server in advance. The created format was shared among all system users. For the Android terminal users, the created format was downloaded to their terminals. The participants accessed the server with the client applications and logged in using their own accounts.

The participants walked around the study area and recorded the location and height of Kawazu-zakura trees based on visual observation. When the participants found a Kawazu-zakura tree, they opened the created format and input the obtained data. When the participants plotted a point on the map, the coordinate forms were automatically supplied with the point's location.

After the experiment, we employed a questionnaire to elicit users' impressions of the system. The question items related to the overall impression of the experiment, the advantages of the proposed system, and the points to be improved. They were open-ended questions.

6.2. Results

The experiment was completed smoothly. About 500 Kawazu-zakura trees were plotted in the study area within four hours. The participants were able to use the system simultaneously with their client terminals. The transmission and processing delays of the system were slight. Although sometimes it was difficult for some participants to distinguish tree species, they encountered no problems when operating the developed system, because they had only to plot points and input tree heights. The stored data were confirmed with a WMS request and exported as a KML file.

With the network access to the server, the participants were able to access the results in real-time. They could see the data that they had stored and also the data that other participants had stored.

The experiment results were updated automatically. However, while the network access to the server was unavailable, the input form cannot be updated with newly stored data. Thus, once all the participants activated the network services and the locally stored data were uploaded, it emerged that some data overlapped. This is because a participant plotted certain trees already plotted by another participant.

Figure 12 shows the relationship between the number of data and the data size stored in the system. The stored data consist of the point ID (4 bytes), coordinates (16 bytes), and height (4 bytes) per point. The data size increases in accordance with the number of data points. In this case study, 500 points data occupy 10 kilobytes in the system.

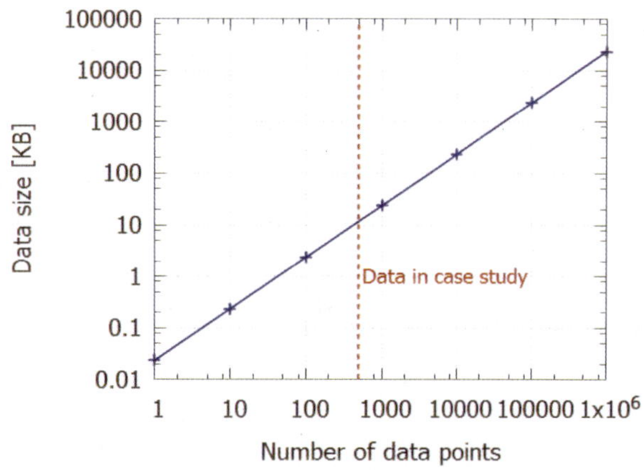

Figure 12. Data size stored in the system.

6.3. Discussion

Here, we discuss the advantages and limitations of the proposed FPSS using the results of the questionnaire.

The results of this application study suggest that a sensing experiment can be performed efficiently with the proposed framework. The data structure and use case of the framework were suitable for participatory sensing services. Multiple fieldworkers could use the same format that was created specifically for the experiment, using various types of client terminals. The system simplified the operation procedure of the sensing and it was easy to use. Because the progress of the experiment was updated in real-time, participants could see how other participants were proceeding with the experiment, and compare their results with those obtained by other participants.

The open source GIS-based implementation was sufficient for the test experiment. The incorporation of web GIS architecture and CMS improved the usability, and the implemented system was easy to use for all participants. If the data size and work load of the system become very large, the system architecture should be reconsidered.

The limitation of the FPSS was the data overlap; some points that a participant plotted without the network service overlapped points plotted by others. This overlap problem is inevitable with this type of system. This is because the obtained data are not synchronized with those of other participants until the network service becomes available. This problem can be prevented by putting a mark on the trees that have already been plotted on the map.

7. Conclusions

This paper introduced a web GIS framework for participatory sensing service based on the comprehensive analysis of the requirements. The proposed FPSS enables an integrated deployment of spatial data capture, storage, and data management functions. In various types of participatory sensing experiments, users can collect and manage spatial data in a unified manner. It can be used across multiple scales; an individual, a group, or a community.

This feature is realized by the optimized system architecture and use case based on the general requirements for participatory sensing. We defined that participatory sensing consists of three steps: (A) Preparation; (B) Data-acquisition; and (C) Application. The sequence of the use case of the proposed framework consists of three steps, namely, (1) Format creation; (2) Data input; and (3) Export, and they correspond to the three steps of participatory sensing. The architecture of the proposed framework consists of the spatial database, the GIS server, the PSS module, and the client application. Then, we developed an open source GIS-based implementation of the proposed framework, which can overcome financial difficulties which are one of the major problems of deploying sensing experiments.

With the prototype, we confirmed that participatory sensing experiments can be performed efficiently with the proposed FPSS. Multiple fieldworkers could easily use the same format that was created specifically for the experiment. They could share the state of the sensing in real-time. If the network service was unavailable, the obtained data were stored in their client terminals and uploaded immediately once the network service became available.

We believe that this work will contribute to the progress and deployment of GIS technologies for participatory sensing. A quantitative evaluation of usability is a future task.

Acknowledgments: This work was supported by JSPS KAKENHI Grant Number JP15H02722.

Author Contributions: Yu Nakayama organized the proposed concepts and wrote the paper; Kazuhiko Nakamura, Hitoshi Saito, and Rui Fukumoto implemented the prototype and performed the experiments

Conflicts of Interest: The authors declare no conflict of interest.

References

1. Lane, N.D.; Miluzzo, E.; Lu, H.; Peebles, D.; Choudhury, T.; Campbell, A.T. A survey of mobile phone sensing. *IEEE Commun. Mag.* **2010**, *48*, 140–150.
2. Khan, W.Z.; Xiang, Y.; Aalsalem, M.Y.; Arshad, Q. Mobile phone sensing systems: A survey. *IEEE Commun. Surv. Tutor.* **2013**, *15*, 402–427.
3. Wang, L.; Zhang, D.; Wang, Y.; Chen, C.; Han, X.; M'hamed, A. Sparse mobile crowdsensing: Challenges and opportunities. *IEEE Commun. Mag.* **2016**, *54*, 161–167.
4. Ma, H.; Zhao, D.; Yuan, P. Opportunities in mobile crowd sensing. *IEEE Commun. Mag.* **2014**, *52*, 29–35.
5. Burke, J.A.; Estrin, D.; Hansen, M.; Parker, A.; Ramanathan, N.; Reddy, S.; Srivastava, M.B. Participatory sensing. *Center Embed. Netw. Sens.* **2006**. Available online: http://escholarship.org/uc/item/19h777qd#page-5 (accessed on 8 April 2017).
6. Estrin, D. Participatory sensing: Applications and architecture [internet predictions]. *IEEE Int. Comput.* **2010**, *14*, 12–42.
7. Sheng, X.; Xiao, X.; Tang, J.; Xue, G. Sensing as a service: A cloud computing system for mobile phone sensing. In Proceedings of the 2012 IEEE Sensors, Taipei, Taiwan, 28–31 October 2012; pp. 1–4.
8. Brouwers, N.; Langendoen, K. Pogo, a middleware for mobile phone sensing. In Proceedings of the 13th International Middleware Conference, Montreal, QC, Canada, 03–07 December 2012; Springer: New York, NY, USA, 2012; pp. 21–40.
9. Zavala-Romero, O.; Ahmed, A.; Chassignet, E.P.; Zavala-Hidalgo, J.; Eguiarte, A.F.; Meyer-Baese, A. An open source Java web application to build self-contained web GIS sites. *Environ. Model. Softw.* **2014**, *62*, 210–220.
10. Agrawal, S.; Gupta, R.D. Development and Comparison of Open Source based Web GIS Frameworks on WAMP and Apache Tomcat Web Servers. *Proc. Int. Arch. Photogramm. Remote Sens. Spat. Inf. Sci.* **2014**, *4*, 1–5.
11. Tiwari, A.; Jain, D.K. Geospatial Framework For Dengue using Open Source Web GIS Technology. In Proceedings of the Joint International Workshop of ISPRS WG VIII/1 and WG IV/4 on Geospatial Data for Disaster and Risk Reduction, Hyderabad, India, 21–22 November 2013; pp. 21–22.
12. Chakraborty, D.; Sarkar, D.; Agarwal, S.; Dutta, D.; Sharma, J.R. Web Based GIS Application using Open Source Software for Sharing Geospatial Data. *Int. J. Adv. Remote Sens. GIS* **2015**, *4*, 1224–1228.
13. Nakayama, Y.; Mori, S. FOSS4G Based Mobile Web-GIS for Field Survey in Natural Environmental Studies. In Proceedings of the 9th International Conference on Ubiquitous Intelligence & Computing and 9th

International Conference on Autonomic & Trusted Computing (UIC/ATC), Fukuoka, Japan, 4–7 September 2012; pp. 707–712.

14. Reddy, S.; Burke, J.; Estrin, D.; Hansen, M.; Srivastava, M. A framework for data quality and feedback in participatory sensing. In Proceedings of the 5th International Conference on Embedded Networked Sensor Systems, Sydney, Australia, 4–9 November 2007; ACM: New York, NY, USA, 2007; pp. 417–418.
15. Guo, B.; Yu, Z.; Zhou, X.; Zhang, D. From participatory sensing to mobile crowd sensing. In Proceedings of the 2014 IEEE International Conference on Pervasive Computing and Communications Workshops (PERCOM Workshops), Budapest, Hungary, 24–28 March 2014; pp. 593–598.
16. Fakoor, R.; Raj, M.; Nazi, A.; Di Francesco, M.; Das, S.K. An integrated cloud-based framework for mobile phone sensing. In Proceedings of the First Edition of the MCC Workshop on Mobile Cloud Computing, Helsinki, Finland, 17 August 2012; pp. 47–52.
17. Kazemi, L.; Shahabi, C. A privacy-aware framework for participatory sensing. *ACM SIGKDD Explor. Newsl.* **2011**, *13*, 43–51.
18. Christin, D.; Roßkopf, C.; Hollick, M.; Martucci, L.A.; Kanhere, S.S. Incognisense: An anonymity-preserving reputation framework for participatory sensing applications. *Pervasive Mob. Comput.* **2013**, *9*, 353–371.
19. Cardone, G.; Foschini, L.; Bellavista, P.; Corradi, A.; Borcea, C.; Talasila, M.; Curtmola, R. Fostering participaction in smart cities: A geo-social crowdsensing platform. *IEEE Commun. Mag.* **2013**, *51*, 112–119.
20. Cardone, G.; Cirri, A.; Corradi, A.; Foschini, L. The participact mobile crowd sensing living lab: The testbed for smart cities. *IEEE Commun. Mag.* **2014**, *52*, 78–85.
21. Cardone, G.; Cirri, A.; Corradi, A.; Foschini, L.; Maio, D. MSF: An efficient mobile phone sensing framework. *Int. J. Distrib. Sens. Netw.* **2013**. Available online: http://journals.sagepub.com/doi/full/10.1155/2013/538937 (accessed on 8 April 2017).
22. Guo, B.; Wang, Z.; Yu, Z.; Wang, Y.; Yen, N.Y.; Huang, R.; Zhou, X. Mobile crowd sensing and computing: The review of an emerging human-powered sensing paradigm. *ACM Comput. Surv. (CSUR)* **2015**, *48*, 7.
23. Pournajaf, L.; Xiong, L.; Sunderam, V.; Goryczka, S. Spatial task assignment for crowd sensing with cloaked locations. In Proceedings of the IEEE 15th International Conference on Mobile Data Management (MDM), Brisbane, Australia, 15–18 July 2014; Volume 1, pp. 73–82.
24. Foremski, P.; Gorawski, M.; Grochla, K.; Polys, K. Energy-efficient crowdsensing of human mobility and signal levels in cellular networks. *Sensors* **2015**, *15*, 22060–22088.
25. Liu, Y.; Niu, J.; Liu, X. Comprehensive tempo-spatial data collection in crowd sensing using a heterogeneous sensing vehicle selection method. *Pers. Ubiquitous Comput.* **2016**, *20*, 397–411.
26. Zavala-Romero, O.; Chassignet, E.P.; Zavala-Hidalgo, J.; Velissariou, P.; Pandav, H.; Meyer-Baese, A. OWGIS 2.0: Open source Java application that builds web GIS interfaces for desktop and mobile devices. In Proceedings of the 22nd ACM SIGSPATIAL International Conference on Advances in Geographic Information Systems, Dallas/Fort Worth, TX, USA, 4–7 November 2014; ACM: New York, NY, USA, 2014; pp. 311–320.
27. Nakayama, Y.; Nakamura, K.; Saito, H.; Fukumoto, R. Open cafe system: Development and application of FOSS4G package in natural environmental field. *Theory Appl. GIS* **2011**, *19*, 17–23. (In Japanese with English Abstract)
28. Iwasaki, N.; Sprague, D.S.; Koyanagi, T.; Furuhashi, T.; Yamamoto, S. Development of the historical agro-environment browsing system constructed by FOSS4G. *Theory Appl. GIS* **2009**, *17*, 83–92. (In Japanese with English Abstract)
29. OpenStreetMap. Available online: http://www.openstreetmap.org/ (accessed on 1 Feburuary 2017).

8

Ophiolitic Remnants from the Upper and Intermediate Structural Unit of the Attic-Cycladic Crystalline Belt (Aegean, Greece): Fingerprinting Geochemical Affinities of Magmatic Precursors

Christina Stouraiti [1,*], Iakovos Pantziris [1], Charalampos Vasilatos [1], Christos Kanellopoulos [1,2], Panagiotis Mitropoulos [1], Panagiotis Pomonis [1], Robert Moritz [2] and Massimo Chiaradia [2]

1 Department of Geology and Geoenvironment, National and Kapodistrian University of Athens, Panepistimiopolis, Ano Ilissia, 15784 Athens, Greece; ipantziris@gmail.com (I.P.); vasilatos@geol.uoa.gr (C.V.); ckanellopoulos@gmail.com (C.K.); pmitrop@geol.uoa.gr (P.M.); ppomonis@geol.uoa.gr (P.P.)

2 Department of Earth Sciences, University of Geneva, Rue des Maraîchers 13, 1205 Genève, GE, Switzerland; Robert.Moritz@unige.ch (R.M.); Massimo.Chiaradia@unige.ch (M.C.)

* Correspondence: chstouraiti@geol.uoa.gr

Abstract: The ophiolitic rocks of the Attic-Cycladic crystalline belt are considered of key importance for understanding the Mesozoic tectonic evolution of the Aegean region. Unresolved questions concern their tectono-stratigraphic relationships across the region. The mode of occurrence of the Cycladic ophiolites varies, as they appear as: (a) dismembered blocks (olistoliths) within the supra-detachment units of Paros and Naxos; (b) mélange formations in the upper structural unit of western Samos and Skyros; and (c) meta-ophiolitic mélange in the Cycladic Blueschist Unit (CBU) from central Samos. The trace element geochemistry and Pb-Sr-Nd isotopes of the mafic ophiolitic rocks indicate four geochemical groups: (a) the upper unit metabasites from Paros and western Samos (Kallithea) display an evolved basaltic composition (Mg# 40.2–59.6), with low Zr/Nb values (5–16) and high Ce/Y values (1.3 to 2.6) compared to MORB, indicating island-arc tholeiite affinities; (b) Naxos upper unit metabasalts show spider diagrams patterns indicating ocean island basalt (OIB-type) affinities; (c) Central Samos metagabbros (CBU) are primitive rocks with Back-Arc Basin basalt affinities; (d) the Skyros metadolerites and Tinos (Mt Tsiknias) and S. Evia (CBU) metagabbros, cluster as a separate geochemical group; they exhibit high MgO values (>10 wt %), very low TiO_2 values (0.1–0.2 wt %), Y and Yb, and depleted trace element N-MORB normalized patterns, similar to volcanic rocks formed in modern oceanic fore-arc settings, such as boninites. A combination of the Pb- and Sr-isotopic compositions of Cycladic metabasites indicate that the Pb and Sr incorporated in the Cycladic ophiolites correspond to mixtures of magmatic fluids with seawater ($^{206}Pb/^{204}Pb$ = 18.51–18.80; $^{207}Pb/^{204}Pb$ = 15.59–15.7; $^{208}Pb/^{204}Pb$ = 39.03–39.80 and initial $^{87}Sr/^{86}Sr80$ = 0.705–0.707). Furthermore, peridotite relicts from Samos, Paros, and Naxos—irrespective of the structural unit—display chemical affinities of ocean floor peridotites formed in a supra-subduction zone. The characteristics of harzburgite relicts in Cycladic serpentinites and Skyros indicate a highly residual nature of the mantle source. Geochemical data from this study shed further light on the complex structure of the oceanic lithosphere from which the Cycladic ophiolites originated.

Keywords: Aegean ophiolites; Sr-Nd-Pb isotopes; MORB, BABB; supra-subduction zone

1. Introduction

Ophiolitic remnants are widespread in the Cyclades, southern Aegean, occurring in the upper portion of the tectono-stratigraphic stack (Figure 1) [1–6]. The mode of occurrence of ophiolitic associations varies in different parts of the Cyclades (Figure 1b,c) [7–9]. The typical mode of occurrence is within low-pressure metamorphosed mélanges in the Upper Cycladic Unit, as on the islands of Anafi [5], Ikaria [10], and Samos (Kallithea nappe) [11], interpreted as subduction–accretion units of Cretaceous age [12] (Figure 1c). In contrast to that is the centrl Cyclades ophiolitic upper unit, which is limited to the occurrence of serpentinized peridotites and minor mafic lithotypes [4]. In the central Cyclades islands of Paros and Naxos, supra-detachment basins associated with metamorphic core complexes enclose olistoliths of the latter lithologies overlain by Miocene and Pliocene sedimentary rocks [12,13]. Additionally, highly sheared serpentinites at the base of these basins mark the low angle detachment faults of Naxos-Paros, and their tectonic position is interpreted as representing a very weak horizon during progressive exhumation [14] (Figure 1c,d).

The initial age of formation of ophiolitic rocks of the upper unit of the central Cyclades is unknown, although in other Cycladic islands and Crete both Jurassic and Cretaceous age ophiolitic rocks have been reported [5,10,15]. Additionally, dismembered metaophiolites occur in the intermediate structural unit enclosed as blocks within mafic mélange units of the Cycladic Blueschists Unit (CBU) [4,6,15–17] (Figure 1b). Based on the geochemical characteristics and structural appearance, these metaophiolitic rocks in the Cycladic islands (e.g., Samos, Syros, Tinos: [15]) have been correlated to the Jurassic ophiolites of the Pindos basin (western ophiolite belts) [5]. An important question arising is whether the late Cretaceous ophiolite remnants from some Cycladic Upper Unit occurrences were high-level parts of a long (~25 Ma) accretion–subduction process of which the Eocene blueschist ophiolite remnants are deep-level representatives [4].

In the Hellenides orogenic belt a major Tethyan-type ophiolite emplacement occurred in mid- to late Jurassic times [18]. This ophiolite belt ('Eohellenic' ophiolites) is interpreted as originating from a Mesozoic Neo Tethyan oceanic basin, the Pindos Ocean, and subsequently thrust northeastwards onto the Pelagonian passive continental margin (Figure 1; [19,20]). Ophiolite remnants in the upper structural units of the Cycladic islands and Crete have been considered as the southernmost continuation of this belt, based mainly on their tectonic position and geochemical characteristics [5] (Figure 1b). However, it is not yet well understood whether all the Cycladic ophiolites correlate with the Middle Jurassic ophiolites of the Pindos ocean ("Eohellenic nappe") in western Greece [21], or if some occurrences correlate with the Upper Cretaceous (70–95 Ma) ophiolites of western Turkey, and Cyprus [18]. Ophiolite rocks from a mélange formation of Skyros and other northern Aegean islands are interpreted, according to their tectonic position, as possible Cretaceous age obducted ophiolite from the internal Vardar ocean [5,22].

Based on their tectonic setting, an ophiolite is emplaced either from down-going oceanic lithosphere via subduction–accretion, or from the upper plate in a subduction zone through trench–continent collision [23,24]. In the case of the Aegean ophiolites, both modes of emplacement occur, indicating a change of the tectonic setting in the Aegean region through the Mesozoic time [25–29].

In this paper, we examine in detail the chemical compositions of mafic and associated ultramafic rocks from the structurally upper unit of Paros-Naxos, Samos, and Skyros, as well as minor occurrences of metamafic rocks from the migmatitic gneiss dome of Naxos (Figure 2). Moreover, we make a comprehensive overview of published geochemical data of ophiolitic lithologies from typical ophiolite exposures of the upper structural unit (UU) of Tinos [15] and Ikaria [10], and compare them with the metamorphosed ophiolitic mélange units (CBU) from Tinos, southern Evia, Andros, Samos [4,15,28], and Naxos [29], in order to place the new data into a regional context.

With these new whole-rock geochemical data that include Sr-Nd-Pb isotopic data for metabasites and ultramafics, we aim to: (a) present a basic geochemical characterization from the ophiolite occurrences of different Aegean islands (Cycladic islands and Skyros); and (b) recognize distinguishing chemical characteristics in order to make regional correlations.

This research paper is based on sampling and whole-rock (major and trace element) analytical data collected by [30] and augmented later on by isotopic data on selected rock samples.

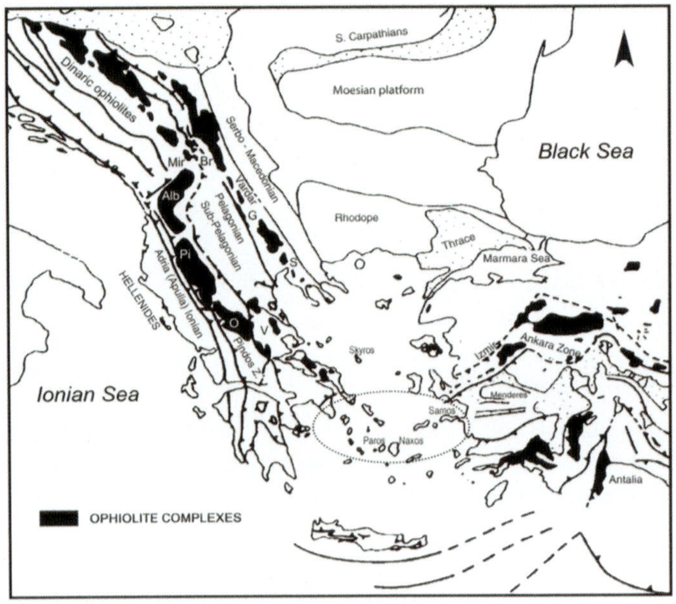

(**a**)

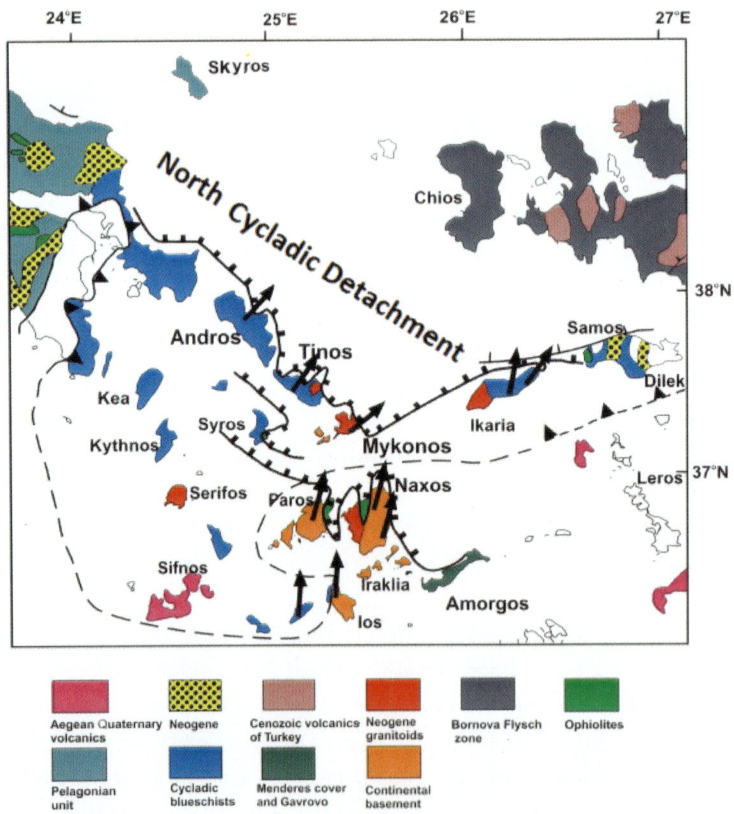

(**b**)

Figure 1. *Cont.*

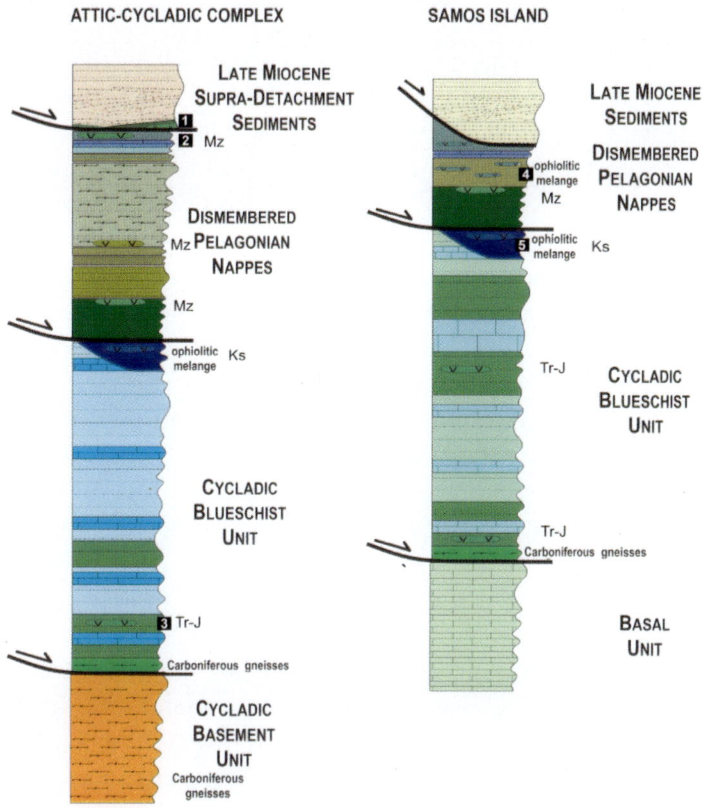

(c)

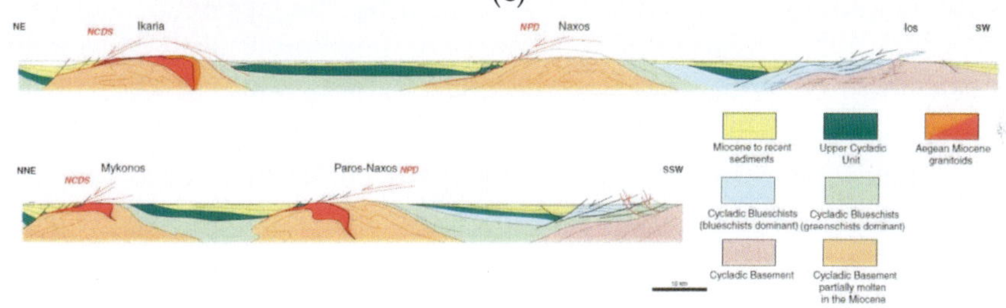

(d)

Figure 1. (**a**) Distribution of Mesozoic ophiolite complexes in the Hellenides and its northward and eastward continuation in the Dinarides and the Taurides, respectively. The tectonic sketch map is modified after [7,8]. Abbreviations: Alb = Albanian, BR = Brezovica, G = Guevgueli, Mir = Mirdita (Albania), O = Othris, Pi = Pindos, S = Sithonia, V = Vourinos. (**b**) Simplified geological map of the Attic-Cycladic Crystalline Belt, showing the major tectonic units. Distribution of the Cycladic Blueschist Unit (CBU), the Cycladic Basement rocks, and the Pelagonian unit within the ACC is also shown (modified after [12] and references therein). (**c**) Schematic tectono-stratigraphic sequence of the main Cycladic units after [1,12–14] and Samos nappe piles after [19,26]. The stratigraphic location of the ophiolitic rocks is marked with numbers on the columns: 1—Naxos and Paros dismembered supra-detachment ophiolites; 2—Skyros melange; 3—Naxos amphibolites from the CBU; 4—western Samos (Kallithea) mélange; 5—central Samos meta-ophiolitic mélange (CBU). Age abbreviations: Mz—Mesozoic, Ks—Cretaceous; Tr-J—Triassic-Jurassic. (**d**) Structural cross-sections across the Cycladic metamorphic belt showing the main structural units and the location of ophiolitic rocks, modified after [9].

2. Geological Setting

Earlier studies in the Attic-Cycladic crystalline belt (Figure 1b) have described two main structural units with different tectono-metamorphic histories (e.g., [1,2,31]). The Upper Unit (or Upper Cycladic Unit) forms the hanging-wall to the Cycladic extensional detachments. It consists of a heterogeneous sequence of Permian to Tertiary sediments, Mesozoic ophiolites (Figure 1c) and pre-Eocene metamorphic rocks transgressively covered by Late Cretaceous unmetamorphosed carbonates, as well as Late Cretaceous medium-pressure/high-temperature rocks and granitoids (e.g., [3,9,32,33]. This upper unit contains mélanges with metaigneous blocks and tectonic slabs (ranging from 1 m to several hundred meters long) that are enclosed in an ultramafic or metasedimentary matrix (e.g., [17,31,34]).

The structurally intermediate unit is equivalent to the Cycladic Blueschist Unit (CBU), which is the dominant tectonic unit in the Cyclades. The CBU consists of remnants of pre-Alpine crystalline basement and a stack of tectonic subunits comprised of a metamorphosed volcano–sedimentary succession (e.g., [31]). The CBU underwent regional eclogite- and blueschist-facies metamorphism during Late Cretaceous to Eocene compression (M1 metamorphic event) [18,35,36]. The Blueschist facies event was followed by a late Oligocene–Miocene, mostly greenschist-facies overprint (M2) with peak conditions of ~450–500 °C and 4–9 kbar [2;35]. In Paros and Naxos M2 reached amphibolite-facies conditions (e.g., [37]). The CBU is considered equivalent to the Pindos zone and to rocks directly underlying the Pelagonian basement.

In the CBU there are widespread tectonic slices of ophiolites, intercalated with marbles, metapelites, mafic volcanites and flysch [3]. These exposures include the metamorphic rocks in Andros, Tinos, Sifnos, Syros, Naxos and, further north, the Lavrion Blueschist Unit [38] and the Styra-Ochi unit of southern Evia [34]. In most studies these units are considered equivalents of the Pindos ocean and some are correlative with sub-ophiolitic mélanges (Figure 1b).

The age of meta-ophiolitic fragments (olistoliths) from the Upper Unit of the Attic-Cycladic crystalline belt and the Late Miocene supra-detachment sedimentary units is considered to be pre-Late Cretaceous according to the Cretaceous metamorphic ages of those units [32,33]. However, an overview of published radiometric data by [39] indicates that both Jurassic and Cretaceous meta-ophiolites occur in the HP mélanges of the CBU, but it is still questionable whether those age groups occur within the same mélange formations or at different lithostratigraphic and/or tectonic levels (Figure 1c).

2.1. Local Geology

On Paros, ophiolitic rocks occur in the Marmara nappe, which in places are transgressively covered by Barremian age limestones (Figure 1c). This transgression is considered to represent the timing of emplacement of ophiolites from the Pindos ocean over the Pelagonian continental margins (Eohellenic event) [40]. The lower part of the Marmara Nappe (Supra-detachment units of map, Eastern Paros in Figure 2a), consists mainly of serpentinites, which preserve relicts of peridotite and only minor blocks of metabasites (metagabbros, Figure 2a).

On Naxos, mafic-ultramafic rock associations from the Upper unit occur as: (a) blocks and slabs of variable size (hundreds of meters) of serpentinized peridotites containing lenses of metagabbros (Galanados, Figure 2b); (b) small occurrences of metagabbros found along the detachment plane; and (c) slices of brecciated metadolerites enclosed in a matrix of sandstone, conglomerate and chert (Agios Thaleleos, Figure 2b). Additionally, metamorphosed and foliated ultramafic rocks mark the contact between the upper and lower structural units [16]. These metamorphosed ultramafic rocks occur as elongated bodies, and often they are marked by the formation of monomineralic zones at their contact with the surrounding gneisses and schists. Serpentinites from two stratigraphic horizons are sampled in this study: (a) the main horizon of meta-ultrabasic bodies of Naxos dome at the boundary of leucogneiss-core with the Lower Series [29], from the localities of Agia and Kourounohori; and (b) the Upper (supra-detachment) Unit serpentinites from the localities of Galanados and Agios Thaleleos (Figure 2b). The aim was to evaluate the similarities/differences between samples from different structural units of the same islands.

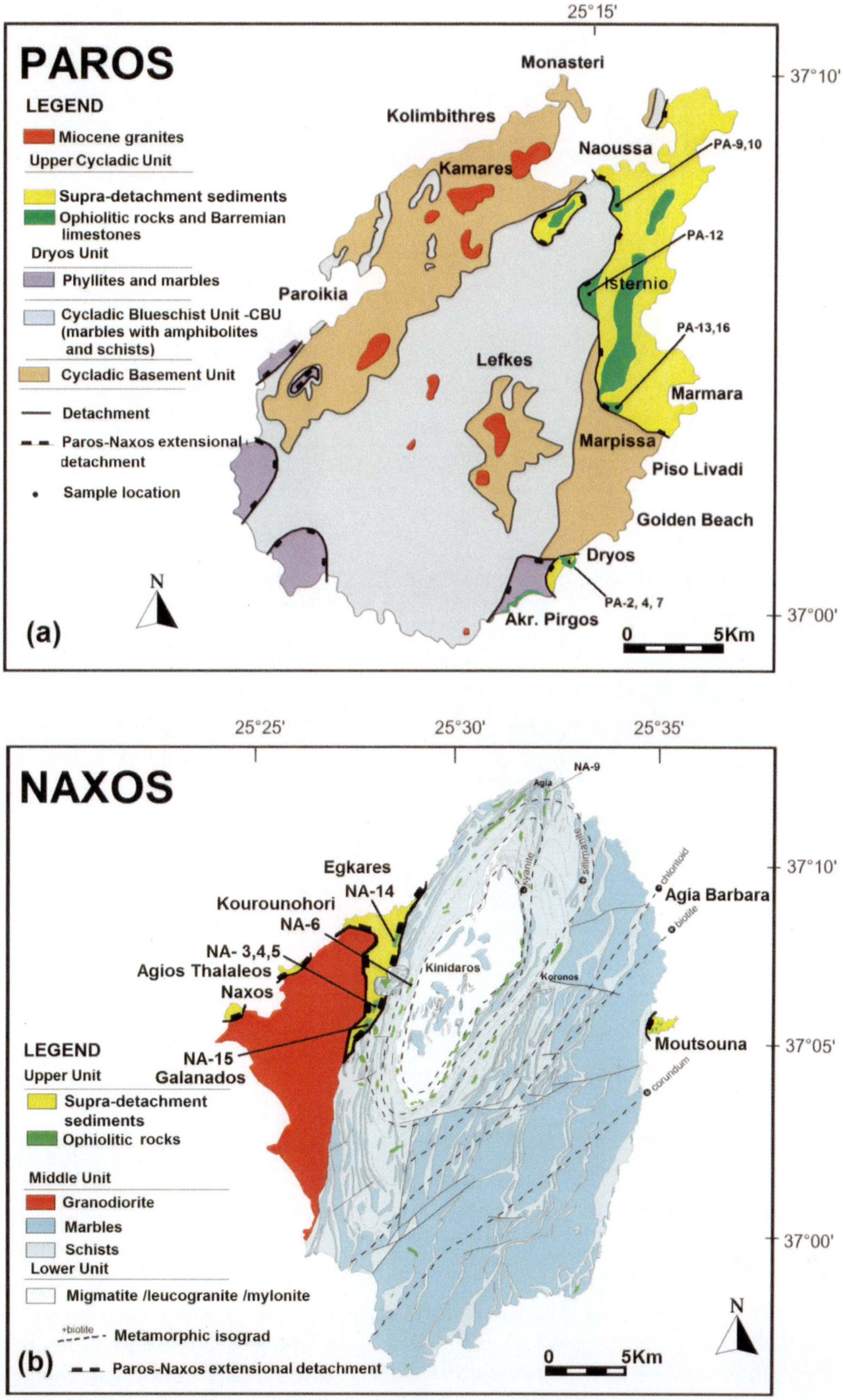

Figure 2. *Cont.*

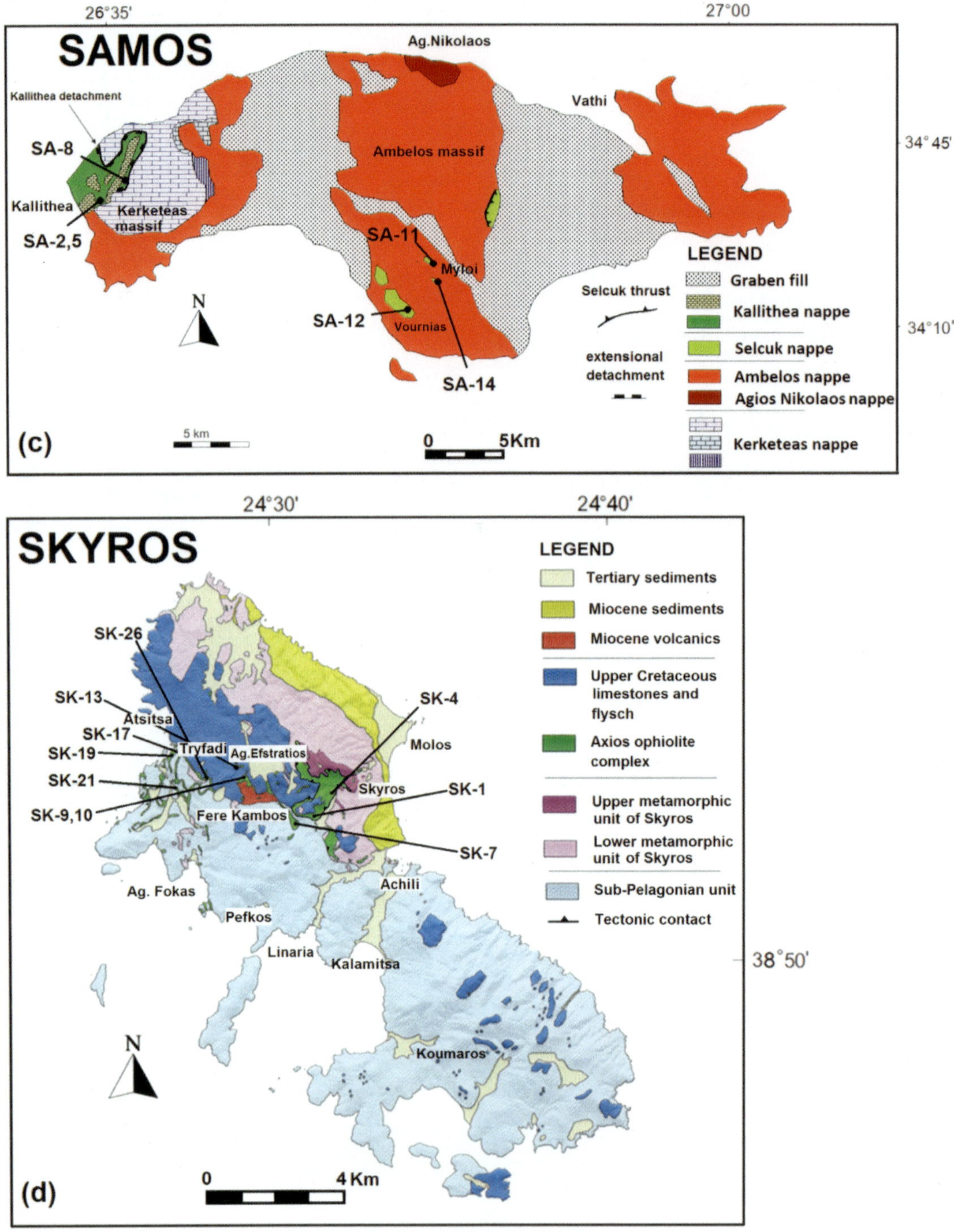

Figure 2. Detailed geological maps of the (**a**) Paros (after [14]); (**b**) Naxos by [8,25]; detailed geological maps of the (**c**) Samos (modified after [26,40]); (**d**) Skyros by [27]). Numbers indicate sample locations.

On Samos island six subunits have been distinguished by previous authors (Figures 1c and 2c). These include the metamorphosed Selçuk, Ampelos, Agios Nikolaos, and Kerketas nappes, and the uppermost Kallithea nappe in western Samos [26,40], Figure 2c. An ophiolitic mélange has been described at the uppermost unit of Kallithea nappe (western Samos, Figure 2c). This unit is described as chaotic mixture of pillow basalts, relict peridotites, serpentinites, conglomerates, radiolarites, clastic sediments, and limestones containing ammonitico rosso, having a thickness of 200 to 300 m [11]. The size of mafic and ultramafic blocks within this chaotic formation ranges from few centimeters to tens of meters. Moreover, a metamorphosed mélange is described in the Selçuk nappe, at central Samos (Figure 2c). The Selçuk nappe has been correlated to the upper most part of the Cycladic Blueschist Unit [40]. For the purpose of this study we have sampled ophiolitic occurrences from the Upper Unit (Kallithea nappe) as well as central Samos metaophiolitic mélange.

On Skyros island, central Aegean (Figure 2d), an ophiolitic mélange overthrusts the Palaeozoic–Middle Jurassic Pelagonian basement rocks and is considered as part of the Eohellenic nappe [5]. K-Ar dating on a gabbroic rock (diorite) from Skyros mélange determined a Lower Cretaceous age (125 $\pm$ 8 Ma), which has been assigned to the formation age of the magmatic protolith [5]. Ophiolite lithologies outcrop in the central–northern part of the island (Figure 2d) and consist of highly deformed serpentinite, massive gabbroic rocks, dolerite dykes, and massive basaltic/andesitic lavas and ophicalcite [41]. In the framework of this study, selected mafic and ultramafic rocks from the Skyros mélange are compared with the structurally correlative Upper Cycladic Unit ophiolitic rocks.

3. Analytical Techniques

Thirty-three rock samples (13 metabasites and 20 serpentinites) were analyzed for major and trace elements. Sample locations are provided in Supplementary Table S1. Of these, 11 samples—selected as representative for each lithological type—were analyzed for Sr and Nd isotope compositions. Lead isotope ratios were measured on representative samples from Naxos, Paros, and Samos metabasic rocks. Lead isotope analyses of the metaperidotites samples were not possible to be determined due to the very low U and Pb concentrations in these samples, except for two samples from Naxos (NA-6, NA-14).

Major and trace element concentrations of the rock samples were determined by the XRF technique at Leicester University (UK). Major elements were determined on glass beads fused in a gold–platinum crucible at 1100 °C from ignited powders, and trace elements were determined on pressed powder pellets (see Supplementary Materials). Major elements were determined using Rh anode excitation, and the trace elements Sc, V, Cr, Ni, Zn, Ga, Rb, Sr, Y, Zr, Nb, La, Ce, Nd, and Ba, on pressed powder briquettes, using either Rh or W excitation, both on the Phillips PW1600 X-ray fluorescence spectrometer (Phillips electronics, Eindhoven, The Netherlands). Details of operating conditions and techniques for this instrument are given in [42]. Although the concentrations of some of the trace elements (particularly Nb, La, Ce, and Th) were near or below the limit of detection by XRF techniques in many samples, a number of key, relatively immobile, elements were able to be determined with sufficiently high precision by XRF methods to enable clear deductions concerning the petrogenesis of the Cycladic ophiolitic rocks.

Radiogenic isotopic compositions of Sr, Nd and Pb were measured on 12 whole rock (six metaperidotite and six metabasites) samples from Naxos, Paros, Samos, and Skyros at the University of Geneva, Switzerland. Between 100 and 120 mg of whole rock powder were dissolved during seven days in Savillex® Teflon (Savillex, Minnetonka, MN, USA) vials using 4 mL of concentrated HF and 1 mL of HNO_3 14 M, at a temperature of 140 °C and with the help of ultrasonication for 30 min twice a day. Subsequently, samples were dried and re-dissolved for three days (also with 30 min ultrasonication twice a day) in 3 mL of HNO_3 14 M, and dried again. Sr, Nd, and Pb were then separated using cascade columns with Sr-Spec, TRU-Spec, and Ln-Spec resins, respectively, according to a protocol modified from [43]. Finally, the material was redissolved in 2% HNO_3 solutions and ratios were measured using a Thermo Neptune PLUS Multi-Collector ICP-MS (Thermo Fisher Scientific, Waltham, MA, USA) in static mode. Ratios used to monitor internal fractionation were: $^{88}Sr/^{86}Sr = 8.375209$ for the $^{87}Sr/^{86}Sr$ ratio, $^{146}Nd/^{144}Nd = 0.7219$ for the $^{143}Nd/^{144}Nd$ ratio, and $^{203}Tl/^{205}Tl = 0.418922$ for the three Pb ratios (a Tl standard was added to the solution). External standards used were SRM987 ($^{87}Sr/^{86}Sr = 0.710248$, long-term external reproducibility: 10 ppm), JNdi-1 ($^{143}Nd/^{144}Nd = 0.512115$; [44] long-term external reproducibility: 10 ppm), and SRM981 [45] for Pb (long-term external reproducibility of 0.0048% for $^{206}Pb/^{204}Pb$, 0.0049% for $^{207}Pb/^{204}Pb$ and 0.0062% for $^{208}Pb/^{204}Pb$). $^{87}Sr/^{86}Sr$, $^{143}Nd/^{144}Nd$ and Pb isotope ratios were further corrected for external fractionation (due to systematic differences between measured isotope ratios and those proposed for SRM987, JNdi-1 and SRM981, by [44–46] respectively]) by −0.039‰, +0.047‰ and +0.5‰ amu, respectively. Interferences at masses 84 (^{84}Kr), 86 (^{86}Kr) and 87 (^{87}Rb) were corrected by monitoring ^{83}Kr and ^{85}Rb, ^{144}Sm interference on ^{144}Nd was monitored on the mass ^{147}Sm and corrected by using a $^{144}Sm/^{147}Sm$ value of 0.206700, and ^{204}Hg interference on ^{204}Pb was corrected by monitoring ^{202}Hg.

All the analyzed metabasites have low U/Pb (ranging from 0.1 to 0.5) as well as Th/Pb ratios, and therefore no corrections due to time-integrated decay of U and Th were made because the addition of radiogenic lead is considered to be insignificant (Table 1). An exception to this is Naxos metabasite NA-3, which displays an elevated Th/Pb value of 1.3, and the serpentinite NA-14, which is characterized by slightly more radiogenic compositions, suggesting a higher U/Pb ratio.

Table 1. Whole-rock major and trace element compositions of metabasites and ultramafic rocks from the Aegean ophiolites.

SAMPLE	**NA-3**	**NA-4**	**NA-5**	**NA-7**	**NA-9**	**NA-15**	**PA-9**	**PA-10**	**SA-2**	**SA-5**
Rock Type wt %	**Meta Basalt (UU)**	**Meta Basalt/ Andesite (UU)**	**Meta Basalt/ Andesite (UU)**	**Amphi- Bolite (CBU)**	**Amphi- Bolite (CBU)**	**Meta Gabbro (UU)**	**Meta Gabbro (UU)**	**Meta Gabbro (UU)**	**Meta Basalt (UU)**	**Meta Basalt (UU)**
SiO_2	42.00	53.49	54.53	50.21	56.6	45.94	50.41	52.19	38.88	49.04
TiO_2	1.79	1.41	0.35	0.21	0.10	1.40	0.61	0.51	0.69	1.92
Al_2O_3	17.11	13.19	12.75	7.33	3.56	15.00	14.36	16.36	15.99	17.01
Fe_2O_{3T}	9.59	8.62	4.75	5.59	7.51	8.89	7.75	7.49	8.57	11.17
MnO	0.14	0.14	0.16	0.07	0.29	0.10	0.15	0.17	0.12	0.12
MgO	3.26	5.27	2.30	25.87	19.17	4.66	5.21	4.54	6.39	3.39
CaO	11.4	11.07	10.06	6.09	12.20	11.55	11.05	11.03	12.99	4.62
Na_2O	2.14	1.06	1.24	0.08	0.17	3.20	1.99	2.79	2.18	6.44
K_2O	4.13	0.16	4.55	0.04	0.16	0.73	1.98	2.05	1.53	0.20
P_2O_5	0.49	0.25	0.10	0.01	0.01	0.54	0.17	0.42	0.13	0.01
LOI	8.27	5.78	9.20	5.21	1.46	7.48	6.51	2.60	12.69	6.76
TOTAL	100.32	100.44	100.00	100.71	100.22	99.49	100.21	100.57	100.15	100.72
Mg#	40.23	54.76	48.95	90.16	83.48	50.93	57.10	54.55	59.62	37.54
Trace elements (ppm)										
Sc	30	29	28	14	24	37	32	29	40	30
V	236	246	104	81	65	217	178	187	254	194
Cr	240	171	40	1212	1643	416	144	131	125	83
Co	45	42	21	57	71	46	34	33	34	30
Ni	130	90	32	1550	1035	225	42	45	40	32
Cu	10	50	39	bdl	4	49	88	42	18	3
Zn	100	67	55	37	144	64	95	73	57	77
Ga	19	16	11	8	6	14	13	15	15	15
Rb	79	2	142	bdl	8	15	56	62	39	22
Sr	206	395	204	12	33	216	299	262	152	231
Y	31	22	19	13	5	27	24	21	12	20
Zr	161	119	100	10	7	144	82	124	35	53
Nb	34	22	8	4	1	31	5	11	6	9
Ba	456	16	449	19	22	147	676	429	136	105
Pb	3	2	18	3	7	2	17	14	4	5
Th	4	2	11	2	bdl	4	3	9	2	3
U	1	1	2	1	1	1	2	2	bdl	1
La	19	17	21	4	bdl	24	11	32	11	15
Ce	47	29	37	3	2	39	30	54	16	28
Nd	26	16	19	6	2	20	14	23	8	11
Zr/Nb	4.7	5.4	12.5	2.5	7	4.6	16.4	11.2	5.8	5.9
Ce/Y	1.5	1.3	1.9	0.2	0.4	1.4	1.2	2.5	1.3	1.4
Nb/Y	1.09	1.0	0.42	0.3	0.2	1.14	0.2	0.52	0.5	0.45
SAMPLE	**SA-8**	**SA-11**	**SK-1**	**NA-6**	**NA-14**	**PA-2**	**PA-4**	**PA-7**	**PA-12**	**PA-13**
Rock Type wt %	**Meta Basalt (UU)**	**Meta Gabbro (CBU)**	**Meta Dolerite (UU)**	**Serp. Harzb. (CBU)**	**Serp. Harzb. (UU)**	**Harzb. (UU)**	**Harzb. (UU)**	**Harzb. (UU)**	**Harzb. (UU)**	**Harzb. (UU)**
SiO_2	46.06	44.93	50.35	43.13	40.66	39.86	40.71	39.93	41.80	39.76
TiO_2	0.68	1.04	0.13	0.04	0.01	0.02	0.02	0.02	0.05	0.01
Al_2O_3	16.74	18.80	16.70	1.74	0.9	1.44	1.66	1.43	2.68	0.85
Fe_2O_{3T}	8.57	8.01	4.24	8.47	8.86	8.15	8.26	7.87	8.37	7.91
MnO	0.14	0.15	0.06	0.13	0.07	0.07	0.07	0.06	0.13	0.10
MgO	6.29	9.09	10.04	42.58	36.76	36.29	36.98	37.15	34.12	37.60
CaO	10.21	11.61	12.82	1.61	0.76	0.13	0.06	0.04	1.99	0.97

Table 1. *Cont.*

SAMPLE	SA-8	SA-11	SK-1	NA-6	NA-14	PA-2	PA-4	PA-7	PA-12	PA-13
Rock Type wt %	Meta Basalt (UU)	Meta Gabbro (CBU)	Meta Dolerite (UU)	Serp. Harzb. (CBU)	Serp. Harzb. (UU)	Harzb. (UU)	Harzb. (UU)	Harzb. (UU)	Harzb. (UU)	Harzb. (UU)
Na_2O	1.18	1.73	1.47	bdl	bdl	0.27	0.03	0.04	bdl	bdl
K_2O	2.55	1.59	0.44	bdl	bdl	0.04	0.01	0.01	bdl	bdl
P_2O_5	0.18	0.24	0.01	0.09	0.07	bdl	bdl	bdl	0.01	0.01
LOI	7.54	3.47	3.60	1.66	11.86	12.59	12.31	12.37	10.96	13.33
TOTAL	100.15	100.55	99.87	99.45	99.95	98.87	100.11	98.94	100.05	100.47
Mg#	59.24	69.20	82.42	90.87	89.15	89.81	89.86	90.33	88.97	90.39
Trace elements (ppm)										
Sc	27	27	39	8	6	4	8	8	9	5
V	172	224	168	50	37	27	41	44	60	39
Cr	261	162	1159	2895	3932	1935	2634	2799	2525	1789
Co	35	43	36	100	112	63	88	81	95	95
Ni	72	244	210	2306	2312	2000	2564	2394	2119	2502
Cu	5	16	80	18	5	8	5	20	20	6
Zn	145	64	17	40	37	37	45	49	40	51
Ga	23	19	8	2	2	1	1	bdl	1	bdl
Rb	101	62	8	bdl	2	bdl	bdl	2	bdl	bdl
Sr	50	259	116	5	6	20	6	5	12	4
Y	12	30	5	bdl	1	1	1	2	2	1
Zr	67	100	20	bdl	bdl	bdl	bdl	2	2	bdl
Nb	55	2	<0.5	bdl	bdl	1	bdl	bdl	bdl	bdl
Ba	299	893	38	bdl	4	4	6	3	11	3
Pb	10	7	1	bdl	1	2	1	3	1	2
Th	3	bdl	bdl	bdl	bdl	bdl	bdl	bdl	bdl	bdl
U	1	1	1	bdl	1	bdl	1	1	1	3
La	10	2	2	bdl	bdl	bdl	3	bdl	2	bdl
Ce	27	8	5	bdl	bdl	bdl	bdl	bdl	bdl	bdl
Nd	13	8	3	bdl	bdl	bdl	bdl	bdl	2	bdl
Zr/Nb	13.4	50	50			1				
Ce/Y	2.25	0.26	1							
Nb/Y	4.58	0.06	0.08							

SAMPLE	PA-16	SA-12	SA-14	SK-4	SK-7	SK-9	SK-10	SK-13	SK-17	SK-19
Rock Type wt %	Harzb. (UU)	Olivin Pyrox. (CBU)	Olivin Pyrox. (CBU)	Harzb. (UU)	Harzb. (UU)	Lherz. (UU)	Harzb. (UU)	Harzb. (UU)	Harzb. (UU)	Harzb. (UU)
SiO_2	36.70	42.35	39.25	36.68	40.00	37.75	38.30	42.60	40.30	40.95
TiO_2	0.02	0.02	0.11	0.35	0.02	0.02	0.02	0.01	bdl	0.01
Al_2O_3	0.69	1.54	4.82	8.61	1.27	1.29	1.14	0.43	0.32	0.88
Fe_2O_{3T}	8.27	8.52	12.18	7.35	8.43	8.09	11.26	6.76	8.49	7.75
MnO	0.11	0.11	0.11	0.13	0.14	0.13	0.18	0.07	0.30	0.14
MgO	37.39	36.84	32.48	35.44	36.04	35.55	34.88	36.68	36.58	36.18
CaO	1.03	0.03	0.37	0.91	1.39	2.90	1.00	0.04	0.80	0.28
Na_2O	bdl	bdl	bdl	bdl	bdl	bdl	Bdl	bdl	bdl	bdl
K_2O	bdl	bdl	bdl	bdl	bdl	bdl	Bdl	bdl	bdl	bdl
P_2O_5	0.01	0.01	0.01	0.12	0.01	0.01	0.01	0.01	0.01	0.01
LOI	13.19	11.26	10.81	12.09	11.72	13.45	13.21	11.95	13.20	11.95
TOTAL	99.32	100.56	100.05	100.59	99.31	99.10	100.00	100.45	100.00	100.07
Mg#	89.95	90.02	84.07	90.52	89.43	89.69	85.98	91.48	89.51	90.24
Trace elements (ppm)										
Sc	4	4	7	6	6	13	33	3	7	34
V	7	34	87	90	45	49	32	16	36	5
Cr	3226	3367	2940	2014	2824	2542	2542	2213	3314	1578
Co	107	82	86	99	94	123	102	84	98	103
Ni	2720	2175	2218	1754	2285	2507	2375	2108	2452	2131
Cu	5	9	53	2	2	10	6	2	6	4
Zn	38	60	61	49	36	26	31	25	35	21
Ga	bdl	2	4	8	1	bdl	bdl	bdl	bdl	bdl
Rb	bdl	2	bdl	2	bdl	bdl	bdl	bdl	bdl	bdl
Sr	47	3	3	12	2	4	12	5	2	12
Y	1	4	3	7	1	2	1	1	bdl	1
Zr	bdl	2	3	82	bdl	2	bdl	bdl	bdl	bdl
Nb	bdl	bdl	bdl	6	bdl	bdl	1	bdl	bdl	bdl
Ba	3	bdl	7	3	4	5	11	22	bdl	18
Pb	1	1	1	3	1	2	1	1	3	2
Th	bdl	bdl	bdl	3	bdl	bdl	bdl	bdl	bdl	bdl
U	bdl	bdl	bdl	2	bdl	bdl	2	1	bdl	1
La	bdl	2	2	7	bdl	bdl	bdl	bdl	2	bdl
Ce	bdl	5	bdl	22	bdl	bdl	bdl	bdl	bdl	bdl
Nd	bdl	2	bdl	11	bdl	bdl	3	bdl	bdl	2
Zr/Nb				13.6			1			
Ce/Y		1.25		3.14						

Table 1. *Cont.*

SAMPLE	SK-21	SK-23	SK-26	Samos	Skyros	N-MORB	E-MORB	IAT [3]	OIB
Rock Type wt %	Olivin Webst. (UU)	Harzb. (UU)	Harzb. (UU)	Meta Gabbro [1] (CBU)	(SGBR) Meta Gabbro [2]				
SiO_2	39.83	40.59	38.68	49.90	50.37			51.03	
TiO_2	0.01	0.01	0.02	0.90	0.23	1.27	1.00	0.82	2.86
Al_2O_3	0.47	0.99	1.27	16.30	17.06			15.76	
Fe_2O_{3T}	9.00	7.49	10.19	6.40	5.67			11.76	
MnO	0.09	0.13	0.14	0.10	0.11			0.21	
MgO	30.72	38.86	37.45	8.00	9.00			6.30	
CaO	5.31	0.17	0.07	11.0	9.98			10.72	
Na_2O	bdl	bdl	bdl	3.60	2.80			2.40	
K_2O	bdl	bdl	bdl	0.50	0.81	0.072	0.25	0.18	1.44
P_2O_5	0.01	0.01	0.01	0.10	0.02	0.116	0.142	0.13	0.618
LOI	13.73	12.17	11.93	3.60	3.80				
TOTAL	99.06	100.31	99.67	100.40	99.96				
Mg#	87.11	91.13	87.92	71.28	76.06				
Trace elements (ppm)									
Sc	11	6	7	34.8	41				
V	19	37	44	207.3	177				
Cr	3359	2938	3104	366.7	520			100	
Co	100	87	105	27.3	31				
Ni	2388	2234	2826	151.7	111			14	
Cu	14	1	5	20	15				
Zn	19	33	32	40	7			79	
Ga	bdl	bdl	1						
Rb	bdl	bdl	bdl	15	11	0.6	5	3	31
Sr	21	1	1	179	193	90	155	133	660
Y	1	bdl	bdl	19	10	28	22	18	29
Zr	bdl	bdl	bdl	53	32	74	73	29	280
Nb	bdl	bdl	bdl	1.1	0.1	2.3	8.3	1	48
Ba	bdl	3	4	108	34	6.3	57	57	350
Pb	2	2	bdl	bdl	1.4	0.3	0.6		3.2
Th	bdl	bdl	bdl	0.1		0.1	0.6		4
U	1	bdl	bdl	0.1	0.1		0.2		1
La	bdl	bdl	bdl	2.5	1.4	2.5	6.3	1.59	37
Ce	bdl	bdl	bdl	7	2.3	7.5	15	4.99	80
Nd	bdl	bdl	bdl	6.6	1.4	7.3	9	4.39	38.5
Yb				1.76	1.22	3.1	2.4	1.84	2.2
Ce_N/Yb_N				1.04	0.49	0.63	1.64	0.71	9.55
Zr/Nb				48.1	320	32.2	8.8	29	5.8
Ce/Y				3.14	0.2	0.2	0.7	0.3	2.7
Nb/Y				2.8	3.2	2.6	3.3	1.6	9.7

bdl.: Below the detection limit (see Supplementary Material for detection limits). Total iron given as Fe_2O_{3T} and LOI is loss on ignition. Mg# = 100*MgO/(MgO + FeOt) with oxides in molar proportions. Mg# calculated for all samples on anhydrous basis. Locations: NA—Naxos. PA—Paros. SA—Samos. SK—Skyros. Rock type abbreviations: basalt/and.—basaltic andesites, Amphib.—Amphibolite; Harzb.—Harzburgite; Olivin Pyrox.—Pyroxenite; Lherz.—Lherzolite; Olivin Webst.—Websterite. [1] Central Samos metagabbros from CBU ($n = 5$), data from [15]; [2] Skyros metagabbro (SGBR), data from [41]; [3] IAT—Island Arc Tholeiite, South Sandwich island arc [47]. Ce_N/Yb_N normalized values to Chondrite CI and average elemental composition of MORB and OIB from [48].

4. Petrography

4.1. Naxos

The metaophiolitic outcrops of Naxos comprise mafic and ultramafic rocks. Mafic rocks are classified as (a) metagabbros and (b) metavolcanics. Ultramafic rocks comprise dunites and serpentinites.

4.1.1. Amphibolites

The metagabbros (CBU) are now amphibolites and display mainly granoblastic to lepidoblastic medium-grained textures, and occasionally relicts of igneous texture are preserved. They consist of 30–60 vol % amphibole, 20–30 vol % strongly altered plagioclase, 10–20 vol % chlorite, 15–20 vol % epidote, and accessory magnetite, quartz, and carbonates. Amphibole typically occurs as needles and slender prisms of actinolite, which locally display orientation. In places, slender prisms of actinolite and fibres of tremolite and chlorite surround porphyroblasts of relict plagioclase (NA-9, Figure 3a).

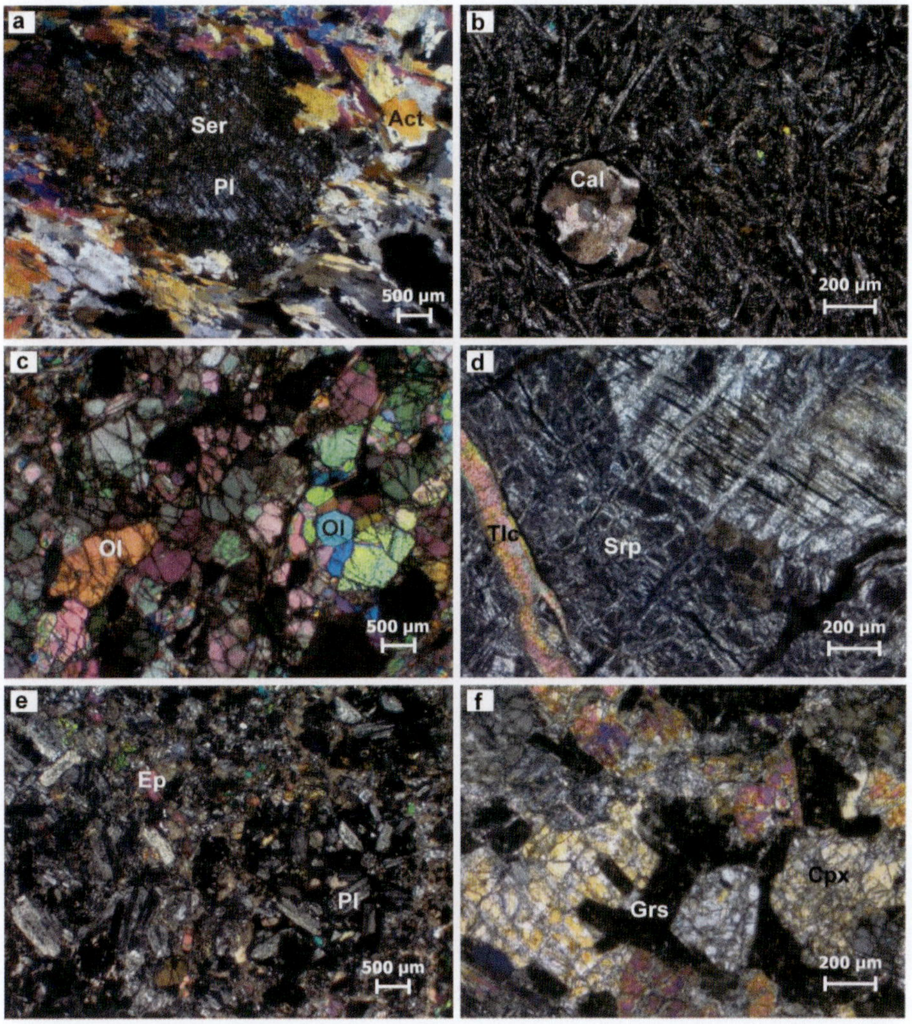

Figure 3. Photomicrographs (XPL) of: (**a**) sericitized plagioclase porphyroblast surrounded by slender prisms of actinolite (NA9 amphibolite, Naxos); (**b**) intersertal texture with amygdule filled with secondary calcite and iron oxides (NA3 spilite, Naxos); (**c**) equigranular polygonal olivine crystals forming a mosaic texture in dunite (NA6 dunite, Naxos); (**d**) mesh texture and bastite pseudomorph in serpentinite (PA12 serpentinite, Paros); (**e**) porphyritic intersertal texture in meta-andesite (SA8 meta-andesite, Samos); (**f**) blasto-subophitic texture in rodingite (SK1 rodingite, Skyros). (Act: Actinolite, Cal: Calcite, Cpx: Clinopyroxene, Ep: Epidote, Grs: Grossular, Ol: Olivine, Pl: Plagioclase, Ser: Sericite, Srp: Serpentine, Tlc: Talc).

Plagioclase laths are totally replaced by a cloudy microcrystalline aggregate of sericite, albite, calcite, epidote, and clinozoisite, forming pseudomorphs due to sericitization and saussuritization.

4.1.2. Metavolcanics

Metavolcanics of Naxos consist of spilites and metabasalts (prasinites). Spilites are characterized by relict intersertal, spherulitic, variolitic, and amygdaloidal texture, consisting of 50–70 vol % albite, 10–20 vol % chlorite, 5–10 vol % calcite, 5–10 vol % epidote, devitrified glass, and quartz. Spherulite shows radiating aggregates of acicular plagioclase altered to albite, while variolitic textures are fan-like altered plagioclase needles within glassy groundmass. Locally, rounded and elliptical amygdules that are 0.5–1 mm in diameter (Figure 3b), are filled mainly with calcite, while zeolite, epidote and secondary quartz are presented as minor constituents. Iron oxides, occurring in the groundmass and along fractures, are late magmatic minerals. Some late veins of quartz cut across the rocks.

Metabasalts show fine-grained granoblastic to lepidoblastic texture, consisting of clinozoisite, epidote, chlorite, albite, actinolite, white mica, calcite, titanite, and quartz. Late veinlets of quartz and feldspars crosscut the rocks.

4.1.3. Dunites and Serpentinites

Dunites show a rather constant mineralogical composition, comprising mainly of olivine with subordinate orthopyroxene and spinel. These fine-grained rocks display mosaic and fine equigranular textures composed of small, polygonal and occasionally unstrained olivine neoblasts (Figure 3c). Spinel is dark red and display anhedral grains heterogeneously distributed in the olivine matrix. Locally, spinel grains exhibit narrow rims of magnetite. The degree of serpentinization varies between 10% and 30%.

Serpentinites display in general mesh texture and consist essentially of serpentine minerals, mainly antigorite with minor chlorite, calcite and talc. The only preserved primary phase is Cr-spinel, which occasionally is totally replaced by magnetite. Pseudomorphic prismatic plates of bastite (lizardite pseudomorphs after pyroxenes), which sometimes mimic the cleavage of earlier pyroxene, are common.

4.2. Paros

The collected metaophiolitic mafic and ultramafic rock samples from Paros are classified as: (a) metagabbros and (b) serpentinites.

4.2.1. Metagabbros

The metagabbros display relict granular to granoblastic textures and consist of 30–40 vol % plagioclase, 20–30 vol % amphibole, 5–10 vol % chlorite, ~5 vol % epidote, minor titanite, and accessory magnetite, quartz, and carbonates. Plagioclase laths are replaced by a microcrystalline aggregate of sericite, albite, calcite, epidote and clinozoisite, forming pseudomorphs due to sericitization and saussuritization. Amphibole typically occurs as thin prisms and fibres of actinolite and tremolite, replacing clinopyroxene. Relicts of clinopyroxene occur as inclusions within some amphibole prisms, or more rarely as large xenomorphic crystals up to 5 mm in diameter, surrounded by continuous thin rings of actinolite and chlorite. These metagabbros show better preserved igneous textures relative to the amphibolites of Naxos, and have not acquired a visible metamorphic fabric.

4.2.2. Serpentinites

Serpentinites exhibit mesh and locally hourglass textures. They consist essentially of serpentine minerals, mainly antigorite with minor chlorite, calcite and talc. The only preserved primary phases are relicts of olivine and Cr-spinel, which occasionally is totally replaced by magnetite. Pseudomorphic prismatic plates of bastite (lizardite pseudomorphs after pyroxenes), which sometimes mimic the cleavage of earlier pyroxene, are common (Figure 3d). Sparsely, veinlets of calcite, magnesite and talk crosscut the serpentinites.

4.3. Samos

The ophiolitic rock samples from Samos are classified as: (a) metagabbros; (b) metavolcanics and (c) serpentinized harzburgites—serpentinites.

4.3.1. Metagabbros

Metagabbros are medium- to fine-grained rocks with poikiloblastic texture. They contain 15–35 vol % plagioclase, 10–20 vol % clinozoisite, 10–15 vol % clinopyroxene, 10–15 vol % amphibole, 5–10 vol % chlorite and <5 vol % quartz. Plagioclase laths are partly replaced by a microcrystalline matrix of sericite, albite, calcite, epidote, and clinozoisite, forming pseudomorphs due to sericitization and saussuritization. Amphibole typically occurs as thin prisms and fibres of actinolite and tremolite, replacing clinopyroxene. Clinopyroxene relicts occur as xenomorphic crystals surrounded by rims of actinolite. Clinozoisite forms idiomorphic crystals, whereas epidote is xenomorphic. Locally, igneous textures are well preserved and there is not a pervasive metamorphic fabric presented.

4.3.2. Metavolcanics

Metavolcanics are classified as metabasalts and metandesites. Metandesites display mainly porphyritic and interintersertal textures. They contain 40–60 vol % plagioclase, 5–10 vol % clinopyroxene, 5–10 vol % chlorite, 2–7 vol % amphibole and minor epidote, calcite, and quartz. Plagioclase form idiomorphic sericitized porphyrocrysts that are set in a fine-grained to microcrystalline matrix composed of allotriomorphic clinopyroxene relicts, epidote crystals, iron oxides, chlorite and devitrified glass (Figure 3e). Sporadic patches of interstitial quartz are common. Elongated veins filled with secondary minerals appear occasionally crosscutting the rock.

4.3.3. Serpentinized Harzburgites, Serpentinites

Serpentinized harzburgites show porphyroclastic texture with olivine constituting 50–70 vol %, orthopyroxene 15–30 vol %, and spinel up to 5 vol %. Olivine forms porphyroclasts as well as smaller polygonal neoblasts (~0.2 mm). The orthopyroxene forms isolated large crystals with lobate boundaries and often displays diopside exsollution lamellae, kink bands, and gliding parallel to (001). Minor clinopyroxene forms neoblastic grains in the recrystallized matrix. Dark-brown spinel forms disseminated anhedral crystals with lobate boundaries. The degree of serpentinization varies between 30% and 60%.

Serpentinites exhibit mesh textures. They consist essentially of serpentine minerals, mainly antigorite and chrysotile with minor chlorite, calcite, and talc. Occasionally relicts of olivine and Cr-spinel are presented. Pseudomorphic prismatic plates of bastite (lizardite pseudomorphs after pyroxenes), which sometimes mimic the cleavage of earlier pyroxene, are common. Sparsely, veinlets of calcite, serpentine, and talc crosscut the serpentinites.

4.4. Skyros

The collected metaophiolitic mafic and ultramafic rock samples from Skyros are classified as: (a) rodingites and (b) serpentinites.

4.4.1. Rodingites

Rodingites mainly display a blasto-subophitic texture with local relicts of primary subophitic texture, which is strongly indicative of a doleritic protolith (Figure 3f). They contain 30–50 vol % clinopyroxene, 20–30 vol % chlorite, 10–20 vol % garnet with minor calcite, titanite, and ilmenite. Clinopyroxene and chlorite formed probably after actinolite but they retain a crystal habit of an older clinopyroxene. Garnet (grossular) occurs in between the clinopyroxene grains and clearly pseudomorphs original plagioclase (Figure 3f). It comprises weakly birefringent, frequently tiny, micron-size crystals. Its small size and anisotropic character make it difficult to distinguish from vesuvianite, which also commonly appears in rodingites. Fe–Ti oxides with a dusty appearance are dispersed throughout the samples.

4.4.2. Serpentinites

The serpentinites are composed essentially of antigorite, chrysotile, and lizardite, with subordinate amounts of carbonates, chromite, magnetite, magnesite, talc, tremolite, and chlorite. Antigorite is the dominant mineral and occurs as interpenetrating and interlocking elongate plates. Sometimes, the antigorite flakes and blades are arranged in roughly parallel aggregates. Serpentine, which developed as an alteration after olivine, usually has a mesh texture while that formed after orthopyroxene has a bastite texture. These textures indicate dunite and harzburgite parent rocks.

The opaque minerals in the studied serpentinites are represented mainly by dark-brown spinel and magnetite. Spinel occurs as disseminated subhedral crystals and/or irregular grains. Alteration of spinel along grain borders and cracks is very common. Magnetite occurs as anhedral grains in the interstitial spaces between the serpentine minerals or as veinlets.

5. Whole-Rock Geochemistry

The chemical compositions of analyzed samples are presented in Table 1. Sample locations are indicated in Figure 2a–d. Standard geochemical diagrams have been utilized in order to classify the investigated former magmatic rocks (Figure 4).

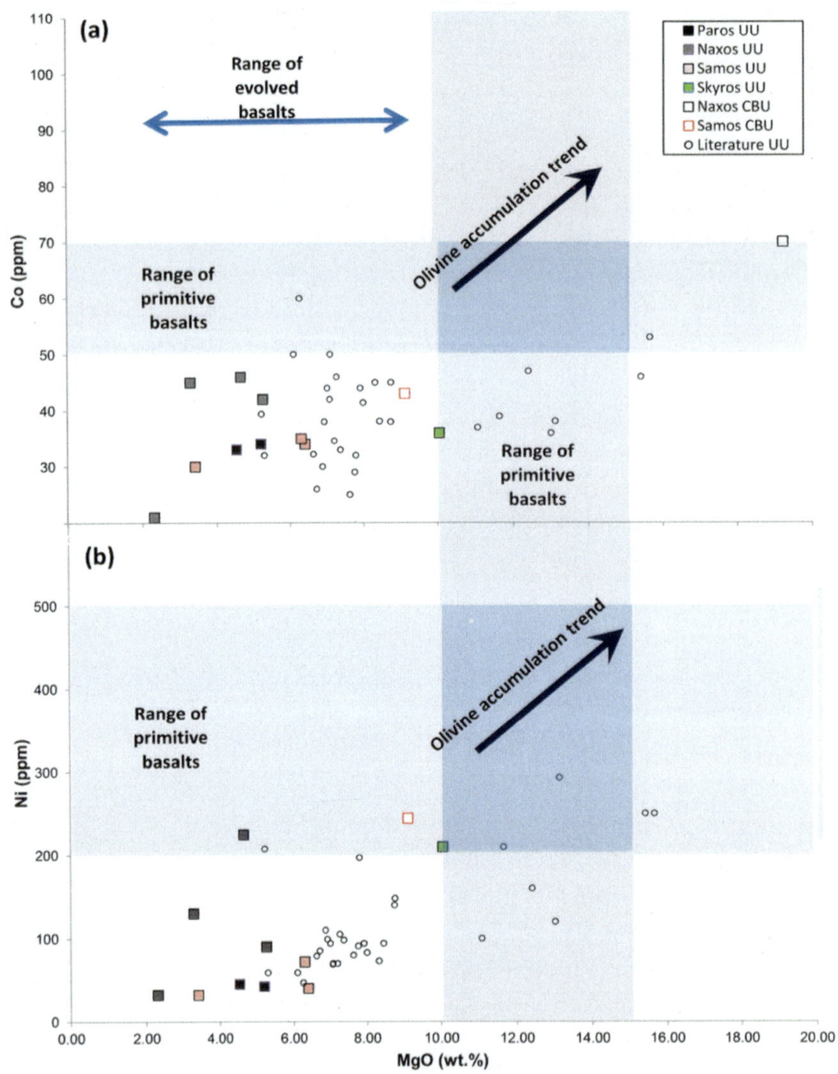

Figure 4. (**a**) MgO vs. Co and (**b**) MgO vs. Ni plot of Aegean metabasites. Upper Unit (UU) ophiolitic metabasites from the literature are also plotted: Ikaria ophiolitic mélange [10]; Tinos ophiolitic upper unit [15] (Mt. Tsiknias). Range expected for magmas in equilibrium with their mantle source: Ni: 200–500 ppm, Co: 50–70 ppm (from [49]). Naxos amphibolites (open squares) plot beyond the compositional range of the Figure 4a,b.

5.1. Upper Unit Metabasites

Studied ophiolitic metabasites from the Cycladic islands and Skyros span basalts, gabbros, and minor basaltic andesites and dolerites, ranging from 39 to 54.5 wt % SiO_2. Samples from the Cycladic islands Paros, Naxos, and western Samos (Kallithea) include moderately to highly evolved basaltic rocks with 2.3 to 6.4 wt % MgO and Mg-number (Mg#) values from 40.2 to 59.6 (Figure 4) (Table 1). Only Skyros metabasites include primitive basaltic composition rocks with 10 wt % MgO at 50 wt % SiO_2, Ni > 200 ppm but Co < 50 ppm, thus indicating early olivine fractionation (Table 1) (Figure 4). TiO_2 contents of the relatively more primitive basaltic rocks (Paros: PA-9; Samos: SA-8, SA-2) are low, i.e., 0.6 to 0.7 wt % when compared to N- and E-type MORB compositions, and resemble typical island arc tholeiites or some Back-Arc Basin basalts, e.g., East Scotia (Table 1) [47,50]. In fact, there is a large range of TiO_2 content in the studied rocks that can be attributed to various processes, including fractionation and post-magmatic alteration.

A common feature in all the mafic rocks is the high potassium concentration. Even the less evolved rocks of gabbroic composition from Paros, SE Naoussa, contain up to 2 wt % K_2O (Table 1). Moreover, the Pb concentrations of all the Cycladic metabasites exceed 1 ppm (i.e., 1–17 ppm), which is high for typical MORB [51]. Notwithstanding, values up to 3 ppm may be normal for Back-Arc Basin basalts, e.g., in lavas from Southern Mariana Trough [52], yet most Cycladic rocks display extremely high values, up to 18 ppm, which clearly indicate secondary alteration (Table 1). The petrographic characteristics of the Cycladic metabasites (UU) document ocean-floor hydrothermal alteration (hydration), especially associated with extensive sericitization of plagioclase, which justifies the mobility of alkalis and lead. This is further attested in the spread of $^{207}Pb/^{206}Pb$ diagram (see Section 5.4).

Few detailed studies have been made of the Cycladic upper unit metabasites and the discussion in this section is based on the dataset of [15] on Tinos metagabbros (UU) and Ikaria ophiolitic rocks (Faros mélange) [10]. In order to elucidate the petrogenesis of the Cycladic rocks, we rely on the relatively immobile key elements. The Total Akali-Silica (TAS) proxy diagram of [51] is considered to provide an effective discrimination of altered ophiolitic rocks, avoiding the wide dispersion of the IUGS TAS [53,54]. The Nb/Y ratio is used as a proxy for total alkalis ($Na_2O + K_2O$) and Zr/Ti as a proxy for SiO_2. In Figure 5 most Cycladic metabasites span the field of tholeiiteic basalts and basaltic andesites. A group of samples from the Naxos Upper Unit (metabasalts, basaltic andesites) and two amphibolites from Ikaria plot in the field of alkali basalts (Nb/Y > 1). Sample SA-8 (meta-andesite) plot as an outlier due to extremely high Nb/Y value (7.85), indicating extreme alkalinity of the magma and possibly a different magmatic type.

High precision Zr-Y determinations are a particularly sensitive method of distinguishing different magma batches in a magmatic suite. Both elements remain essentially incompatible within basaltic magma during fractional crystallization. Hence, a single magma batch will retain an almost constant Y/Zr ratio [42]. In Figure 6a there is a reasonable linear correlation between Zr and Y but the Zr/Y ratio is variable. The majority of upper unit metabasites exhibit mainly a tholeiitic affinity (one linear trend), whereas a small group of samples from Naxos shows a different linear trend with higher Zr/Y (e.g., 5.2–5.4). Such relationship cannot be attributed to crystallization of a single magma batch. Similar geochemical systematics have been reported from the sheeted dyke suite and pillow lavas of Troodos complex [42] and basalts from modern Back-Arc Basins [55].

The V–Ti plot (Figure 6b) of [56] is used for further tectonic-magmatic classification of ophiolites from suprasubduction zones, and is considered here to investigate the main ophiolite complexes of the Western Ophiolite Belt (Pindos: [57]; E. Othrys: [58]; Gevgeli: [59]). This diagram (Figure 6b) also shows the spectrum of magmatic affinities recognised above: (a) MORB tholeiites and (b) island arc tholeiite affinities (Paros and western Samos-Kallithea); (c) boninitic affinities (high-MgO, low-TiO_2 andesites and gabbros) displayed by Skyros.

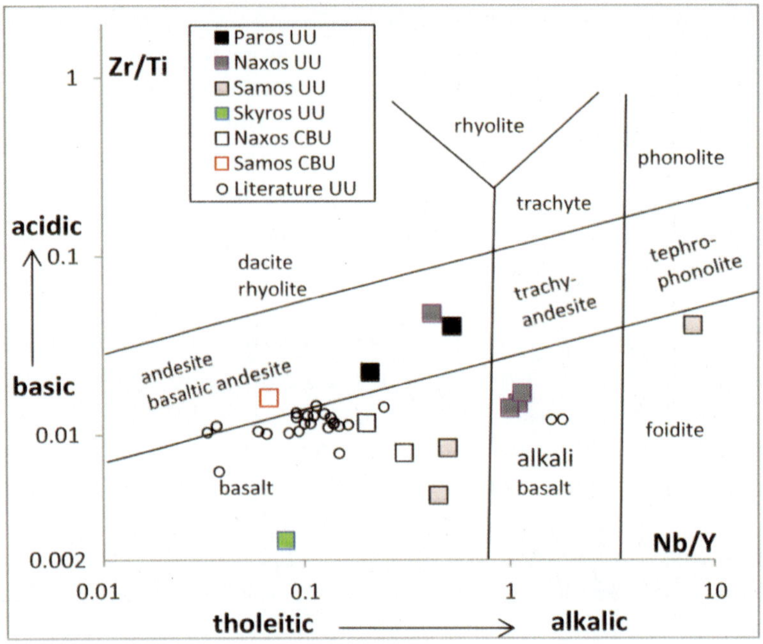

Figure 5. Classification of Aegean ophiolitic metabasites using the immobile element-based TAS proxy diagram (modified after [53,54]). Literature data for Upper Unit metabasites are plotted for comparison (see text).

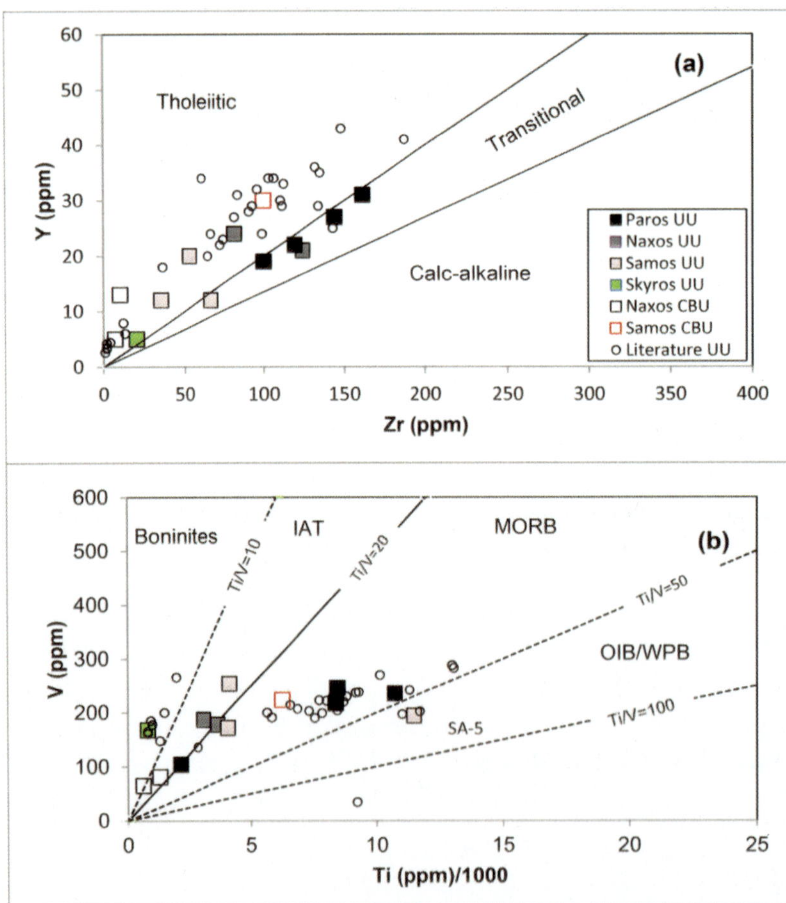

Figure 6. (**a**) Zr vs. Y and (**b**) Ti vs. V [56] tectonic-magmatic discrimination of Cycladic ophiolitic metabasites.

5.2. Naxos Amphibolites and Central Samos Metagabbro (CBU)

Two amphibolite samples from the Naxos gneiss dome (NA-7: Kourounohori, NA-9: Agia) show very low TiO_2 contents (TiO_2: 0.1–0.2 wt %) and extremely high MgO (19–27 wt %) for mafic rocks, and very low Al_2O_3 at $SiO_2 \geq 52\%$ (Table 1). According to the extreme composition of this group of rocks and considering the microscopic observations of thin sections, NA-9 ampibolite is interpreted as a metamorphosed tholeiitic basalt, whereas NA-7 is probably a metasomatised cummulate rock, in particular a former pyroxenite (NA-7). The bulk-rock geochemical characteristics of these amphibolites are comparable to amphibolites from metamorphic soles beneath the ultramafic ophiolites in northern Evia [8]. A metagabbro (SA-11) from a metamorphosed ophiolitic mélange in central Samos-Myli, displays primitive basaltic composition with 9 wt % MgO at 50.3 wt % SiO_2. Ni and Co contents of the latter sample is lower than 250 ppm and 50 ppm, respectively, thus indicating early olivine fractionation [49] (Figure 4a,b). In Figure 6, SA-11 metagabbro plots in the field of MORB tholeiites whereas Naxos amphibolites plot in the field of island arc tholeiites.

5.3. Multi-Element Geochemical Patterns

For representation of multi-element spider diagram patterns (Figure 7), only immobile elements are included and ordered in increasing compatibility, in the standard form after [51]. Typical compositions of basaltic rocks from modern well defined settings are shown for direct comparisons. In general, relatively immobile elements under seafloor hydrothermal alteration of oceanic crust are considered the High-Field-Strength Element (HFSE) group (e.g., Zr, Nb, Ta, Ti), the REEs, Mg, Y, and especially the transition metals (Sc, V, Cr, Fe, Co, Ni, and Zn) [60,61]. On the contrary, the Large Ion Lithophile Elements (LILE) K, Rb, Ba, Sr, and U are considered readily mobile under low-temperature alteration of ocean floor basalts and basaltic glass, and they are commonly added through the process of the formation of clay [62,63].

In these diagrams, N-type MORB [48] is selected as a normalized factor because the resulting flat N-MORB pattern is a useful reference regarding tectonic setting.

The Cycladic samples can be classified into three geochemical groups (Figure 7a–g):

(a) Paros metagabbros (Figure 7a) (SE Naoussa—Upper Unit) and Samos (Kallithea nappe) metavolcanites (Figure 7b) display a Zr-Y flat pattern and higher Th/Y ratio relative to MORB or OIB; these characteristics are typical of modern arc lavas from complex tectonic settings (C-MORB, Taitao lavas, Chile; [51] and references therein). The gradient of the patterns for the Paros-W. Samos samples is even higher than that for back-arc basin basalts (BABB) tholeiites (Figure 7a,b). Moreover, the $(La/Nb)_N$ ratios, which range from 1.2 to 2.35 (Figure 6b), emphasize the HFSE depletion of this type of magmas relative to LREE, which is another characteristic feature of arc magmas [64,65].

(b) Sample SA-11 (metagabbro from CBU), from central Samos, displays a flat pattern typical for MORB (Figure 7b). Following the definition of N-type MORB of [66], this sample exhibits Ce/Nb > 2.5 but $Ce_N/Yb_N = 1$ (Table 1) and, combined with initial Sr > 0.7035, it is assigned to an enriched MORB category (E-MORB) rather than N-MORB. However, comparing sample SA-11 with a complete dataset from [15] (Figure 7f) it becomes clear that there is a whole range of compositions typical of BABB, similar to Troodos ophiolitic lavas [51].

(c) Trace element patterns of Naxos metabasaltic rocks (NA-3, NA-15) share many characteristics with oceanic island basalts (OIB) (Figure 7c): low Zr/Nb (i.e., ~5) relative to N-MORB [67], high Nb/Y, and the negative gradient in the Ti–Yb(Y) part of the diagram, which are characteristics of enriched ocean-island basalts (OIB). This is distinct from the flat patterns for Ti–Yb(Y) that characterize both N- and EMORB compositions.

(d) Naxos amphibolites (CBU) also display a high Th/Y ratio similar to Paros and western Samos metabasalts; however, they also have a strong depletion for most of the trace elements LREE and HFSE, compared to typical MORB patterns (Figure 7c).

(e) Skyros metagabbro shows a typical pattern of boninitic rocks (Figure 7d) from modern oceanic arc environments, such as the fore-arc volcanics of the Mariana-Izu-Bonin [68]. One metabasalts and a metagabbro from the same area also show similar boninitic affinities (data from [41], Table 1).

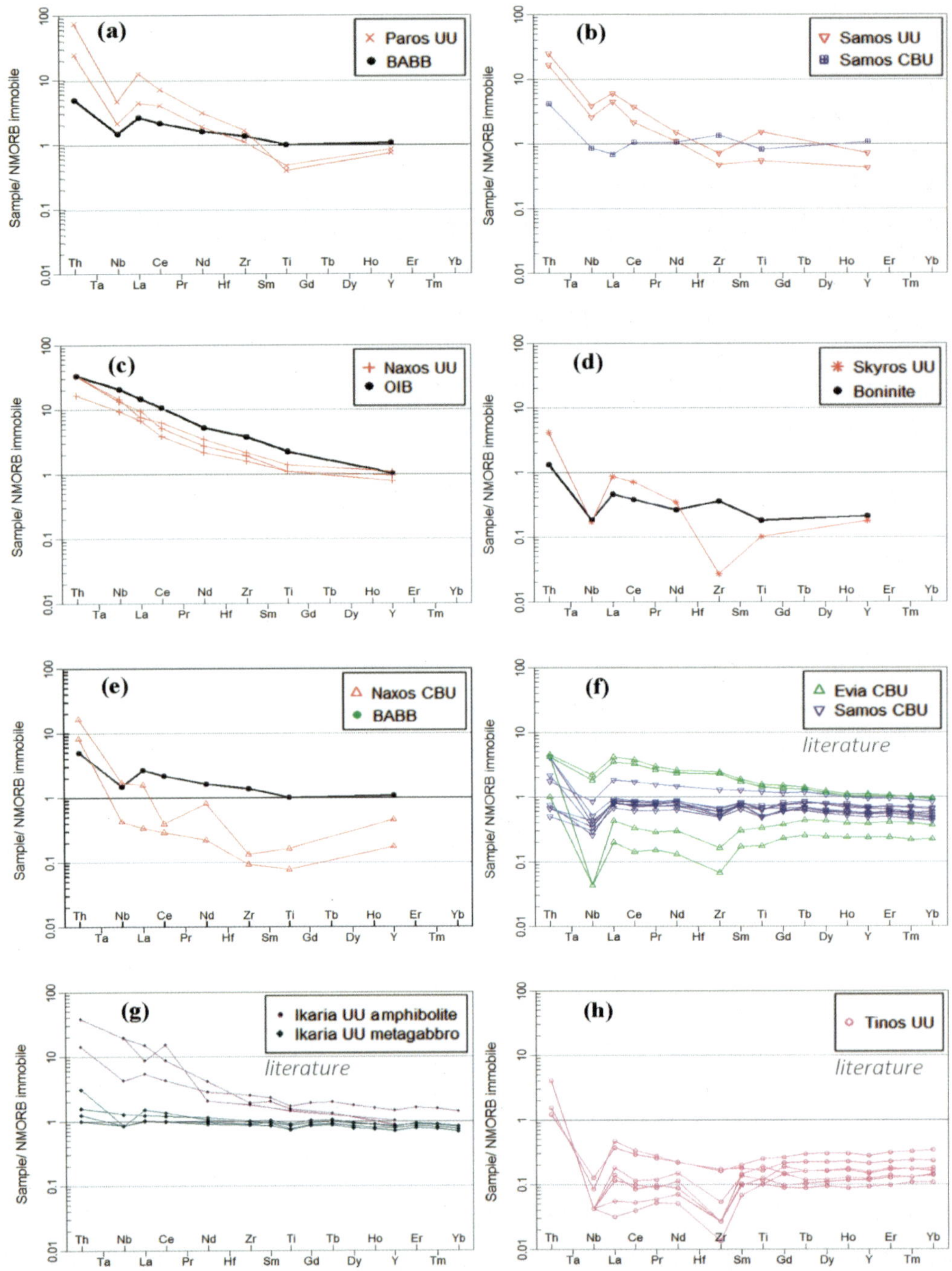

Figure 7. Incompatible, immobile element patterns (spidergrams) for Aegean ophiolites along with some typical basaltic compositions from different tectonic setting. Data sources, see [51] and references in the text; BABB from [52]; Boninite from [66]. Normalisation is to N-MORB [51]. (**a**) Paros; (**b**) western Samos-Kallithea nappe and central Samos (CBU) metagabbros; (**c**) Naxos metabasalts (UU); (**d**) Skyros metadolerite (SK-1); (**e**) Naxos amphibolites (CBU) from the flanks of the gneiss dome; (**f**) central Samos CBU) and southern Evia (CBU) metagabbros from [15]; (**g**) Ikaria upper unit (Faros and Kefala units from [10]); (**h**) Tinos-Mt Tsiknias (Upper Unit) metagabbros from [15].

5.4. Upper Unit Serpentinites and Serpentinized Peridotites

Major and trace element contents of 20 representative samples from Cyclades and Skyros have been analyzed and the results are listed in Table 1.

a. Cycladic serpentinites

Based on the olivine-orthopyroxene-clinopyroxene CIPW normative classification, most of the Cycladic serpentinite protoliths are assigned to harzburgites and minor olivine orthopyroxenite whereas only one sample from Skyros (SK-9) is assigned to the lherzolite category. However, published data for metaperidotites from the main horizon of Naxos gneiss dome [29] (Katzir et al., 1999) display a broader compostional range from lherzolites to harzburgites and rare dunites, whose variability is not shown in the upper unit peridotites of Naxos possibly due to the scattered occurrence of the upper unit serpentinites (Figure 8). In major element variation diagrams the Mg# is used as a fractionation index (Figure 9a–c). The Mg# values are high in all the Cycladic rocks, ranging from 84 to 91.5. SiO_2, Al_2O_3, and CaO display negative correlation trends defined by the Paros-Naxos peridotites, which are similar to those found for the main horizon of metaperidotites from Naxos. These trends have been attributed to a successive depletion process caused by melt extraction from an original lherzolitic mantle source as indicated in Figure 9d [29]. In the binary diagram of MgO/SiO_2 vs. Al_2O_3/SiO_2 (Figure 9d), the serpentinized peridotites and serpentinites of our study plot below the "successive magmatic depletion" line, as expected for moderately serpentinized peridotitic rocks [69–71], which is attributed to restricted loss of Mg during seafloor weathering after serpentinization [71,72]. The Aegean serpentinites are comparable to peridotites of the western ophiolite belt (e.g., east Othris [73].

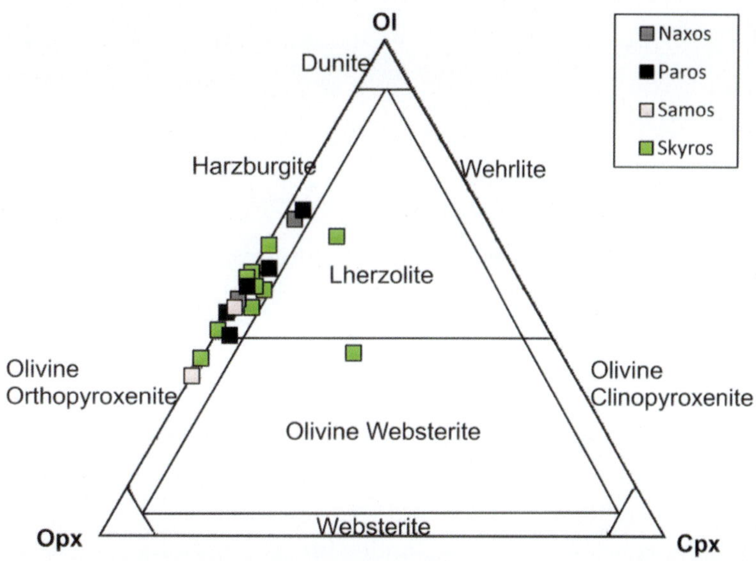

Figure 8. Spinel peridotite norm compositions of the ultramafic rocks from the Cyclades and Skyros plotted on the ultramafic classification diagram of IUGS.

Compositional variation of the Cycladic serpentinites show a dominant chemical group with average low Al_2O_3 (0.3–1.5 wt %), TiO_2 (0.01–0.04 wt %) and CaO (0.03–1.52 wt %), and low V (5–49 ppm), coupled with high Ni (2318 ppm), Co (81–123 ppm), and Cr (1935–3932 ppm), indicating a highly residual nature of the mantle source; they are comparable to the composition of harzburgites from Pindos ophiolite complex [74] and mantle wedge serpentinites from oceanic fore-arc settings [75] (Figure 9). A subordinate group from Paros-Naxos displays a relatively less refractory character testified by the higher Al_2O_3 (1.79–2.86 wt %) and V (50–60 ppm), and lower Co (96–103 ppm) and Cr (2398–2677 ppm) concentrations. Two serpentinite samples, one from Samos (SA-14) and one from

Skyros (SK-4), display unusually high Al_2O_3 values of 4.82 and 8.61, respectively (Figure 9b). Based on thin section observations and mineral analysis data, these values can be attributed to extensive chlorite intergrowth in serpentine.

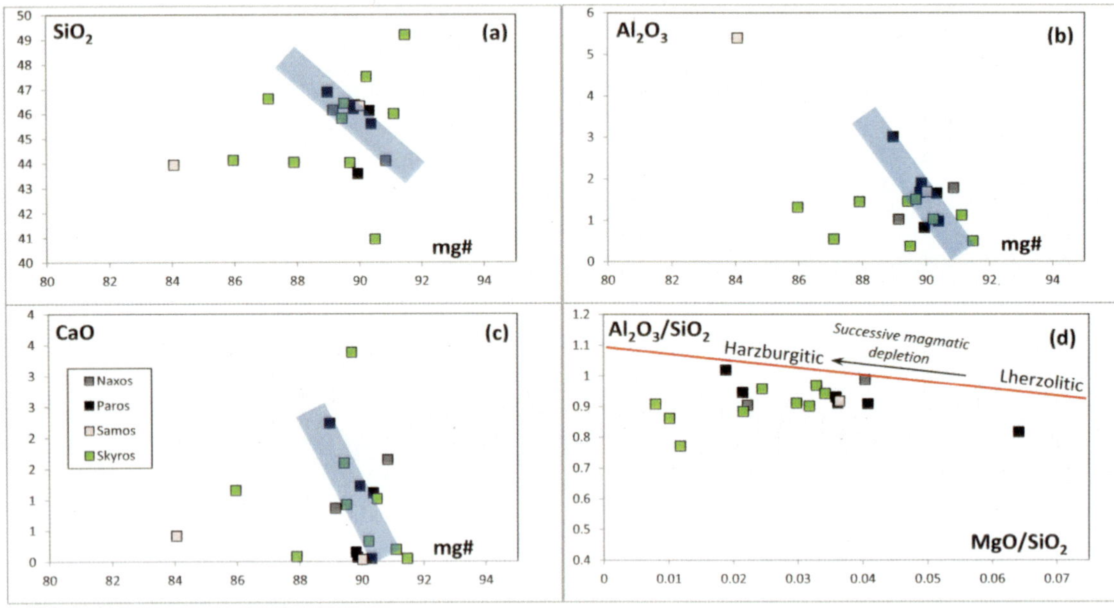

Figure 9. Variation diagrams of (**a**) SiO_2; (**b**) Al_2O_3 and (**c**) CaO vs. Mg# as fractionation index for the Aegean ultramafic rocks; (**d**) Binary diagram of Mg/SiO_2 vs. Al_2O_3/SiO_2 with serpentinized peridotites and serpentinites. Successive magmatic depletion line according to [69–71]. The shaded field shows the main magmatic trend defined by Paros samples. Skyros serpentinites display the largest scatter, due to alteration effects.

b. Skyros serpentinites

Skyros serpentinites are classified mostly as hartzburgites, except from one sample as lherzolite and another as olivine websterite (Figure 8). Skyros ultramafic rocks display the highest Mg# (up to 91.5). Ni and Cr abundances are as high as 2720 ppm and 3932 ppm, respectively.

5.5. Sr-Nd Results

Sr and Nd isotope ratios were determined for 12 samples. The three peridotite samples failed to produce Nd isotopic data due the very low content of REEs (<1 ppm). The results are summarized in Table 2 and presented in Figure 10a,b. Initial Sr and εNd values have been recalculated at 160 Ma, which corresponds to the available mean ages of the Pindos Jurassic ophiolite, and at 80 Ma, for the case of metaigneous rocks from the CBU [15,76–78]. The Cycladic ophiolites exhibit initial $^{143}Nd/^{144}Nd$ values lower than 0.5130, which is the lower limit value of most N-type middle ocean ridge basalts (N-type MORB), and higher initial $^{87}Sr/^{86}Sr$ ratios than 0.7035 [66]. These isotopic values, if primary, suggest that the source of these rocks was isotopically modified from an enriched mantle component (OIB). Generally, the initial Sr-Nd isotopic ratios of Cycladic metabasites display a large scatter considering the number of samples analyzed. In the Sr-Nd isotopic diagram of Figure 10a, a main, horizontal trend is recognized. This trend is defined by the metamorphosed gabbros of central Samos (CBU), and sample SA-11 (Myli location, this study) also plot along this trend together with a serpentinite from Samos and the Tinos (UU) metagabbros (Mt. Tsiknias) [15]. This trend is characterized by a shift towards higher $^{87}Sr/^{86}Sr$ (ISr80 = 0.7062 to 0.708) at almost constant initial $^{143}Nd/^{144}Nd$ (εNd80 = +6.54 to +7.48), calculated for a presumed protolith age of 80 Ma. The Skyros metadolerite (SK-1) also plots together with the Samos and Tinos metabasites. In Figure 10b initial isotopic ratios are recalculated at 160 Ma which is the approximate age of the Pindos ophiolite complex ([77]; and references therein]). In the latter diagram, Paros (PA-9) and Naxos (NA-3) upper unit metabasites plot close to Avdella mélange gabbros, and define a separate group showing elevated initial Sr (0.70413–0.70563) and distinctly lower εNd (+4.13–+0.32) isotopic values compared to the Samos field. Sr and Nd isotope compositions of metagabbros from Andros CBU (calculated for 160 Ma) can be clearly distinguished from the other Cycladic metabasites (Paros, Naxos, Tinos) samples by even higher $^{87}Sr/^{86}Sr$ = 0.70613 to 0.70694 ratios and lower εNd = −3.77 to −1.54 values (Figure 10b), strongly indicating significant crustal involvement from terrigenous sediment or arc crust [15,28].

Upper Unit ophiolites of Paros and Naxos show a strong coupling of the two isotopic systems towards more radiogenic Sr and less radiogenic Nd isotopic ratios. This second trend could possibly be attributed to the combined effect of contamination of the source from a subduction-derived component (hydrous fluids or melt) and seafloor hydrothermal alteration.

5.6. Lead Isotope Geochemistry

Lead isotopes, in combination with other isotopic systems (Sr-Nd), have been proven as an important source of information on the source reservoir of magmatic rocks [79]. In this study, lead isotope compositions were measured on samples from metavolcanites and metagabbros from the Upper and intermediate Units of the Cyclades. Lead isotope compositions of the metabasites are given in Table 3.

The analyzed metabasites from the Cyclades (metagabbros, metabasalts, and metabasaltic andesites) display a wide range of lead isotope compositions for a relatively small number of samples, especially regarding $^{207}Pb/^{204}Pb$ ratios ($^{206}Pb/^{204}Pb$ = 18.40–21.53; $^{207}Pb/^{204}Pb$ = 15.56–15.77; $^{208}Pb/^{204}Pb$ = 38.32–39.6) [80] (Figure 11a,b). Similarly, high whole-rock $^{207}Pb/^{204}Pb$ ratios (Figure 11a) have been reported for the hydrothermally altered parts of the Troodos ophiolite complex (greenschists and epidosites) [80]. In the ^{207}Pb-^{206}Pb and ^{208}Pb-^{206}Pb diagrams, the Cycladic metabasites display a spread beyond the normal MORB field and Troodos glasses [80] and plot within the field of seawater, attesting the effect of seawater alteration on the Cycladic metabasites compositions. Sample NA-14 (serpentinized harzburgite) has a high $^{206}Pb/^{204}Pb$ ratio, probably due to time-integrated development of radiogenic lead caused by a higher U/Pb ratio compared to the other samples (not plotted in Figure 11).

Table 2. Rb-Sr and Sm-Nd isotope data of representative ophiolitic metabasites and serpentinites from the Aegean islands of Paros, Naxos, Samos, and Skyros.

Sample No.	Rock Type	Rb	Sr	$^{87}Rb/^{86}Sr$	$^{87}Sr/^{86}Sr$ (m)	2σ	$^{87}Sr/^{86}Sr$ (80 Ma)	$^{87}Sr/^{86}Sr$ (160 Ma)
Metabasites								
NA-3	Metabasalt	79	206	1.1098	0.706656	0.000004	0.705395	0.704132
NA-9	Amphibolite	8	33	0.7018	0.710557	0.000009	0.709759	0.708961
PA-9	Metagabbro	56	299	0.5420	0.706858	0.000015	0.706242	0.705625
SA-8	Metandesite	101	50	5.8454	0.705731	0.000004	0.707109	0.692435
SA-11	Metagabbro	62	259	0.6928	0.707224	0.000004	0.706437	0.705648
SK-1	Metadorerit.	8	116	0.1996	0.707165	0.000006	0.706938	0.706711
Serpentinites								
NA-6	Dunite	1	5	0.5262	0.707540	0.000013	0.706942	0.706343
NA-14	Serpentinite	2	6	0.9045	0.708138	0.000011	0.707109	0.706078
PA-7	Serpentinite	2	5	1.0584	0.708772	0.000016	0.707569	0.706364
PA-16	Serpentinite	1	47	0.0610	0.709102	0.000004	0.709033	0.708964
SA-12	Serpentinite	2	3	1.6571	0.709911	0.000100	0.708034	0.706155
SK-4	Serpentinite	2	12	0.6212	0.708550	0.000023	0.707847	0.707143

Table 2. *Cont.*

Sample No.	Rock Type	Sm	Nd	$^{147}Sm/^{144}Nd$	$^{143}Nd/^{144}Nd$ (m)	2σ	εNd (CHUR) (0 Ma)	$^{143}Nd/^{144}Nd$ (80 Ma)	$εNd_{(CHUR)}$ (80 Ma) *	$^{143}Nd/^{144}Nd$ (160 Ma)	$εNd_{(CHUR)}$ (160 Ma)
Metabasites											
NA-3	Metabasalt	6	26	0.136990	0.512786	0.000005	2.91	0.512715	3.52	0.512644	4.13
NA-9	Amphibolite	bdl	2	-	0.512406	0.000020	−4.51	-	-	-	-
PA-9	Metagabbro	3	14	0.136890	0.512592	0.000004	−0.90	0.512520	−0.29	0.512449	0.32
SA-8	Metandesite	2.8	13	0.130253	0.512689	0.000003	1.01	0.512622	1.69	0.512554	2.37
SA-11	Metagabbro	3	8	0.253171	0.513000	0.000008	7.07	0.512868	6.49	0.512735	5.91
SK-1	Metadorerit.	bdl	3	-	0.512932	0.000018	5.75	-	-	-	-
Serpentinites											
NA-6	Dunite	n.d	n.d	-	-	-	-	-	-	-	-
NA-14	Serpentinite	n.d	n.d	-	-	-	-	-	-	-	-
PA-7	Serpentinite	n.d	n.d	-	0.512710	0.000062	1.41	-	-	-	-
PA-16	Serpentinite	n.d	n.d	-	-	-	-	-	-	-	-
SA-12	Serpentinite	bdl	2	-	0.512990	0.000041	6.87	-	-	-	-
SK-4	Serpentinite	2	11	0.104920	0.512240	0.000005	−7.76	0.512185	−6.83	0.512130	−5.89

n.d: not detected, bdl: below detection limit. * 80 Ma is the reported age of Samos metagabbros from [15]. $^{87}Rb/^{86}Sr$ and $^{147}Sm/^{144}Nd$ isotopic ratios were calculated on the basis of whole rock Rb/Sr and Sm/Nd ratios (concentrations from XRF analysis), using the applications of GCDKit 4.0 software [81] and equations of [82]. Initial isotopic ratios and epsilon Nd values were calculated using the plug-in modules of GCDKit software.

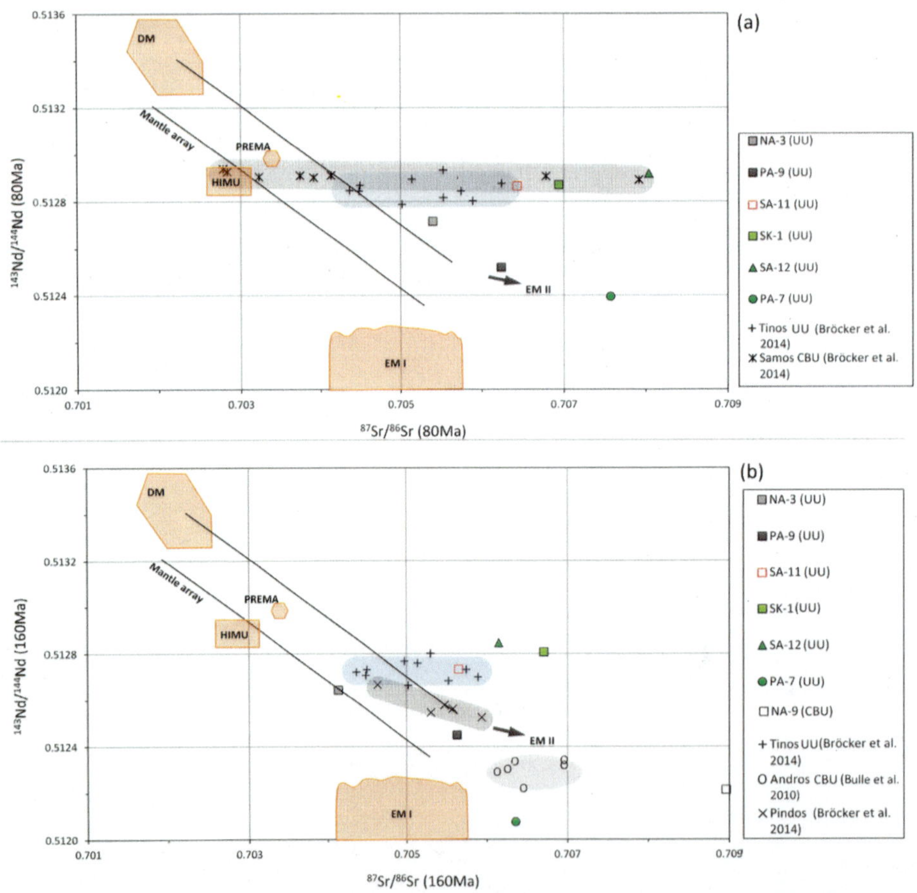

Figure 10. Sr–Nd isotope compositions of mafic rocks from the Upper Units of Naxos, Paros, Samos (Kallithea), and Skyros and Naxos amphibolites, also showing data from the HP-mélange blocks from central Samos, Tinos, and a Pindos-Avdella melange [15,28]. (**a**) $^{87}Sr/^{86}Sr$ vs. $^{143}Nd/^{144}Nd$ plot for age recalculation t = 80 Ma; schematic fields of comparison indicated by abbreviations: N-MORB = Normal type Mid-Ocean Ridge Basalt; OIB = Ocean-Island basalt; IAB = Island Arc Basalt; (**b**) $^{87}Sr/^{86}Sr$ vs. $^{143}Nd/^{144}Nd$ plot with ratios calculated for t = 160 Ma. Terrestrial mantle reservoirs: DMM, HIMU, PREMA, EM-I, EM-II from [69].

Table 3. Lead isotope compositions of ophiolitic metabasites and associated serpentinites from the Aegean islands (NA—Naxos, SA—Samos).

Sample No.	NA-3	NA-9	SA-8	SA-11	NA-6	NA-14
Rock Type	**Metabasalt**	**Amphibolite**	**Metabasalt**	**Metabasalt**	**Serp. Harzb.**	**Serp. Harzb.**
$^{206}Pb/^{204}Pb$	18.802	18.816	18.670	18.703	18.518	21.534
$^{207}Pb/^{204}Pb$	15.625	15.771	15.607	15.665	15.594	15.838
$^{208}Pb/^{204}Pb$	39.039	39.065	38.807	38.799	38.419	38.326

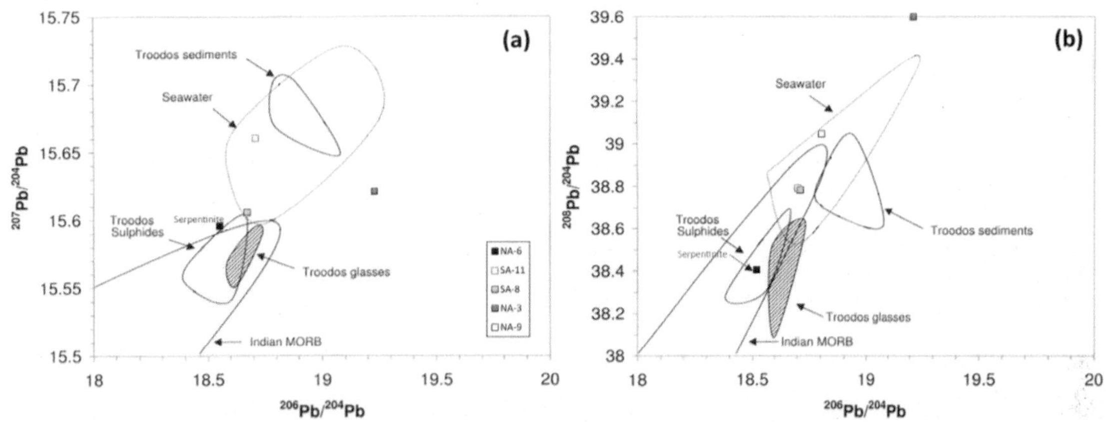

Figure 11. (**a**) $^{208}Pb/^{204}Pb$ versus $^{206}Pb/^{204}Pb$ and (**b**) $^{207}Pb/^{204}Pb$ versus $^{206}Pb/^{204}Pb$ plots showing the isotopic compositions of the Cycladic metabasites. Fields for Troodos glasses, sulfides, sediments, and seawater form [80] and references therein.

6. Discussion

6.1. Determining the Effect of Alteration on the Cycladic Metabasites

In general, the Cycladic metabasites exhibit K enrichment when compared to gabbroic rocks from recent oceanic crust. For instance, normal MORB compositions do not exceed 0.17 wt % K_2O (e.g., [79]. This enrichment in K_2O is attributed to an addition of potassium (plus Rb and Pb) due to low-T seafloor alteration reactions in basalts, which is commonly observed within the oceanic crust ([83]; and references therein]). Sericitization of plagioclase is a common feature in all the mafic lithotypes (metagbbros and metavolcanites). Similar K-enrichments are reported in gabbroic rocks and dolerite dykes from southern Aegean ophiolites (Crete, Rhodes, and Karpathos [21]).

Moreover, the $^{87}Sr/^{86}Sr$ ratio is considered a very powerful indicator of hydrothermal alteration in the oceanic crust, whereby fresh modern crust is unradiogenic ($^{87}Sr/^{86}Sr$ = 0.7025–0.703) and modern seawater is radiogenic (0.7092) [79]. Based on microscopic observations (e.g., Figure 3a) and the formation of secondary minerals, all the Cycladic upper unit samples show evidence of low-temperature seafloor hydrothermal alteration at greenschist facies. Sr-Nd isotopic variations of Samos (and Tinos, data from [28]) samples display a characteristic shift towards higher $^{87}Sr/^{86}Sr$ (up to 0.708) at almost constant initial $^{143}Nd/^{144}Nd$ (i.e., εNd_{80} = +6.54 to +7.48) (Figure 10). This shift indicates a seawater contribution at hydrothermal conditions of upper extrusive oceanic crust, and is a common feature in Mediterranean ophiolite complexes (e.g., Troodos: [83]). Previous authors [83] suggested that interchange of Jurassic and Cretaceous seawater Sr with ocean crust Sr occurs during hydrothermal circulation and may buffer the Sr isotope composition of seawater. Nevertheless, seawater $^{87}Sr/^{86}Sr$ has changed with time; thus, before 80 My, seawater was less radiogenic (0.7076) than today and has increased since then [84,85]. The value of 0.708 is the maximum value of $^{87}Sr/^{86}Sr$ found in the Cycladic metabasites. In this context, the initial $^{87}Sr/^{86}Sr$ of the metabasites reflects the mixing ratio of seawater-derived Sr and basalt Sr present in a sample. Thus, the significance of Sr isotope values for determining the characteristics of the magmatic protolith and tectonic setting is limited. From Figure 11a,b it is clear that the Pb-isotopic ratios for Samos and Naxos metabasites show wide variation, from close to MORB into the seawater field. The elevated $^{207}Pb/^{206}Pb$ and the large scatter of values, which overlaps with the seawater, also suggest that the Pb incorporated in the Cycladic metabasites is derived from the seawater. The positive correlation between $^{206}Pb/^{204}Pb$ and $^{208}Pb/^{204}Pb$ for all the Cycladic rocks (serpentinite-metagabbros-metabasalts) is attributed to the mixing relationships between the magmatic and seawater end-member components (hydrothermal system).

6.2. Comparisons of the Upper Unit and CBU Ophiolites of the Broader Aegean Region

The geochemical characteristics of the Upper Unit ophiolites reveal significant differences between the Cycladic rocks (Paros-Naxos-Western Samos) and Skyros:

(a) Random occurrence of ophiolitic lithologies characterizes the Cycladic mélange units, whereas the Skyros ophiolite consists of more coherent slices of metabasites with essentially boninitic affinities.

(b) Basalts with Ocean Island type (OIB) affinities and alkaline characteristics are recognized in the Cycladic islands (Naxos-Ikaria), but are also common elsewhere in the Cycladic belt and the mainland ophiolite complexes: the amphibolite soles of central–northern Evia ophiolite [8] and Ikaria [10], as well as the main Western Ophiolite Belt (Avdella mélange-Pindos and Koziakas: [86]). These rocks display extreme HFSE enrichments relative to N-type MORB (high Ti, P and Nb, $Zr/Y \sim 5$, $Nb/Y > 1$). Alkaline basalts (AB) found in ophiolites mainly include rocks formed at seamounts (OIB) [87]. They mainly occur as tectonic slices or blocks incorporated into mélanges, that is, as accreted fragments of oceanic terranes. Contrary to the Pindos peridotites, the Cycladic rocks exhibit variable enrichments in alkali earth elements (Sr, Rb, and Ba). These enrichments are a common feature of the Cycladic serpentinites from mélange units, which have been interpreted in different ways: oceanic hydrothermal alteration, synmetamorphic fluid–rock interaction, igneous differentiation, or some combination of these processes (e.g., S. Evia, Syros: [17,34]). Due to the random variation in these mobile elements (Sr, Rb, and Ba) we consider these enrichments a result of ocean floor hydrothermal alteration.

(c) The peridotite relicts in Cycladic ophiolites display considerable chemical and petrographic variation and a general refractory nature. Overall their composition correlates with the two groups distinguished by [74] for the ultramafic rocks of the Pindos complex. The observed elemental variations are typical of serpentinites corresponding to harzburgites from mantle wedge settings (Supra Subduction Zones—SSZ [75]).

There is a scarcity of ages for the formation of metaophiolites from the upper unit of the Cycladic islands (Paros, Naxos, and western Samos) and the broader Aegean region. In most cases, available ages are based on isotopic data for metamorphic soles and on paleontological data of closely associated sediments (e.g., [18] and references therein). Due to that fact, regional correlations between the remnant ophiolitic occurrences exposed in the mélange units of the CBU and the upper unit are difficult to establish. Only for a few ophiolite occurrences ion probe U–Pb zircon ages have been reported [15,76,77]. In terms of geochemical characteristics, previous work revealed mid-ocean ridge to island arc affinities (e.g., metagabbros and glaucophanites) from Syros, Tinos, and adjacent islands, and this observation has been interpreted to indicate a back-arc setting [88,89].

6.3. Regional Correlation and Significance

A regional correlation of Cycladic metabasites from upper unit ophiolite mélanges and CBU mélanges is attempted (a total of 60 analyses) in order to examine possible clustering of compositions. Published data include the supra-detachment mélange units of Ikaria (Faros and Kefala: [10]), Tinos Upper Unit (Mt Tsiknias: [15]), which can be directly correlated to the low-pressure ophiolitic mélange units of the present study. The metaophiolitic mélanges (within CBU) include S. Andros (Late Jurassic), S. Evia (undated), central Samos-Selcuk nappe, and Tinos CBU (Late Createous) [15,28]).

A combination of discriminant plots have been used to examine the diversity of magma types and refine their tectonic setting. In the Zr vs. Zr/Y discriminant plot (Figure 12a) [60], the Zr/Y ratio is used against Zr as a fractionation index, to investigate long term source heterogeneities and distinguish between basalts erupted in within-plate settings, mid-ocean ridges or volcanic arcs. In Figure 12a the Zr vs. Zr/Y systematics (Figure 12a) suggests that the distribution of geochemical groups is random between the upper unit and CBU. Three magmatic types are recognized: normal MORB, Volcanic Arc Basalts (VAB) and Within Plate Basalts (WPB). Skyros and Tinos (Upper Unit) metagabbros (Figure 12a) define a separate field and plot together with Tethyan ophiolite boninitic lavas, late dykes and the Mariana Fore Arc Basalts fall [57].

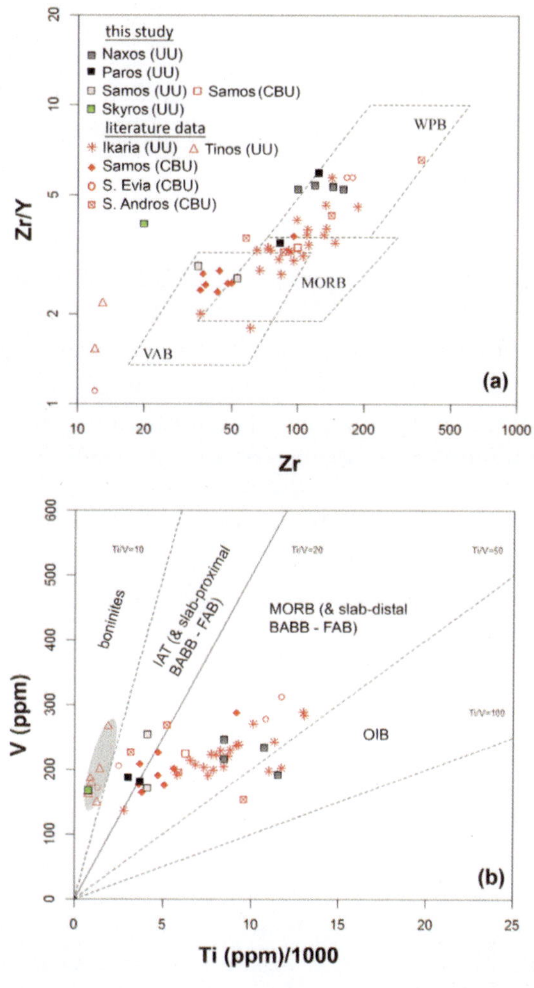

Figure 12. Tectonomagmatic discrimination of metabasites (gabbros, basaltic andesites, and dolerites) of Aegean ophiolites: (**a**) Zr vs. Zr/Y [60], and (**b**) Ti/V discriminant plot of [56] modified in field names after [50]. Abbreviations in (**a**): VAB = Volcanic Arc Basalt, WPB = Within-Plate Basalt and MORB = Mid-Ocean Ridge Basalt; in (**b**): OIB = Ocean Island Basalt, IAT = Island Arc Tholeiite; FAB = Fore-Arc Basalt, BABB = Back-Arc Basin Basalts. Symbols: black = this study, red = published data (data from Samos, Tinos Upper Unit, S. Evia: [15]; Andros: [28]; Ikaria: [10]).

True MORB samples are only a few, the majority of Cycladic metabasites plot in the transitional VAB/MORB field. From the WPB group, only Naxos metabasalts, and two amphibolite samples from Ikaria (Faros) [10], can be geochemically classified as alkali basalt: Nb/Y > 1 and high P, Ba, Sr and Ce_N/Yb_N, higher than E-MORB (i.e., >2) these geochemical features rather resemble OIB-type (see Table 1). Alkali basalts are very commonly associated with tholeiitic rocks in many Mesozoic Tethyan ophiolite mélanges in the Hellenides, and were mostly interpreted as remnants of intra-oceanic islands (OIB) or sea-mounts in the Pindos ocean (Pindos-Avdella: [74], Koziakas: [86]). Alkali basalts in modern oceanic setting are associated with fractures zones, seamounts, and aseismic ridges [50].

The V–Ti discriminant plot of [56] is considered to provide further information for distinguishing between ophiolites already recognized as having a suprasubduction zone origin [51]. Based on Ti vs. V relationships (Figure 12b), the Skyros and Tinos (Upper Unit) metagabbros have very low Ti and exhibit arc-like Ti/V <10. These rocks plot in the same area of late stage boninitic basalts and basaltic andesite dykes of the Pindos-Mirdita ophiolite complexes [57,74]. In this diagram (Figure 12b) Paros and Andros and part of Samos metagabbros exhibit arc-like Ti/V = 10–20; however, most of Cycladic rocks plot within the broad field of mid-ocean ridge basalts (MORB) and back-arc basin basalts (BABB): Ti/V = 20–50.

Previous work on the HP mélanges from the NW Cycladic islands (Syros: [32]; Sifnos: [89]) revealed a similar complicated range of magmatic types in metagabbros and glaucophanites: mid-ocean ridge to island arc affinities. This observation has been interpreted to indicate a transition from subduction initiation to back-arc spreading magmatism (e.g., Tinos, Syros: [28,88]; Sifnos: [89]).

6.4. Tectonic Setting

Extensive geochronological dating (U-Pb on zircons) and geochemical studies have focused on metagabbros and felsic gneisses within mélange units [15,17,28,90,91] and the marble-schist sequences of the CBU [76,91]. Previous studies have shown, on the basis of δO^{18} and εHf values of zircons, that Late Cretaceous igneous zircons (Syros and Tinos: [76,87,90,91]) are characterized by MORB-like affinities, suggesting that the Late Cretaceous magmas were produced by partial melting of depleted MORB mantle. Fu et al. [91] also pointed out the significant crustal component in 160 Ma metagabbros from Andros, implying that some Jurassic gabbroic rocks of the Hellenides are not part of SSZ-type (supra-subduction zone) ophiolites that are common elsewhere along the margin of the Pelagonian zone (Western Ophiolite belt).

So far, both harzburgite compositions and those of metabasites (gabbros and dolerites) are linked by the fact that their chemical affinities share many characteristics with supra-subduction zone setting magmatism [66,92].

7. Conclusions

The results of this study can be summarized as follows:

(a) Metavolcanics (dolerites, basaltic andesites, minor basalts) and metagabbros from all the upper unit ophiolitic exposures exhibit wide ranges of major element composition (SiO_2: 39–56.6 wt %; Mg-number: 37.5–69.2 and high K_2O i.e., >0.2 wt %), indicating derivation from evolved magmas.

(b) The petrographic, major and trace element and Sr-isotope data show that all the studied rocks (metagabbros, metavolcanics, and peridotites) underwent ocean floor alteration. Microscopic observations corroborate that low-temperature alteration during the generation of the ocean crust affected the upper unit metavolcanites and metagabbros.

(c) Based on the immobile- element TAS proxy classification diagram, Ti/V, Zr/Y vs. Zr discriminant plots and spidergram patterns of N-MORB-normalized compositions, three magmatic groups are recognized among the upper unit (UU) ophiolitic metabasites: (i) metabasites from Paros and western Samos (Kallithea melange) display affinities of island arc tholeiites (IAT); (ii) Naxos basaltic rocks are highly enriched in incompatible elements and display an OIB type affinity, representing an oceanic island setting; (iii) Metagabbros from the northern Skyros ophiolitic-mélange (Fere-Kambos) and Tinos (Upper Unit—Mt. Tsiknias) exhibit boninitic affinities indicating a fore-arc setting.

(d) Central Samos metagabbros of the Cycladic Blueschists Unit (CBU) are primitive rocks with Back-Arc Basin affinities.

(e) Most peridotite relicts from Samos, Paros, and Naxos—irrespective of the structural unit—display common chemical affinities similar to ocean floor peridotites formed in a supra-subduction zone environment. Major and trace element characteristics of harzburgite relicts in Cycladic serpentinites and Skyros indicate a highly residual nature of the mantle source, comparable to the mantle wedge serpentinites from oceanic fore-arc settings, in accordance with previous studies.

(f) A combination of Pb and Sr isotopic characteristics of the Cycladic metabasites shows the importance of seawater hydrothermal alteration on the composition of the studied metaophiolitic rocks. Trace element patterns of Central Samos metabasites (CBU) and Nd isotopes suggest affinities of Back-Arc Basin Basalts (BABB).

Acknowledgments: The University of Leicester provided the XRF and microprobe analytical facilities. We thank the authors of the GCDkit software (Vojtěch Janušek, Colin M. Farrow and Vojta Erban) for providing the free software with which geochemical plots were prepared. National and Kapodistrian University of Athens is acknowledged for supporting this work. We would like to thank Konstantinos Soukis for valuable information and discussions on the structural relationships of the ophiolite exposures with the main tectonic units in the Cyclades. Two anonymous reviewers are kindly thanked for their constructive criticism that led to the improvement of previous versions of this manuscript.

Author Contributions: This paper is partly based on unpublished data from the PhD thesis of Iakovos Pantziris, National and Kapodistrian University of Athens, Geology Department; the research work was designed and supervised by professor Panagiotis Mitropoulos who conceived all the experimental work; Iakovos Pantziris performed the sampling and most of the whole rock analytical experiments; Christos Kanellopoulos, Robert Moritz and Massimo Chiaradia performed the isotopic analyses, at the University of Geneva. Panagiotis Pomonis accomplished the petrography work; Charalampos Vasilatos, Christina Stouraiti and Panagiotis Pomonis analyzed the data and made the graphical presentation and the interpretation of the results; Christina Stouraiti wrote the paper.

Conflicts of Interest: The authors declare no conflict of interest.

References

1. Dürr, S.; Altherr, R.; Keller, J.; Okrusch, M.; Seidel, E. The median Aegean crystalline belt: Stratigraphy, structure, metamorphism, magmatism. In *Alps, Apennines, Hellenides*; Closs, H., Roeder, D.H., Schmidt, K., Eds.; IUGS Report; Schweizerbart: Stuttgart, Germany, 1978; Volume 38, pp. 455–477.
2. Altherr, R.; Kreuzer, H.; Wendt, I.; Lenz, H.; Wagner, G.A.; Keller, J.; Harre, W.; Höhndorf, A.A. Late Oligocene/Early Miocene high temperature belt in the Attic–Cycladic belt in the Attic–Cycladic crystalline complex (SE Pelagonian, Greece). *Geol. Jahrb.* **1982**, *23*, 97–164.
3. Patzak, M.L.; Okrusch, M.; Kreuzer, H. The Akrotiri unit on the island of Tinos, Cyclades, Greece: Witness to a lost terrane of Late Cretaceous age. *Neues Jahrb. Geol. Paläontol.* **1994**, *194*, 211–252.
4. Katzir, Y.; Matthews, A.; Garfunkel, Z.; Schliestedt, M. The tectonometamorphic evolution of a dismembered ophiolite (Tinos, Cyclades, Greece). *Geol. Mag.* **1996**, *133*, 237–254. [CrossRef]
5. Pe-Piper, G.; Piper, D.J.W. *The Igneous Rocks of Greece. The Anatomy of an Orogeny*; Schweizerbart: Stuttgart Germany, 2002; 573p.
6. Soukis, K.; Stockli, D. Structural and thermochronometric evidence for multi-stage exhumation of southern Syros, Cycladic islands, Greece. *Tectonophysics* **2013**, *595–596*, 148–164.
7. Robertson, A.H.F.; Shallo, M. Mesozoic-Tertiary evolution of Albania in its regional Eastern Mediterranean context. *Tectonophysics* **2000**, *316*, 197–254. [CrossRef]
8. Gartzos, E.; Dietrch, V.J.; Migiros, G.; Serelis, K.; Lymperopoulou, T. The origin of amphibolites from the metamorphic soles beneath the ultramafic ophiolites in Evia and Lesvos (Greece) and their geotectonic implication. *Lithos* **2009**, *108*, 224–242. [CrossRef]

9. Jolivet, L.; Menant, A.; Sternai, P.; Rabillard, A.; Arbaret, L.; Augier, R.; Laurent, V.; Beaudoin, A.; Grasemann, B.; Huet, B.; et al. The geological signature of a slab tear below the Aegean. *Tectonophysics* **2015**, *659*, 166–182. [CrossRef]
10. Pe-Piper, G.; Photiades, A. Geochemical characteristics of the Cretaceous ophiolitic rocks of Ikaria island Greece. *Geol. Mag.* **2006**, *143*, 417–429. [CrossRef]
11. Pomonis, P.; Hatzipanagiotou, K. Petrography and geochemistry of relict peridotites of the Kallithea-Drakei area (W. Samos). *Bull. Geol. Soc. Greece* **1998**, *32*, 215–224.
12. Jolivet, L.; Brun, J.-P. Cenozoic geodynamic evolution of the Aegean. *Int. J. Earth Sci.* **2010**, *99*, 109–138. [CrossRef]
13. Malandri, C.; Soukis, K.; Maffione, M.; Özkaptan, M.; Vassilakis, E.; Lozios, S.; van Hinsbergen, D.J.J. Vertical-axis rotations accommodated along the Mid-Cycladic lineament on Paros Island in the extensional heart of the Aegean orocline (Greece). *Lithosphere* **2016**. [CrossRef]
14. Bargnesi, E.A.; Stockli, D.F.; Mancktelow, N.; Soukis, K. Miocene core complex development and coeval supradetachment basin evolution of Paros, Greece, insights from (U–Th)/He thermochronometry. *Tectonophysics* **2013**, *595–596*, 165–182.
15. Bröcker, M.; Löwena, K.; Rodionov, N. Unraveling protolith ages of meta-gabbros from Samos and the Attic–Cycladic Crystalline Belt, Greece: Results of a U–Pb zircon and Sr–Nd whole rock study. *Lithos* **2014**, *198–199*, 234–248.
16. Katzir, Y.; Garfunkel, Z.; Avigad, D.; Matthews, A. The geodynamic evolution of the Alpine orogen in the Cyclades (Aegean Sea, Greece): Insights from diverse origins and modes of emplacement of ultramafic rocks. In *The Geodynamics of the Aegean and Anatolia*; Taymaz, T., Yilmaz, Y., Dilek, Y., Eds.; Geol. Soc. London Spec. Publ., The Geological Society of London: London, UK, 2007; Volume 291, pp. 17–40.
17. Bröcker, M.; Enders, M. Unusual bulk-rock compositions in eclogite-facies rocks from Syros and Tinos (Cyclades, Greece): Implications for U-Pb zircon geochronology. *Chem. Geol.* **2001**, *175*, 581–603. [CrossRef]
18. Robertson, A.H.F. Overview of the genesis and emplacement of Mesozoic ophiolites in the Eastern Mediterranean Tethyan region. *Lithos* **2002**, *65*, 1–67. [CrossRef]
19. Papanikolaou, D. Timing of tectonic emplacement of the ophiolites and terrane paleogeography in the Hellenides. *Lithos* **2009**, *108*, 262–280. [CrossRef]
20. Robertson, A.H.F. Late Palaeozoic–Cenozoic tectonic development of Greece and Albania in the context of alternative reconstructions of Tethys in the Eastern Mediterranean region. *Int. Geol. Rev.* **2012**, *54*, 373–454. [CrossRef]
21. Rassios, A.; Smith, A.G. Constraints on the formation and emplacement age of western Greece ophiolites (Vourinos, Pindos and Othris) inferred from deformation structures in peridotites. In *Ophiolites and Ocean Crust: New Insights from Field Studies and the Ocean Drilling Program*; Dilek, Y., Moores, E.M., Elthon, D., Nicolas, A., Eds.; Geological Society of America Special Papers; The Geological Society of America: Boulder, CO, USA, 2000; pp. 473–483.
22. Jacobshagen, V.; Dürr, S.; Kockel, E.; Kopp, K.O.; Kowalczyk, G. Structure and geodynamic evolution of the Aegean region. In *Alps, Appenines, Hellenides*; Closs, H., Roeder, D., Schmid, K., Eds.; E. Schweitzerbartsche Verlagsbuchhandlung: Stuttgart, Germany, 1978; pp. 537–564.
23. Wakabayashi, J.; Dilek, Y. What constitutes' emplacement of an ophiolite? Mechanisms and relationship to subduction initiation and formation of metamorphic soles. In *Ophiolites in Earth History*; Dilek, Y., Robinson, P.T., Eds.; Geol. Soc. London Spec. Publ., The Geological Society of London: London, UK, 2003; Volume 218, pp. 427–447.
24. Dilek, Y.; Furnes, H. Ophiolites and their origin. *Elements* **2014**, *10*, 93–100.
25. Vanderhaeghe, O. Structural development of the Naxos migmatite dome. In *Gneiss Domes in Orogeny*; Whitney, D.L., Teyssier, C., Siddoway, C.S., Eds.; The Geological Society of America: Boulder, CO, USA, 2004; pp. 211–227.
26. Theodoropoulos, D. Geological map of Greece, 1:50,000, Samos Island. Institute of Geology and Mineral Exploration: Athens, Greece, 1979.
27. Papantoniou, G. The Structure and History of the Strike—Slip Fault Zone of Skyros Island, Central—North Aegean. Master's Thesis, National and Kapodistrian University of Athens, Athens, Greece, 2015.
28. Bulle, F.; Bröcker, M.; Gärtner, C.; Keasling, A. Geochemistry and geochronology of HP mélanges from Tinos and Andros, Cycladic Blueschist Belt; Greece. *Lithos* **2010**, *117*, 61–81. [CrossRef]

29. Katzir, Y.; Avigad, D.; Matthews, A.; Garfunkel, Z.; Evans, B.W. Origin and metamorphism of ultrabasic rocks associated with continental margin, Naxos (Cyclades, Greece). *J. Metamorph. Geol.* **1999**, *17*, 301–318. [CrossRef]
30. Pantziris, I. Geochemical—Petrological Study of Basic—Ultrabasic Rocks from the Aegean Islands. Ph.D. Thesis, National and Kapodistrian University of Athens, Athens, Greece, 2000.
31. Okrusch, M.; Bröcker, M. Eclogite facies rocks in the Cycladic Blueschist Belt, Greece: A review. *Eur. J. Mineral.* **1990**, *2*, 451–478.
32. Bröcker, M.; Franz, L. The base of the Cycladic Blueschist Unit on Tinos Island (Greece) re-visited: Field relationships, phengite chemistry and Rb–Sr geochronology. *Neues Jahrb. Mineral. Abh.* **2005**, *181*, 81–93.
33. Bröcker, M.; Franz, L. Dating metamorphism and tectonic juxtaposition on Andros Island (Cyclades, Greece): Results of a Rb–Sr study. *Geol. Mag.* **2006**, *143*, 609–620. [CrossRef]
34. Katzir, Y.; Matthews, A.; Garfunkel, Z.; Evans, B.W. Origin, HP/LT metamorphism and cooling of ophiolitic mélanges in southern Evia (NW Cyclades) Greece. *J. Metamorph. Geol.* **2000**, *18*, 699–718. [CrossRef]
35. Bröcker, M.; Kreuzer, H.; Matthews, A.; Okrusch, M. $^{40}Ar/^{39}Ar$ and oxygen isotope studies of polymetamorphism from Tinos Island Cycladic blueschist belt, Greece. *J. Metamorph. Geol.* **1993**, *11*, 223–240. [CrossRef]
36. Tomaschek, F.; Kennedy, A.K.; Villa, I.M.; Lagos, M.; Ballhaus, C. Zircons from Syros, Cyclades, Greece—Recrystallization and mobilization of zircon during high pressure metamorphism. *J. Petrol.* **2003**, *44*, 1977–2002. [CrossRef]
37. Gautier, P.; Brun, J.P.; Moriceau, R.; Sokoutis, D.; Martinod, J.; Jolivet, L. Timing, kinematics and cause of Aegean extension: A scenario based on a comparison with simple analogue experiments. *Tectonophysics* **1999**, *315*, 31–72. [CrossRef]
38. Baziotis, I.; Mposkos, E. Origin of metabasites from upper tectonic unit of the Lavrion area (SE Attica, Greece): Geochemical implications for dual origin with distinct provenance of blueschist and greenschist's protoliths. *Lithos* **2011**, *126*, 161–173. [CrossRef]
39. Bröcker, M.; Pidgeon, R.T. Protolith ages of meta-igneous and meta-tuffaceous rocks from the Cycladic Blueschist Unit, Greece: Results of a reconnaissance U–Pb zircon study. *J. Geol.* **2007**, *115*, 83–98. [CrossRef]
40. Ring, U.; Gessner, K.; Güngör, T.; Passchier, C.W. The Menderes massif of western Turkey and the Cycladic massif in the Aegean: Do they really correlate? *J. Geol. Soc. Lond.* **1999**, *156*, 3–6. [CrossRef]
41. Karkalis, C.; Magganas, A.; Koutovitis, P. Petrological, mineralogical and geochemical data of Eohellenic ophiolitic rocks in the island of Skyros, Greece. *Bull. Geol. Soc. Greece* **2016**, *50*, 1867–1877.
42. Tarney, J.; Marsh, N.G. Major and trace elements geochemistry of Holes CY-1 and CY-4: Implications for petrogenetic models. In *Cyprus Crustal Study Project; Initial Report, Holes CY-1 and 1a*; Gibson, I.L., Mapas, J., Robinson, P.T., Xenophontos, C., Eds.; Paper No. 90-20; Geological Survey of Canada: Ottawa, ON, Canada, 1992; pp. 133–175.
43. Pin, C.; Briot, D.; Bassin, C.; Poitrasson, F. Concomitant separation of strontium and samarium-neodymium for isotopic analysis in silicate samples, based on specific extraction chromatography. *Anal. Chim. Act.* **1994**, *298*, 209–217. [CrossRef]
44. Tanaka, T.; Togashi, S.; Kamioka, H.; Amakawa, H.; Kagami, H.; Hamamoto, T.; Yuhara, M.; Orihashi, Y.; Yoneda, S.; Shimizu, H.; et al. JNdi-1: A neodymium isotopic reference in consistency with LaJolla neodymium. *Chem. Geol.* **2000**, *168*, 279–281. [CrossRef]
45. Baker, J.; Peate, D.; Waight, T.; Meyzen, C. Pb isotopic analysis of standards and samples using a ^{207}Pb-^{204}Pb double spike and thallium to correct for mass bias with a double-focusing MC-ICPMS. *Chem. Geol.* **2004**, *211*, 275–303. [CrossRef]
46. McArthur, J.M.; Howarth, R.J.; Bailey, T.R. Strontium isotope stratigraphy: LOWESS version 3: best fit to the marine Sr-isotope curve for 0–509 Ma and accompanying look-up table for deriving numerical age. *J. Geol.* **2001**, *109*, 155–169.
47. Luff, I.W. Petrogenesis of the island arc tholeiite series of the South Sandwich islands. Ph.D. thesis, University of Leeds, Leeds, UK, 1982.
48. Sun, S.S.; McDonough, W.F. Chemical and isotopic systematics of oceanic basalts: Implications for mantle composition and processes. In *Magmatism in Ocean Basins*; Saunders, A.D., Norry, M., Eds.; Geol. Soc. London Spec. Publ., The Geological Society of London: London, UK, 1989; Volume 42, pp. 313–345.

49. Merle, R.; Marzoli, A.; Reisberg, L.; Bertrand, H.; Nemchin, A.; Chiaradia, M.; Callegaro, S.; Jourdan, F.; Bellieni, G.; Kontak, D.; et al. Sr, Nd, Pb and Os isotope systematics of CAMP tholeiites from Eastern North America (ENA): Evidence of a subduction-enriched mantle source. *J. Petrol.* **2013**, *55*, 133–180. [CrossRef]
50. Wilson, M. *Igneous Petrogenesis*; Unwin Hyman: London, UK, 1989; p. 232.
51. Pearce, J.A. Immobile element fingerprinting of ophiolites. *Elements* **2014**, *10*, 101–108. [CrossRef]
52. Pearce, J.; Stern, R.; Bloomer, S.; Fryer, P. Geochemical mapping of the Mariana arc-basin system: Implications for the nature and distribution of subduction Component. *Geochem. Geophys. Geosyst.* **2005**, 6. [CrossRef]
53. Pearce, J.A. A user's guide to basalt discrimination diagrams. *Geol. Assoc. Can. Spec. Publ.* **1996**, *12*, 79–113.
54. Floyd, P.A.; Winchester, J.A. Magma type and tectonic setting discrimination using immobile elements. *Earth Planet. Sc. Lett.* **1975**, *27*, 211–218. [CrossRef]
55. Pearce, J.; Stern, R. Origin of Back-Arc Basin Magmas: Trace Element and Isotope Perspectives. In *Back-Arc Spreading Systems: Geological, Biological, Chemical, and Physical Interactions*; Christie, D.M., Fisher, C.R., Lee, S.-M., Givens, S., Eds.; Geophysical Monograph Series, 166; American Geophysical Union: Washington DC, USA, 2006; pp. 63–86.
56. Shervais, J.W. Ti-V plots and the petrogenesis of modern and ophiolitic lavas. *Earth Planet. Sci. Lett.* **1982**, *59*, 101–118. [CrossRef]
57. Whattam, S.A.; Stern, R.J. The subduction initiation rule: A key for linking ophiolites, intra-oceanic forearcs and subduction initiation. *Contrib. Mineral. Petrol.* **2011**, *162*, 1031–1045. [CrossRef]
58. Koutsovitis, P. Gabbroic rocks in ophiolitic occurrences from East Othris, Greece: Petrogenetic processes and geotectonic environment implications. *Mineral. Petrol.* **2012**, *104*, 249–265. [CrossRef]
59. Saccani, E.; Photiades, A. Mid-ocean ridge and supra-subduction affinities in the Pindos ophiolites (Greece): Implications for magma genesis in a forearc setting. *Lithos* **2004**, *73*, 229–253. [CrossRef]
60. Pearce, J.A.; Norry, M.J. Petrogenetic implications of Ti, Zr, Y and Nb variations in volcanic rocks. *Contrib. Mineral. Petrol.* **1979**, *69*, 33–47. [CrossRef]
61. Saunders, A.D.; Tarney, J.; Marsh, N.G.; Wood, D.A. Ophiolites as ocean crust or marginal basin crust: A geochemical approach. In *Ophiolites, Proceedings of the International Ophiolite Symposium*; Panayiotou, A., Ed.; Ministry of Agriculture and Natural Resources, Geological Survey Department: Nicosia, Cyprus, 1980; pp. 193–204.
62. Gills, K.M.; Robinson, P.T. Distribution of alteration zones in the upper oceanic crust. *Geology* **1988**, *16*, 262–266. [CrossRef]
63. Staudigel, H. Chemical fluxes from hydrothermal alteration of the oceanic crust. In *Treatise on Geochemistry*, 2nd ed.; Turekian, K.K., Holland, H.D., Eds.; Elsevier: London, UK, 2014; Volume 4, pp. 583–606.
64. Hastie, A.R.; Ramsook, R.; Mitchell, S.F.; Kerr, A.C.; Millar, I.L.; Darren, F. Geochemistry of Compositionally Distinct Late Cretaceous Back-Arc Basin Lavas: Implications for the Tectonomagmatic Evolution of the Caribbean Plate. *J. Geol.* **2010**, *118*, 655–676. [CrossRef]
65. Perfit, M.R.; Gust, D.A.; Bence, R.; Arculus, J.; Taylor, S. Chemical characteristics of island arc basalts: Implications for mantle sources. *Chem Geol.* **1980**, *30*, 227–256.
66. Saunders, A.D.; Norry, M.J.; Tarney, J. Origin of MORB and chemically-depleted mantle reservoirs: Trace element constraints. *J. Petrol. Spec. Lithosph. Issue* **1988**, *1*, 415–445. [CrossRef]
67. Wood, D.A.; Joron, J.L.; Treuil, M. A re-appraisal of the use of trace elements to classify and discriminate between magma series and different tectonic settings. *Earth Planet. Sc. Lett.* **1979**, *45*, 326–336. [CrossRef]
68. Reagan, M.K.; Ishizuka, O.; Stern, R.J.; Kelley, K.A.; Ohara, Y.; Blichert-Toft, J.; Bloomer, S.H.; Cash, J.; Fryer, P.; Hanan, B.B.; et al. Fore-arc basalts and subduction initiation in the Izu-Bonin-Mariana system. *Geochem. Geophys.* **2010**, *11*, Q03X12. [CrossRef]
69. Zindler, A.; Hart, S. Chemical Geodynamics. *Annu. Rev. Earth Planet. Sci.* **1986**, *14*, 493–571. [CrossRef]
70. Niu, Y. Bulk-rock major and trace element compositions of abyssal peridotites: Implications for mantle melting, melt extraction and post-melting processes beneath mid-ocean ridges. *J. Petrol.* **2004**, *45*, 2423–2458. [CrossRef]
71. Boschi, C.; Bonatti, E.; Ligi, M.; Brunelli, D.; Dallai, L.; D'Orazio, M.; Fróh-Green, G.L.; Tonarini, S.; Barnes, J.; Bedini, R. Serpentinization of mantle peridotites along an uplifted lithospheric section, Mid Atlantic Ridge at 11° N. *Lithos* **2013**, *178*, 3–23. [CrossRef]
72. Snow, J.E.; Dick, H.J.B. Pervasive magnesium loss by marine weathering of peridotite. *Geochim. Cosmochim. Acta* **1995**, *59*, 4219–4235. [CrossRef]

73. Magganas, A.; Koutsovitis, P. Composition, melting and evolution of the upper mantle beneath the Jurassic Pindos Ocean inferred by ophiolitic ultramafic rocks in East Othris, Greece. *Int. J. Earth Sci.* **2015**, *104*, 1185–1207. [CrossRef]
74. Saccani, E.; Photiades, A. Petrogenesis and tectono-magmatic significance of volcanic and subvolcanic rocks in the Albanide-Hellenide ophiolitic mélanges. *Isl. Arc* **2005**, *14*, 494–516. [CrossRef]
75. Deschamps, F.; Godard, M.; Guillot, S.; Hattori, K. Geochemistry of subduction zone serpentinites: A review. *Lithos* **2013**, *178*, 96–127. [CrossRef]
76. Bröcker, M.; Keasling, A. Ionprobe U-Pb zircon ages from the high-pressure/low-temperature mèlange of Syros, Greece: Age diversity and the importance of pre-Eocene subduction. *J. Metamorph. Geol.* **2006**, *24*, 615–631. [CrossRef]
77. Liati, A.; Gebauer, D.; Fanning, C.M. The age of ophiolitic rocks of the Hellenides (Vourinos, Pindos, Crete): First U–Pb ion microprobe (SHRIMP) zircon ages. *Chem. Geol.* **2004**, *207*, 171–188. [CrossRef]
78. Parlak, O.; Karaoğlan, F.; Rızaoğlu, T.; Klötzli, U.; Koller, F.; Billor, Z. U–Pb and ^{40}Ar–^{39}Ar geochronology of the ophiolites and granitoids from the Tauride belt: Implications for the evolution of the Inner Tauride suture. *J. Geodyn.* **2013**, *62*, 22–37. [CrossRef]
79. White, W.M.; Klein, E.M. Composition of the oceanic crust. In *Treatise on Geochemistry*, 2nd ed.; Holland, H.D., Turekian, K.K., Eds.; Elsevier: Amsterdam, The Netherlands, 2014; pp. 457–496.
80. Booij, E.; Bettison-Varga, L.; Farthing, D.; Staudigel, H. Pb-isotope systematics of a fossil hydrothermal system from the Troodos ophiolite, Cyprus: Evidence for a polyphased alteration history. *Geochim. Cosmochim. Acta* **2000**, *64*, 3559–3569. [CrossRef]
81. Janousek, V.; Farrow, C.M.; Erban, V. Interpretation of whole-rock geochemical data in igneous geochemistry: Introducing Geochemical Data Toolkit (GCDkit). *J. Petrol.* **2006**, *47*, 1255–1259. [CrossRef]
82. Faure, G. *Principles of Isotope Geology*, 2nd ed.; John Wiley & Sons: New York, NY, USA, 1986; p. 589.
83. Spooner, E.T.C.; Fyfe, W.S. Sub-sea-floor metamorphism, heat and mass transfer. *Contrib. Mineral. Petrol.* **1973**, *42*, 287–304. [CrossRef]
84. Richter, F.M.; Rowley, D.B.; DePaolo, D.J. Sr isotope evolution of seawater: The role of tectonics. *Earth Planet. Sci. Lett.* **1992**, *109*, 11–23. [CrossRef]
85. Veizer, J. Strontium isotopes in seawater through time. *Annu. Rev. Earth Planet. Sci.* **1989**, *17*, 141–167. [CrossRef]
86. Pomonis, P.; Tsikouras, B.; Hatzipanagiotou, K. Petrogenetic evolution of the Koziakas ophiolite complex, (W. Thessaly, Greece). *Mineral. Petrol.* **2007**, *89*, 77–111.
87. Saccani, E. A new method of discriminating different types of post-Archean ophiolitic basalts and their tectonic significance using Th-Nb and Ce-Dy-Yb systematics. *Geosci. Front.* **2015**, *6*, 481–501. [CrossRef]
88. Seck, H.A.; Kötz, J.; Okrusch, M.; Seidel, E.; Stosch, H.G. Geochemistry of a meta-ophiolite suite: An association of metagabbros, eclogites and glaucophanites on the island of Syros, Greece. *Eur. J. Mineral.* **1996**, *8*, 607–623.
89. Moçek, B. Geochemical evidence for arc-type volcanism in the Aegean Sea: The blueschist unit of Siphnos, Cyclades (Greece). *Lithos* **2001**, *57*, 263–289. [CrossRef]
90. Fu, B.; Paul, B.; Cliff, J.; Bröcker, M.; Bulle, F. O-Hf isotope constraints on the origin of zircons in high-pressure melange blocks and associated matrix rocks from Tinos and Syros, Greece. *Eur. J. Mineral.* **2012**, *24*, 277–287. [CrossRef]
91. Fu, B.; Bröcker, M.; Ireland, T. Zircon U–Pb, O, and Hf isotopic constraints on Mesozoic magmatism in the Cyclades, Aegean Sea, Greece. *Int. J. Earth Sci.* **2015**, *104*, 75–87. [CrossRef]
92. Weaver, B.L.; Wood, D.A.; Tarney, J.; Joron, J.L. Geochemistry of ocean island basalts from the south Atlantic: Ascension, Bouvet, St. Helena, Gough and Tristan da Cunha. In *Alkaline Igneous Rocks*; Fitton, J.G., Upton, B.G.J., Eds.; Geol. Soc. London Spec. Publ., The Geological Society of London: London, UK, 1987; Volume 30, pp. 253–267.

9

Stratigraphic Changes in the Pliocene Carnivoran Assemblage from Hagerman Fossil Beds National Monument, Idaho

Dennis R. Ruez Jr. [1,2,3]

[1] Department of Geological Sciences, The University of Texas at Austin, Austin, TX 78712, USA
[2] Current affiliations: Department of Environmental Studies, University of Illinois at Springfield, Springfield, IL 62703, USA; druez2@uis.edu
[3] Research and Collections Center, Illinois State Museum, Springfield, IL 62703, USA

Academic Editor: Olaf Lenz

Abstract: At least 17 carnivoran taxa occur in the Pliocene Glenns Ferry Formation at Hagerman Fossil Beds National Monument (HAFO), Idaho. This assemblage was examined for stratigraphic changes in species distribution, specimen abundance, and species diversity. Three relatively common mustelids, *Trigonictis cookii*, *Trigonictis macrodon*, and *Mustela rexroadensis*, occur at most stratigraphic levels, but are absent during an interval coinciding with the coolest time segment at HAFO. It is within this gap that two less-common mustelids, *Ferinestrix vorax* and *Buisnictis breviramus*, first appear at HAFO; they persist up-section with the more common mustelids listed above. Specimens of *Borophagus hilli* are restricted to the warm intervals at HAFO, irrespective of the relative abundance of surface water. The other canid at HAFO, *Canis lepophagus*, is more abundant during the dry intervals at HAFO, regardless of the estimated paleotemperature. Most remarkable is the recovery of many taxa impacted by abrupt climate change, although a notable change is the much higher relative abundance of carnivoran species following a return to warm temperatures.

Keywords: Glenns Ferry; Blancan; paleoclimate

---

1. Introduction

Hagerman Fossil Beds National Monument (HAFO) in southern Idaho (Figure 1) is internationally significant because it is one of the richest sources of Pliocene vertebrates. Hundreds of localities within the exposed beds of the Glenns Ferry Formation have produced many thousands of fossil mammals housed at museums across the United States [1]. These localities range in age from about 4.2 to 3.1 Ma [2]. This is the first in a series of notes that document the stratigraphic distribution of fossil mammals at HAFO. Comparisons are here made with the estimated paleoclimate during the Pliocene represented in the Glenns Ferry Formation. This study examines the distribution of carnivorans at HAFO, but the relative paucity of those fossils, compared to some other groups, means that future studies may have important implications for the interpretations below.

There are at least 54 species of mammals at HAFO, including 17 species of carnivorans [1]. Eight species of carnivorans were named on holotypes from HAFO; of these, six are still valid. Publication on these specimens began in the 1930s [3–5], but comprehensive description of the carnivorans from HAFO was completed much later [6]. The taxonomy of these species is reviewed elsewhere with a discussion on their geographic distributions [1].

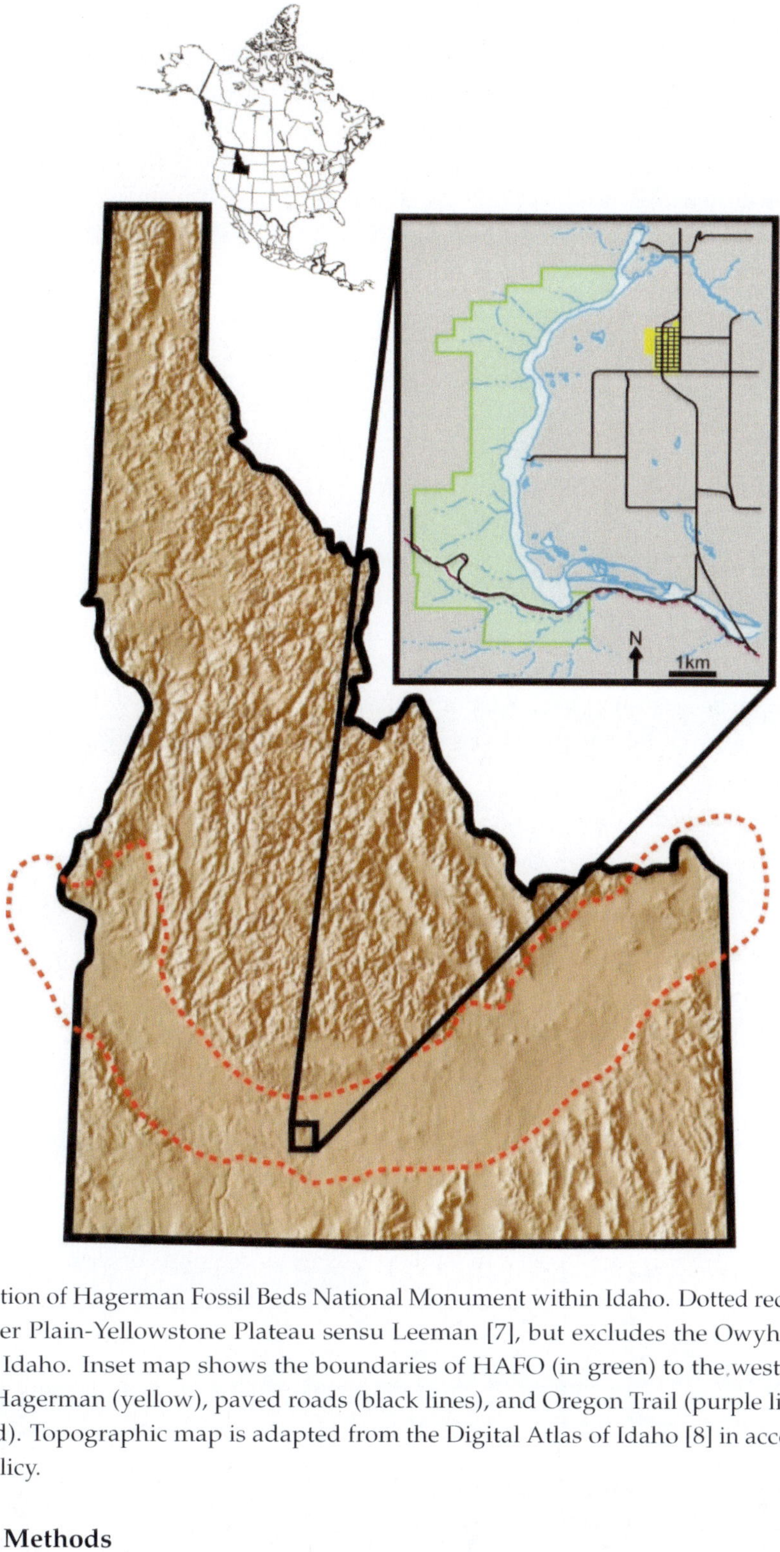

Figure 1. Location of Hagerman Fossil Beds National Monument within Idaho. Dotted red line outlines the Snake River Plain-Yellowstone Plateau sensu Leeman [7], but excludes the Owyhee Plateau in southwestern Idaho. Inset map shows the boundaries of HAFO (in green) to the west of the Snake River, city of Hagerman (yellow), paved roads (black lines), and Oregon Trail (purple lines—dashed where inferred). Topographic map is adapted from the Digital Atlas of Idaho [8] in accordance with their usage policy.

2. Materials and Methods

In spite of the species-level diversity and overall number of mammalian fossils at HAFO, the number of specimens referable to carnivorans is low; only 361 fossils were recognized during the preparation of this manuscript. Only specimens with well-established locality data were included in this study. Additionally, only specimens that could be identified to species, or as *Taxidea* sp. or *Homotherium* sp. (the sparse material does not allow for more specific identification), were used. The 256 specimens used here are listed in Supplementary file (List S1). (Collections of fossils at HAFO have continued since the data in this manuscript were collected, resulting in additional specimens, particularly mustelids [9], and the possible addition of two carnivoran taxa [10].)

Although a Minimum Number of Individuals (MNI) is commonly preferred in analysis of abundance in a fauna (e.g., Lyman [11]), it is not an empirical observational unit such as the number of identifiable specimens (NISP). In fact, the MNI must be calculated from the NISP, and the method involved can include various criteria [12]. In the case of a single locality with abundant fossils, comparisons of MNI instead of NISP can give significantly different results. However, at HAFO there are hundreds of localities, so the MNI would have to be calculated for each locality because it is unlikely that different localities will contain specimens from the same individual. Because the ratio of localities to specimens is so high, about two to one, and no localities are especially abundant with carnivorans, the MNI and NISP of carnivorans at HAFO are similar. In fact, Bjork [6] examined 196 carnivoran fossils, but calculated an MNI of 173. In this particular case, use of NISP instead of MNI has little impact on the results.

Proxy data for paleoecological interpretations for HAFO (Figure 2) follow Ruez [13]. Stratigraphic distribution of specimens is based on placement of localities on the Hagerman Horse Quarry (HHQ) datum (sensu Ruez [2]). This facilitates comparison of sites across HAFO. Specimen and species abundances were plotted at 1-m intervals with a sliding window of 20 m for the stratigraphic span encompassing the majority of localities at HAFO: 900 to 1005 m on the HHQ datum, or about 4.0 to 3.2 Ma. By using a sliding window some detail may be lost, but the pattern will more accurately reflect overall trends and not be as subject to distorting spikes from particularly fossiliferous localities. Further, the sliding window acknowledges that there is difficulty in placing some localities stratigraphically.

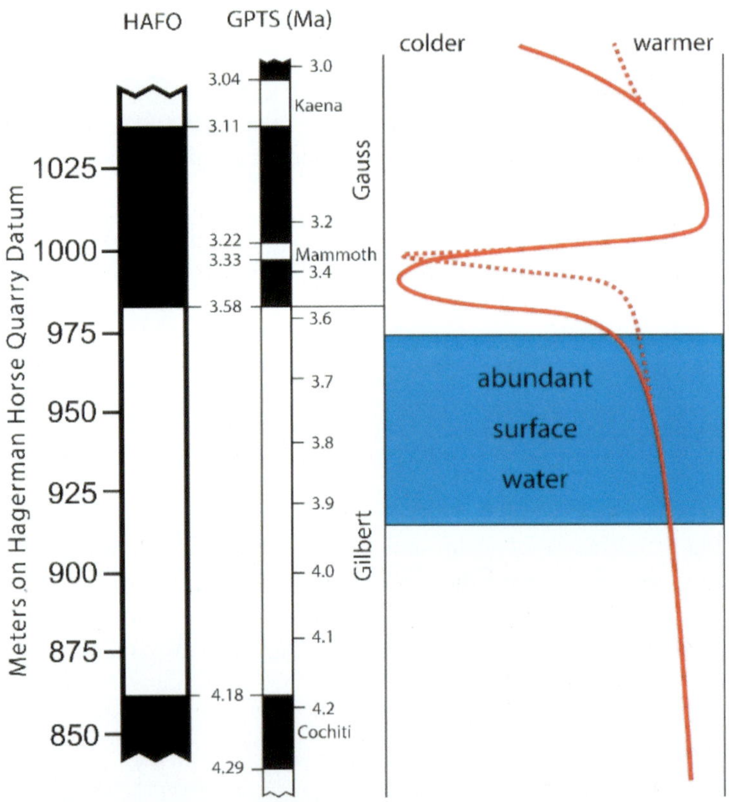

Figure 2. Pliocene paleoecological interpretations at HAFO. Temperature trend is adjusted to the chronology of deposits at HAFO [2]. Solid line indicates the temperature pattern exhibited by data generated from microfossil abundance; the dashed line illustrates where the isotopic data deviate [13]. Note that although the elevations are at a constant interval, the scale changes on the GPTS (after Berggren *et al.* [14]). Dates are in Ma and are not spaced consistently because sediment deposition rates varied; stratigraphic levels are in meters above mean sea level.

Institutional abbreviations: HAFO, Hagerman Fossil Beds National Monument (Hagerman, ID, USA); IMNH, Idaho Museum of Natural History (Pocatello, ID, USA); UMMP, University of Michigan Museum of Paleontology (Ann Arbor, MI, USA); USNM, United States National Museum (Washington, DC, USA).

3. Results

Three mustelids, *Trigonictis cookii*, *Trigonictis macrodon*, and *Satherium piscinarium*, have stratigraphic ranges of more than 100 m (>0.82 myr) each at HAFO; a fourth, *Mustela rexroadensis*, extends for almost 90 m (0.73 myr; Figure 3). The ranges for three of these long-persisting taxa, *T. cookii*, *M. rexroadensis*, and *T. macrodon*, contain significant gaps coinciding with the transition from an interval of abundant surface water to one lacking surface water; that interval also is marked by dramatic cooling. The most abundant mustelid, *S. piscinarium*, does not exhibit such a significant gap, but the number of specimens of this species is reduced within that interval. It is within that interval that two other mustelids, *Buisnictis breviramus* and *Ferinestrix vorax*, first appear at HAFO.

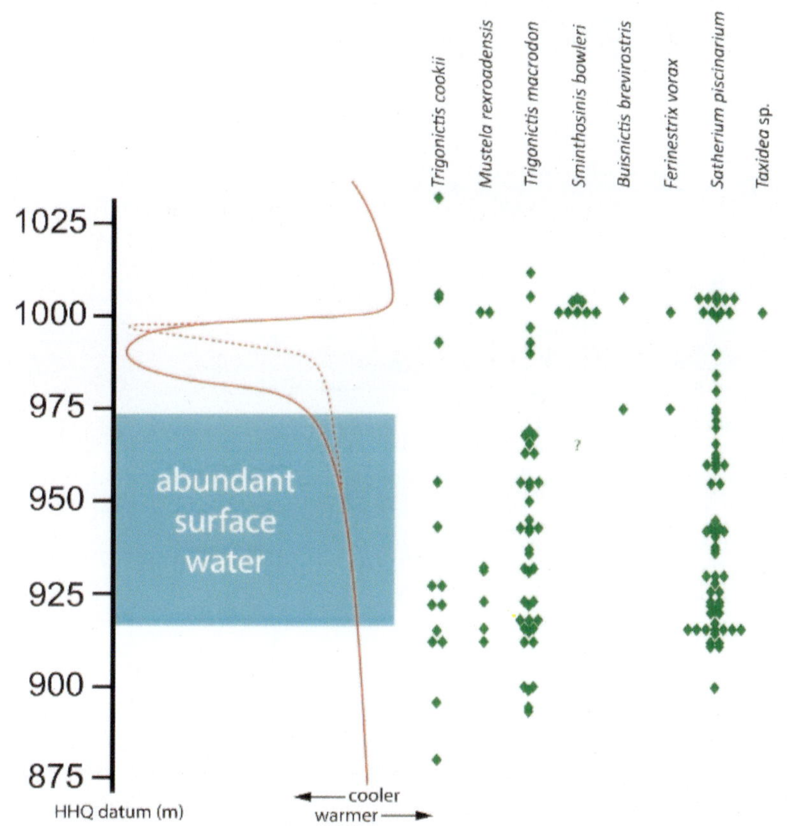

Figure 3. Distribution of mustelids at HAFO. Stratigraphic levels and paleoecologic data follow Figure 2. Each specimen is indicated with a diamond. Question mark under *Sminthosinis bowleri* indicates a specimen referred to that species by Bjork [6], but here considered an indeterminate mustelid.

Sminthosinis bowleri and *Taxidea* sp. first appear at the end of an abrupt warming at HAFO. Another specimen of *S. bowleri*, from about 35 m (~0.45 myr) lower in the Glenns Ferry Formation, was identified previously [6], but this edentulous partial maxilla (UMMP 51681) is also similar in size and morphology to *T. cookii*. That specimen does not have any diagnostic characters allowing for referral to either *S. bowleri* or *T. cookii*.

Two large felids, *Megantereon hesperus* and *Homotherium* sp., are known from only one specimen each with precise stratigraphic data; the single fossil representing *Miracinonyx inexpectatus* (USNM 12613) lacks locality data and is therefore excluded from this analysis. Two small felids, *Puma lacustris* and *Lynx rexroadensis*, are known from HAFO, but they are extremely similar in skeletal morphology. Most elements of these species cannot be distinguished, and unfortunately the ones that have diagnostic characters do not have associated stratigraphic data. Therefore, these two cats are plotted together in Figure 4.

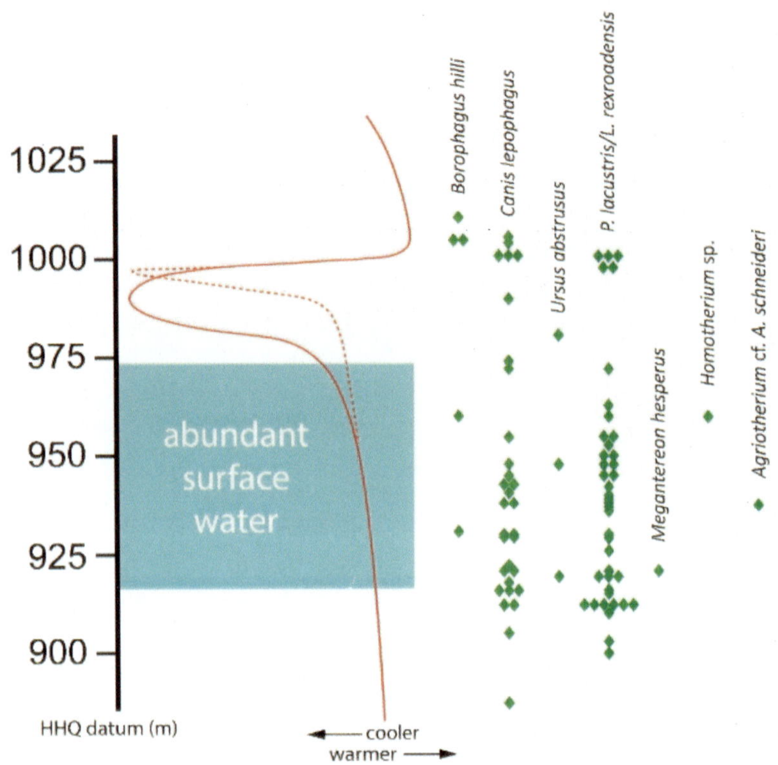

Figure 4. Distribution of large carnivorans (canids, ursids, and felids) at HAFO. Stratigraphic levels and paleoecologic data follow Figure 2. Each specimen is indicated with a diamond.

Large carnivorans (canids, ursids, felids) known from multiple specimens with stratigraphic data all have a long temporal range at HAFO. *Canis lepophagus* and *P. lacustris/L. rexroadensis* have ranges of more than 100 m (>0.82 myr). Even rare species such as *Borophagus hilli*, known from only five fossils with precise field data, and *Ursus abstrusus*, known from only three specimens, have stratigraphic distributions of 80 m (0.69 myr) and 62 m (0.33 myr), respectively.

As in the case of some mustelids, two of the above large carnivorans have gaps in their record that correspond to the end of abundant surface water and the beginning of pronounced cooling. There is a 26 m (0.39 myr) gap in the distribution of *P. lacustris/L. rexroadensis*. A 45 m (0.51 myr) gap occurs in the range of the specimens of *B. hilli*, but again, that species is known from few fossils at HAFO. The range of *Canis lepophagus* does not include this significant gap, but localities within the cool interval only produced a single specimen. Records of *U. abstrusus* occur in intervals both with and without abundant surface water, but all are known prior to the abrupt cooling event at HAFO. The specimens of *Megantereon hesperus*, *Homotherium* sp., and *Agriotherium* cf. *A. schneideri* occur during warm periods of abundant surface water.

Specimen abundance of the four most abundant taxa were examined and compared to the pattern of all carnivorans combined (Figure 5). Overall, the pattern for each species is similar to that for all carnivorans combined. This pattern, in turn, corresponds closely to the temperature trend at HAFO.

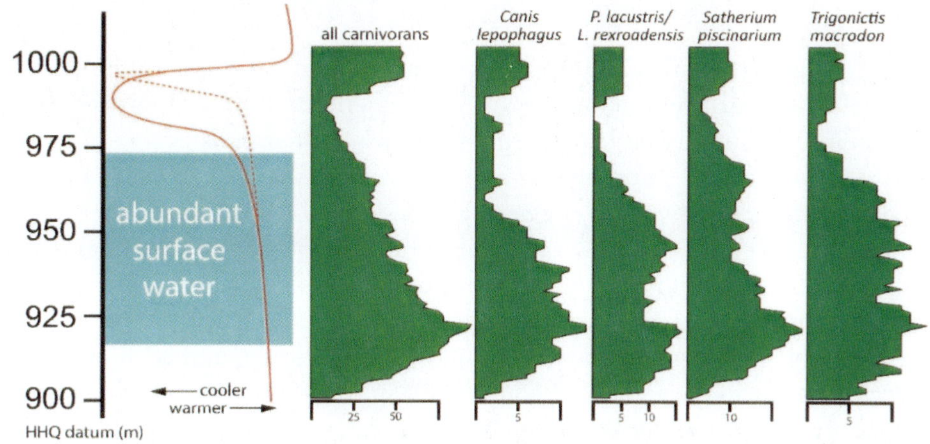

Figure 5. Specimen abundance of the four most abundant carnivoran species at HAFO, compared to data for all carnivorans and paleoecological interpretations. Specimen abundances were plotted at 1 m intervals with a sliding window of 20 m for the stratigraphic span encompassing the majority of localities at HAFO: 900 to 1005 m on the HHQ datum, or about 4.0 to 3.2 Ma. Note that horizontal scales vary.

Specimen-level abundances for *C. lepophagus* and *P. lacustris/L. rexroadensis* decreased during the interval of slow climatic cooling. Patterns for *S. piscinarium* and *T. macrodon* show a delayed response, more closely matching the end of abundant surface water and the onset of rapid temperature decrease. The abundance of *P. lacustris/L. rexroadensis* is also decreased low in the stratigraphic section, near the bottom of the interval of abundant surface water.

The number of carnivoran species is relatively constant throughout most of the HAFO section, until the very top of the section (Figure 6). The pattern is similar, but more pronounced, when the number of carnivoran species is plotted as a proportion of all mammalian species. Both the absolute and relative abundance of carnivoran species have a positive correlation with the estimated temperature trend.

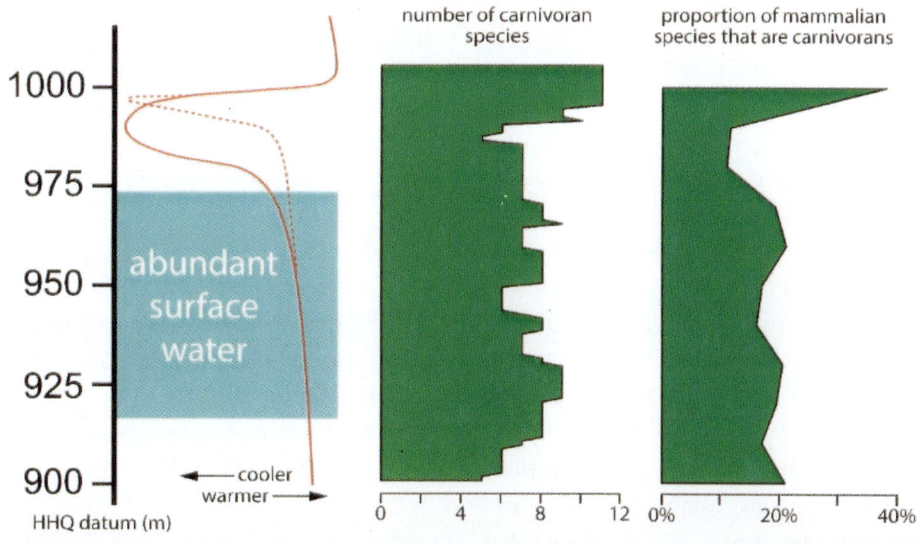

Figure 6. Species abundance at HAFO. Carnivoran species abundances were plotted at 1 m intervals with a sliding window of 20 m for the stratigraphic span encompassing the majority of localities at HAFO: 900 to 1005 m on the HHQ datum, or about 4.0 to 3.2 Ma. Relative abundance of carnivorans was plotted at 10 m intervals with the same 20 m sliding window.

4. Conclusions

Beginning at about the 975 m stratigraphic level there is a faunal change at HAFO that coincides with the end of abundant surface water and a decrease in temperature. At this point, two mustelids, *S. bowleri* and *F. vorax*, first appear at HAFO. Mustelids reach their peak diversity at the warm, dry interval around 1000 m, while the diversity of large carnivorans is highest lower in the section, during an interval of abundant surface water as suggested by sedimentological data [2].

Among the abundant carnivorans, the specimen abundance of *C. lepophagus* begins a sharp decline with an increasing rate of cooling at 950 m; distribution of this species is seemingly unaffected by the presence or absence of abundant surface water. Likewise, *P. lacustris*/*L. rexroadensis* shows a reducing abundance of specimens before the end of the wet interval, and additionally has reduced numbers of specimens low in the section with abundant surface water. The abundance of *T. macrodon* does not decrease markedly until the end of the wet interval, where it actually increases through the continued cooling. Distribution of *T. macrodon* is here suggested as more dependent on the presence of surface water/moisture than temperature. The abundance of *S. piscinarium* appears linked to both temperature and surface moisture. The number of specimens decreases at the end of the wet interval, as in *T. macrodon*, but continues to decrease, matching the temperature curve.

Correlation of specimen and species abundance with estimated temperature at HAFO is not the result of a lack of fossiliferous localities in the cool interval [2]. However, in modern environments warmer temperatures and wetter environments are correlated with increased carnivoran species diversity and higher numbers of individuals [15]; some Quaternary faunas also seem to follow this trend (e.g., Dayan [16]). Species richness in the carnivoran assemblage from HAFO is relatively constant through intervals of varying temperature and levels of aridity, until the sharp increase in temperature near the top of the section where the assemblage becomes more speciose. The study of Pliocene and Pleistocene carnivorans from east Africa documents the highest species diversity during an interval from 3.3 to 3.0 Ma [17]; this range that corresponds to the age of the sediments containing the spike of carnivoran richness at HAFO.

Although the fossil richness of localities throughout HAFO will vary, calculating the relative abundance of carnivoran species as a proportion of all mammalian species adjusts for this difference. The lowest proportion of carnivorans occurred in the dry cool interval, and the highest occurred with the rapid return to warm temperatures. Analysis of Pleistocene mammalian faunas from Italy also showed decreased predator/prey ratios with cooler temperatures [18]; however, unlike HAFO, the ratios in the Italian faunas result from increased herbivore richness during cold times [19]. In contrast to those studies, the proportion of carnivorans in the mammalian fauna of Hayonim Cave in Israel is at both the highest and the lowest during cold intervals [16]. This disparity of results suggests that the correlation of temperature and relative abundance of carnivorans at HAFO may not be the result of impacts on carnivorans, but on other aspects of the faunas.

Changes documented here for HAFO primarily occur in association with the end of abundant surface water and onset of rapid cooling. Even though this interval persisted for about 300 ky until the return of warm temperatures, the carnivoran assemblage proved resilient, at least in part. At least five species were present early at HAFO, disappeared during the rapid cooling, and reappeared afterwards. Specimen abundance also rebounded with the return to warmer temperatures. Three large carnivorans did not reappear after the rapid cooling, but they are known from only a few total specimens.

The carnivoran assemblage at HAFO did differ following the interval of rapid change. In particular, the mustelid diversity greatly increased. For the earliest ~600 ky, there were only four mustelid species represented at HAFO; once the temperature warmed again, eight species of mustelids occurred. This increase is not merely a result of increased overall mammalian diversity at HAFO, because the relative abundance of carnivoran species is double that of earlier times at HAFO. Unfortunately, the distribution of localities does not allow the determination of whether these ecological changes persisted long-term or were a brief response to the rapid warming. There are few deposits of the Glenns Ferry Formation at HAFO that are stratigraphically higher than the rapid warming event. However, future collections may help refine our picture of carnivoran changes with climate at HAFO.

Acknowledgments: This work comprised a portion of my doctoral dissertation, and I appreciate the input of ChrisBell (The University of Texas at Austin (UT)), Greg McDonald (U.S. National Park Service (NPS)), Tim Rowe (UT), Jim Sprinkle (UT), Richard Zakrzewski (Fort Hays State University), and the UT paleontology graduate students for their suggestions and support. Assess to collections at HAFO were facilitated by Phil Gensler (NPS) and Greg McDonald; Bill Akersten (Idaho Museum of Natural Hisotory (IMNH)) and Mary Thompson (IMNH) provided access to materials at IMNH. Funding for my visits to Idaho was provided by The University of Texas at Austin Department of Geological Sciences, Geology Foundation, and Graduate School.

Conflicts of Interest: The author declares no conflict of interest.

Abbreviations

The following abbreviations are used in this manuscript:

GPTS	Geomagnetic Polarity Time Scale
HAFO	Hagerman Fossil Beds National Monument (Hagerman, ID, USA)
myr	Million Years
IMNH	Idaho Museum of Natural History (Pocatello, ID, USA)
UMMP	University of Michigan Museum of Paleontology (Ann Arbor, MI, USA)
USNM	United States National Museum (Washington, DC, USA)

References

1. Ruez, D.R., Jr. Revision of the Blancan (Pliocene) mammals from Hagerman Fossil Beds National Monument, Idaho. *J. Idaho Acad. Sci.* **2009**, *45*, 1–143.
2. Ruez, D.R., Jr. Framework for stratigraphic analysis of Pliocene fossiliferous deposits at Hagerman Fossil Beds National Monument, Idaho. *Rocky Mt. Geol.* **2009**, *44*, 33–70. [CrossRef]
3. Gazin, C.L. New felids from the Upper Pliocene of Idaho. *J. Mammal.* **1933**, *14*, 251–256. [CrossRef]
4. Gazin, C.L. Upper Pliocene mustelids from the Snake River basin of Idaho. *J. Mammal.* **1934**, *15*, 137–149. [CrossRef]
5. Gazin, C.L. Notes on fossil mustelids from the upper Pliocene of Idaho and Texas. *J. Mammal.* **1937**, *18*, 363–364. [CrossRef]
6. Bjork, P.R. The Carnivora of the Hagerman local fauna (late Pliocene) of southwestern Idaho. *Trans. Am. Philos. Soc.* **1970**, *60*, 3–54. [CrossRef]
7. Leeman, W.P. Development of the Snake River Plain-Yellowstone Plateau Province, Idaho and Wyoming: An overview and petrologic model. In *Cenozoic Geology of Idaho*; Bonnichsen, B., Breckenridge, R.M., Eds.; Idaho Department of Lands, Bureau of Mines and Geology: Moscow, ID, USA, 1982; Volume 26, pp. 155–177.
8. Digital Atlas of Idaho. Available online: http://imnh.isu.edu/digitalatlas/index.htm (accessed 30 December 2015).
9. Prassack, K. Current and future research on the paleoecology and taphonomy of Pliocene musteloids from Hagerman Fossil Beds National Monument. In Proceedings of the 10th Conference on Fossil Resources, Rapid City, SD, USA, 13–15 May 2014.
10. Prassack, K. Pliocene musteloid diversity and paleoecology at Hagerman Fossil Beds National Monument. Meeting Program. In Proceedings of the 74th Annual Meeting of Society of Vertebrate Paleontology, Berlin, Germany, 5–8 November 2014.
11. Lyman, R.L. Quantitative units and terminology in zooarchaeology. *Am. Antiq.* **1994**, *59*, 36–71. [CrossRef]
12. Klein, R.G.; Cruz-Uribe, K. *The Analysis of Animal Bones from Archaeological Sites*; University of Chicago Press: Chicago, IL, USA, 1984.
13. Ruez, D.R., Jr. Middle Pliocene paleoclimate in the Glenns Ferry Formation of Hagerman Fossil Beds National Monument, Idaho: A baseline for evaluating faunal change. *J. Idaho Acad. Sci.* **2006**, *42*, 1–16.

14. Berggren, W.A.; Kent, D.V.; Swisher, C.C., III; Aubry, M.-P. A revised Cenozoic geochronology and chronostratigraphy. In *Geochronology, Time Scales and Global Stratigraphic Correlation*; Berggren, W.A., Kent, D.V., Aubry, M.-P., Hardenbol, J., Eds.; Society for Sedimentary Geology (SEPM): Tulsa, OK, USA, 1995; Volume 54, pp. 129–212.
15. Waide, R.B.; Willig, M.R.; Steiner, C.F.; Mittelbach, G.; Gough, L.; Dodson, S.I.; Juday, G.P.; Parmenter, R. The relationship between productivity and species richness. *Annu. Rev. Ecol. Syst.* **1999**, *30*, 257–300. [CrossRef]
16. Dayan, T. Carnivore diversity in the late Quaternary of Israel. *Quat. Res.* **1994**, *41*, 343–349. [CrossRef]
17. Werdelin, L.; Lewis, M.E. Plio-Pleistocene Carnivora of eastern Africa: Species richness and turnover patterns. *Zool. J. Linn. Soc.* **2005**, *144*, 121–144. [CrossRef]
18. Palombo, M.R.; Raia, P.; Giovinazzo, C. Early-middle Pleistocene structural changes in mammalian communities from the Italian Peninsula. In *Early-Middle Pleistocene Transitions: The Land-Ocean Evidence*; Head, M.J., Gibbard, P.L., Eds.; The Geological Society Publishing House: Bath, UK, 2005; Volume 247, pp. 251–262.
19. Mussi, M.; Palombo, M.R. Les carnivores du Pliocène supérieur et du Pléistocène de la peninsula italienne: Quelles implications pour le peuplement humain. *Rev. Paleobiol.* **2005**, *23*, 597–609.

10

Exploitation of Satellite A-DInSAR Time Series for Detection, Characterization and Modelling of Land Subsidence

Roberta Bonì [1,*], Claudia Meisina [1], Francesca Cigna [2], Gerardo Herrera [3,4,5], Davide Notti [1], Stephanie Bricker [2], Harry McCormack [6], Roberto Tomás [3,4,7], Marta Béjar-Pizarro [3,4], Joaquín Mulas [3,4] and Pablo Ezquerro [3]

1 Department of Earth and Environmental Sciences, University of Pavia, Via Ferrata 1, 27100 Pavia, Italy; claudia.meisina@unipv.it (C.M.); davidenotti@gmail.com (D.N.)
2 British Geological Survey, Natural Environment Research Council, Nicker Hill, Keyworth, Nottinghamshire NG12 5GG, UK; francesca.cigna@gmail.com (F.C.); step@bgs.ac.uk (S.B.)
3 Geohazards InSAR Laboratory and Modeling Group, Instituto Geológico y Minero de España (IGME), C/. Alenza 1, 28003 Madrid, Spain; g.herrera@igme.es (G.H.); roberto.tomas@ua.es (R.T.); m.bejar@igme.es (M.B.-P.); j.mulas@igme.es (J.M.); p.ezquerro.martin@gmail.com (P.E.)
4 Unidad Asociada de Investigación IGME-UA de Movimientos del Terreno Mediante Interferometría Radar (UNIRAD), Universidad de Alicante, P.O. Box 99, 03080 Alicante, Spain
5 Earth Observation and Geohazards Expert Group (EOEG), EuroGeoSurveys, the Geological Surveys of Europe, 36–38, Rue Joseph II, 1000 Brussels, Belgium
6 CGG, NPA Satellite Mapping, Crockham Park, Edenbridge Kent TN8 6SR, UK; Harry.McCormack@cgg.com
7 Departamento de Ingeniería Civil, Escuela Politécnica Superior, Universidad de Alicante. P.O. Box 99, 03080 Alicante, Spain
* Correspondence: roberta.boni01@universitadipavia.it

Abstract: In the last two decades, advanced differential interferometric synthetic aperture radar (A-DInSAR) techniques have experienced significant developments, which are mainly related to (i) the progress of satellite SAR data acquired by new missions, such as COSMO-SkyMed and ESA's Sentinel-1 constellations; and (ii) the development of novel processing algorithms. The improvements in A-DInSAR ground deformation time series need appropriate methodologies to analyse extremely large datasets which consist of huge amounts of measuring points and associated deformation histories with high temporal resolution. This work demonstrates A-DInSAR time series exploitation as valuable tool to support different problems in engineering geology such as detection, characterization and modelling of land subsidence mechanisms. The capabilities and suitability of A-DInSAR time series from an end-user point of view are presented and discussed through the analysis carried out for three test sites in Europe: the Oltrepo Pavese (Po Plain in Italy), the Alto Guadalentín (Spain) and the London Basin (United Kingdom). Principal component analysis has been performed for the datasets available for the three case histories, in order to extract the great potential contained in the A-DInSAR time series.

Keywords: A-DInSAR time series; land subsidence; groundwater level change; principal component analysis (PCA)

---

1. Introduction

Land subsidence represents the main response to superficial and deep deformations induced by multiple natural and anthropic phenomena (i.e., vadose zone processes, such as swelling/shrinkage of clay soils, soil consolidation, aquifer compaction, solid and fluid extraction, and load-induced compaction) which take place at different spatio-temporal scales. The impacts of land subsidence can be infrastructural, economic, environmental, and social [1,2], including impacts on the natural heritage, natural surficial drainage system, agricultural activities, building foundations and transportation network, alteration of irrigation network, reduction of aquifer storage, ground failure, and it enhances the risk of floods, affecting the human life and activities. The increasing factors of risk are mainly due to rapid urban development, relatively young alluvium soils, and weak mitigation and adaptation strategies [3]. Therefore, in the pre-mitigation investigation phase, the identification of land subsidence areas and the understanding of driving factors is fundamental in order to adopt suitable land use planning and sustainable management of the available resources. Then, land subsidence investigations are essential to delineate the magnitude and type of deformation related to the temporal evolution of surface displacements (i.e., linear or non-linear), the spatial extension of the affected areas and the mechanism of land subsidence.

When land subsidence evidence is the result of superimposed processes, it is difficult to discern, and to map the different processes and to evaluate the associated triggering factors. Therefore, the complexity of the mechanism recognition requires a multidisciplinary approach including the expertise of engineering geologists, hydrogeologists, and geotechnical engineers. Until now, the scientific community carried out different strategies and a combination of different methods, including field measurements, remote sensing tools, and integrated approaches, to solve the complexity of the problem in many areas of the world [4–8].

Recent advanced ground deformation investigations make use of satellite synthetic aperture radar (SAR) data to examine the mechanisms of land subsidence around the world. In particular, these investigations exploit advanced differential interferometric synthetic aperture radar (A-DInSAR) techniques, which are based on the processing of multiple interferograms derived from a large set (at least 20 images) of SAR images. These techniques allow the retrieval of displacement time series of measuring points over wide areas at millimeter resolution [9–15] and have already been successfully applied to monitoring the evolution of different processes.

In the last two decades, A-DInSAR techniques have experienced major developments, which are mainly related to (i) the progress of satellite SAR data acquired by new missions, such as COSMO-SkyMed and ESA's Sentinel-1 constellations; and (ii) the development of novel processing algorithms. The improvements in the displacement time series (TS) obtained by A-DInSAR techniques need appropriate methodologies to analyze extremely large datasets which consist of very large amounts of measuring points and associated deformation histories with high temporal resolution.

This work contributes to address this crucial aspect, by exploiting the great potential contained in A-DInSAR time series. In this article, the relevant contributions of A-DInSAR time series are analyzed distinguishing the main advantages and limitations for three topics that commonly concern land subsidence investigations:

1. Detection of the magnitude and distribution of land subsidence;
2. Characterization of the mechanisms involved in land subsidence; and
3. Modelling of land subsidence due to groundwater level changes.

In particular, the paper provides insights into A-DInSAR time series capabilities and suitability as valuable supporting tools for these topics through interesting case histories of moderate rates of displacement in flat areas (Oltrepo Pavese, Po Plain, in Italy) and of land subsidence due to groundwater level change (Alto Guadalentín Basin, in Spain and the London Basin, in the United Kingdom). A-DInSAR time series analysis has emerged as a fundamental tool not only to monitor, but also to detect, ground motion areas. An innovative approach based on A-DInSAR time series analysis is presented to detect ground motion areas with significant deformational behaviors, such as linear, non-linear, and seasonal movements. Moreover, displacement time series acquired by multi-temporal SAR sensors allow the extraction of a historical archive of the temporal evolution of land movements. The collection of multi-sensor displacement time series proved to be crucial for the characterization of land subsidence mechanisms, such as transient and inelastic aquifer deformation otherwise undetectable by considering the average velocity. Finally, changes, such as acceleration and deceleration in the displacement rates, were simulated by calibrating a 1-D model using A-DInSAR time series.

2. Data and Methods

In the framework of the Panel on Land Subsidence of the U.S. National Research Council (NRC) (1991) two information needs were recognized [16]:

1. Earth-science data and information on the magnitude and distribution of land subsidence in order to recognize and to assess future problems; and
2. Research on subsidence processes and engineering methods in order to prevent damage.

According to these information needs, the support of A-DInSAR time series for land subsidence investigations is presented here. A multidisciplinary approach was adopted by combining geological, geotechnical, hydrogeological, and A-DInSAR data to gain insight into land subsidence detection, characterization, and modelling through three representative test cases: Oltrepo Pavese, in Italy, Alto Guadalentín Basin, in Spain, and the London Basin, in the United Kingdom (Figure 1).

Figure 1. Location of the study areas.

Regarding the detection of ground motion areas, a novel methodology was implemented in the plain area of the Oltrepo Pavese (Italy) in order to distinguish different components of motion. The characterization of the mechanisms of land subsidence were analyzed in detail in the Alto Guadalentín Basin in Spain, where the highest subsidence rates measured in Europe (>10 cm/year) were recognized as a direct consequence of long-term aquifer exploitation [17]. Finally, the London Basin was chosen as a study area to use the A-DInSAR time series to characterize the hydrological properties of the deposits of the basin and to integrate the A-DInSAR time series in the modelling of ground deformation. The prediction of the ground motion expected in response to the groundwater level variations was performed in the London Basin thanks to the availability of detailed time-series data from the observation borehole network provided by the Environment Agency (around ~200 piezometers), to reconstruct the historical groundwater level changes across the basin for the 1990s and 2000s.

Additional details about the data for each test case are summarized in Table 1.

Table 1. Test case characteristics and available A-DInSAR data.

Test Case	SAR Data	Sensor	Processing Technique	Time Span	Type of Land Subsidence	Area (km^2)
Oltrepo Pavese (Italy)	ERS-1/2 RADARSAT-1	C-band C-band	SqueeSAR™ SqueeSAR™	1992–2000 2003–2010	Natural and anthropic causes	440
Alto Guadalentín Basin (Spain)	ERS-1/2 ENVISAT ALOS-PALSAR COSMO-SkyMed	C-band C-band L-band X-band	StaMPS StaMPS SPN SPN	1992–2000 2003–2007 2007–2010 2011–2012	Anthropogenic subsidence due to groundwater extraction	277
London Basin (United Kingdom)	ERS-1/2 ENVISAT	C-band C-band	IPTA IPTA	1992–2000 2002–2010	Natural and anthropic causes	1360

2.1. The Oltrepo Pavese Test Case

The Oltrepo Pavese (Italy) is the plain sector of the Po River (Figure 2) covering an area of approximately 440 km^2. The Oltrepo Pavese is a representative site of similar geological contexts in the Po Plain, where geohazards, due to natural and anthropic factors, were previously recognized [18]. Severely damaged structures were also observed in this area due to volume changes of clayey soils (shrinkage and swelling during drying and wet periods, respectively). Regarding the land use coverage of Oltrepo Pavese, discontinuous urban fabrics are the most prevalent land use class and 68% of the monitored area is covered by non-irrigated arable land.

The plain of the Oltrepo Pavese is constituted by alluvial quaternary deposits, originated from the action of Apennine streams and of the Po River [19,20]. These deposits overlay Miocene-Pliocene marine substratum, composed of sandy-marls, sandstones, conglomerates, gypsy-marls, and calcareous-marls. Three main geomorphological units were previously distinguished [20,21] in the quaternary deposits (1) Post-würmian alluvial deposits; (2) Würmian-holocenic alluvial deposits; (3) Pre-würmian alluvial deposits.

Post-würmian alluvial deposits are the most recent sediments that are mainly localized close to the Po River. These deposits are composed of sand, sandy silt, and silt, and they contain a shallow phreatic aquifer.

Würmian-holocenic alluvial deposits are made up of alternating sand and gravel, with interbedded clays or argillaceous silt. These deposits contain a shallow phreatic aquifer and deeper aquifers, of both phreatic and confined types. The aquifer geometry is controlled by the deeply buried structures in the tertiary marine basement, consisting of a series of folds and fold-faults. These alluvial deposits are extensively covered by clayey-silty deposits, which act as a seal, limiting water infiltration [22].

Pre-würmian alluvial deposits, located in the southern part of the study area, consist of older fluvial terraces, and are composed of gravel and sand with a silty matrix. In the Pre-würmian unit, the depth of the groundwater level in this unit reaches values up to 20 m depth.

Regarding the Quaternary sediment thickness, a decrease from west to east is observed in the study area. The minimum thickness and the outcrop of the marine substratum is evident in correspondence with the Stradella thrust, where neotectonic activity was previously described [23].

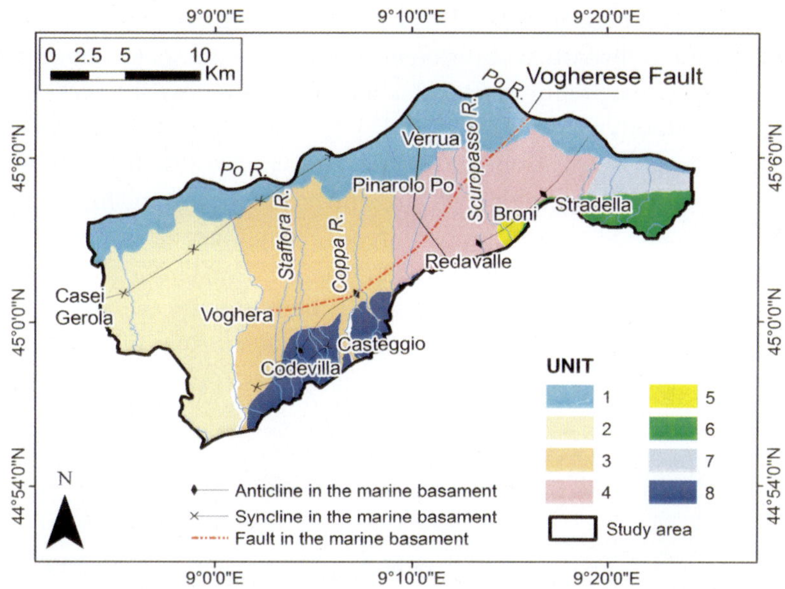

Figure 2. Engineering geological map of the plain area of the Oltrepo Pavese, modified from [24].

The geotechnical properties of the Quaternary deposits in the first 15 m of depth of the Oltrepo Pavese plain are summarized in Figure 2 [24]. Six geotechnical classes of non-cohesive soils (from I1 to I6) and four geotechnical classes of cohesive soils (from C1 to C4) have been distinguished. Therefore, taking into account the geotechnical classification of these soils, eight engineering geological units were introduced as representative of homogeneous geotechnical profiles (Figure 2). Unit 1 represents the Post-würmian alluvial deposit of the Po River, and is mainly constituted of non-cohesive soils (I3). Conversely, units 2, 3, 4, and 7 are Würmian-holocenic alluvial deposits, characterized by cohesive soils. Unit 5 is composed of the alluvial fan of the Scuropasso River, and units 6 and 8 by the Pre-würmian terraced deposits.

The A-DInSAR dataset available for the Oltrepo Pavese test case is composed of SAR images acquired by sensors operating in the C-band (wavelength: 5.6 cm; frequency: 5.3 GHz) onboard the ERS-1, ERS-2, and RADARSAT-1 (RSAT) satellites. The scenes were acquired in the ascending mode, covering the time intervals from 9 July 1992 to 2 August 2000, and from 24 March 2003 to 5 May 2010. The dataset acquired in the descending mode covers the time intervals from 3 April 1992 to 7 January 2001 and from 28 April 2003 to 5 January 2009. The ERS-1/2 images were acquired with a nominal repeat cycle of 35 days, while the RADARSAT images were acquired with a revisiting period of 24 days. These scenes were processed with the SqueeSAR™ technique by Tele-Rilevamento Europa (TRE srl, Milano, Italy). The algorithm allows the extraction of movement measurements, not only from traditional persistent scatterers (PS), such as anthropic structures or rocks, but also from distributed scatterers (DS), such as sparse vegetated areas [10]. This permitted to have a high density of A-DInSAR data over non-urban areas. ERS-1/2 descending and RADARSAT-1 ascending data were exploited to monitor the time intervals, from 1992 to 2000, and from 2003 to 2010 (Figure 3), in order to analyze the datasets with higher spatial and temporal resolutions.

Various geohazard-mapping methodologies were previously implemented by the use of A-DInSAR techniques [25–30]. The problem of detection of land subsidence using A-DInSAR data is compounded by the large-scale analysis of an enormous amount of measuring points. Often, the strategies to detect and to map the land subsidence areas are based on the spatial distribution of the average velocities and on statistic approaches to automatically detect clusters with significant movements. Some difficulties were observed to distinguish ground deformation due to different processes, or in detecting shallow deformations caused by seasonal processes.

In this work, the procedure proposed by [31] is exploited for the identification of land subsidence areas and to disentangle the contribution of different processes to the spatial and temporal distribution of displacement estimated through multisensor A-DInSAR time series. The first step of the performed activity was the decomposition of the vertical and E-W components of motion and the displacement time series (TS) accuracy assessment. Then, different statistical tests, such as the principal component analysis (PCA) and an automated time series classification, were applied. Thereafter, areas with significant movement, so-called "ground motion areas", were mapped. These areas correspond to a cluster of a minimum 3 of PS, with a maximum distance of 50 m, characterized by the same trends (i.e., linear, non-linear, and seasonal). Furthermore, the methodology to detect ground motion areas using the average velocity [25] was implemented in the same area using ERS-1/2 and Radarsat data to assess the results.

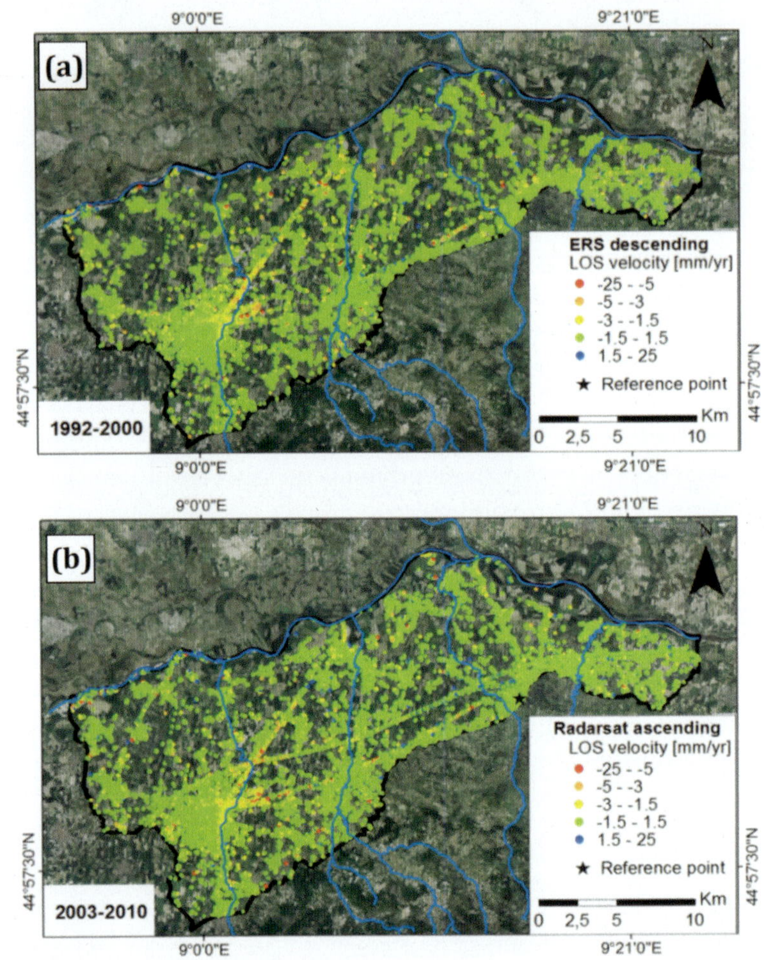

Figure 3. (**a**) Line of sight (LOS) velocity measured by the use of ERS-1/2 descending data (time interval from 1992 to 2000) and (**b**) RADARSAT ascending data (time interval from 2003 to 2010), modified from [31]. The reference points are also reported.

2.2. The Alto Guadalentín Test Case

The Guadalentín Basin is located in Southeast Spain, and has an extension of 277 km^2 (Figure 4). The area was chosen because the highest subsidence rates measured in Europe (>10 cm/year) have been recorded here, as a direct consequence of long-term aquifer exploitation. Urban sites are located in the area, but agriculture is the prevalent land use.

The Guadalentín Basin is underlain by Neogene-Quaternary sediments transported by the Guadalentín River along an intramontane depression located in the eastern part of the Baetic Cordillera, which is an ENE-WSW-oriented alpine orogenic belt resulting from the ongoing convergence of the African and Iberian plates [32]. The Guadalentín is a tributary river of the Segura River which geographically divides the Bajo from the Alto Guadalentín sub-basins, where Lorca City is located (Figure 4). The basin is mainly composed of Quaternary alluvial fan systems that overlap with Tertiary deposits. The Tertiary deposits are composed of conglomerate and calcarenite that outcrop at the border of the basin. The main active fault system of the study area, the NE-SW-oriented Alhama de Murcia Fault (AMF) [33] is represented in Figure 4. The deposits of the basin overlap the Paleozoic metamorphic complexes [34].

The Alto Guadalentín aquifer system is characterized by Plio-Quaternary detrital and alluvial material, including clays, sands, and conglomerates with clay and/or silt matrices; Miocene detritical with conglomerate and sand deposits; and local Triassic carbonate rocks (Figure 4). The lower impermeable limit of the aquifer is composed of Mesozoic marl, and marl with intercalated sand and conglomerates. The aquifer geometry is controlled by the horst and graben structures of the impermeable limit [35]. Since 1960, agricultural advance has led to the exploitation of the aquifer system, which resulted in the aquifer being declared temporarily overexploited in 1987 [36]. Historically, the piezometric level was close to the surface and, as artesian wells were exploited, groundwater drawdown became apparent in 1972 [35].

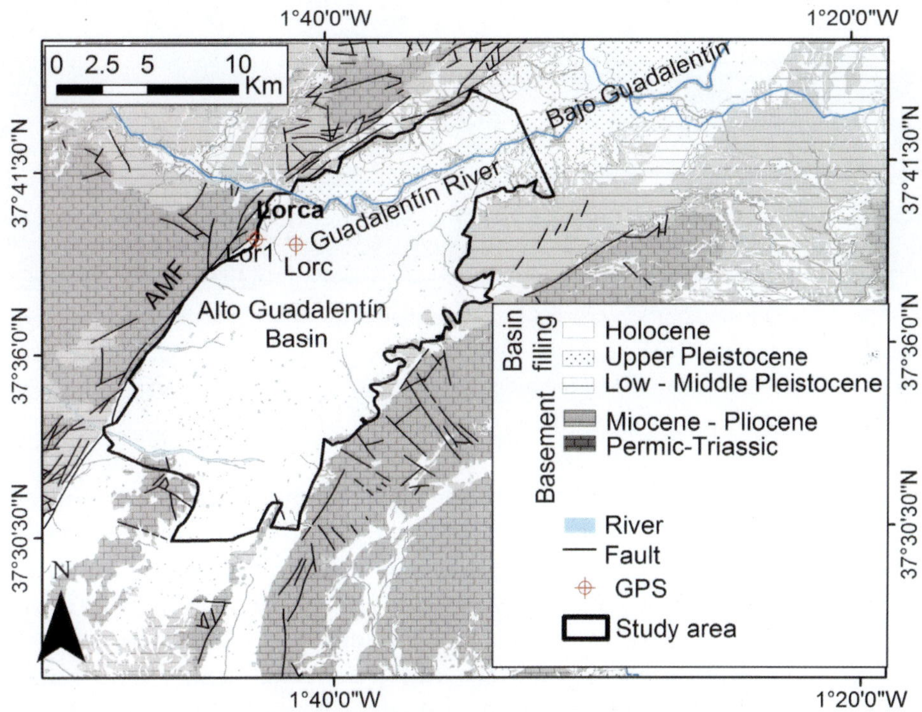

Figure 4. Geological setting of the Alto Guadalentín Basin modified from [37]. The localization of the Global Positioning System (GPS) stations, called Lor1 and Lorc, is also reported.

The A-DInSAR datasets available for the Alto Guadalentín are C-band (ERS-1, ERS-2, and ENVISAT), L-band (ALOS PALSAR), and X-band SAR data (COSMO-SkyMed). ERS-1/2 and ENVISAT scenes were acquired in the descending mode, covering the time intervals from 22 June 1992, to 21 December 2000, and from 15 March 2003 to 15 March 2007. ALOS PALSAR and COSMO-SkyMed scenes were acquired in the ascending mode, covering the time intervals from 19 January 2007 to 14 June 2010, and from 17 May 2011 to 14 October 2012. The SAR images acquired by ERS-1, ERS-2, and ENVISAT sensors were processed using the small baseline approach [13,38]. For each dataset, we first used Delft Object-oriented Radar Interferometric Software (DORIS) [11] to produce the interferograms and then we performed the time series analysis using Stanford Method for Persistent Scatterers

(StaMPS) [15]. Finally, results from both datasets were merged to determine the temporal evolution of displacement over the complete ERS-ENVISAT period. ALOS PALSAR and COSMO-SkyMed dataset were processed using SPN software [12,14]. The use of the A-DInSAR dataset acquired by different sensors using different incidence angles clearly affects the capability to measure the vertical component of the displacement. More precisely, ERS-1/2 and ENVISAT have a 23° incidence angle, which allows the estimation of 92% of vertical displacements, while ALOS and COSMO-SkyMed satellites only detect 83% and 75% of vertical displacements, respectively. As a result, the identification of these geometrical distortions due to the different acquisition angles of the various sensors is essential to perform A-DInSAR analyses. Consequently, LOS velocities have been projected along the vertical direction for each dataset in order to homogenize the datasets (Figure 5). Furthermore, local comparisons were performed with Global Positioning System (GPS) data available for two continuous stations (see the location in Figure 4) located in the study area, demonstrating the high consistency of the vertical motion measurements between the two different surveying techniques. An average absolute error of 4.6 ± 4 mm for the ALOS data and of 4.8 ± 3.5 mm for the COSMO-SkyMed data confirmed the reliability of the projection along the vertical direction [39].

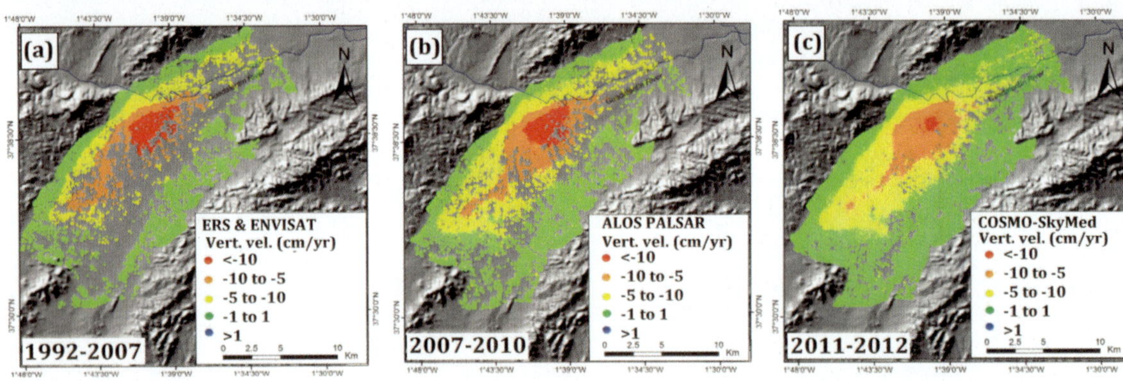

Figure 5. (**a**) Vertical velocity (Vert. vel.) measured by the use of ERS and ENVISAT data (time interval from 1992 to 2007); (**b**) ALOS PALSAR (time interval from 2007 to 2010); and (**c**) COSMO-SkyMed (time interval from 2011 to 2012).

The land subsidence triggered by groundwater exploitation depends on three main factors [4]: the areal and vertical distribution of subsidence-prone materials, their current state of stress, and their stress history. Previous works highlight that the same aquifer system can exhibit different deformation responses, according to the stratigraphic characteristic and the amount of groundwater pumped. Indeed, the complexity of the deformation responses to the applied stress is due to the different composition of the hydrostratigraphic units and the different changing patterns of groundwater levels that the units have experienced [40]. Principal component analysis was performed using the ERS and ENVISAT, ALOS, and COSMO-SkyMed data to analyze the spatio-temporal deformation pattern of each dataset. Then, A-DInSAR and piezometers time series have been compared to give insights about the evolution of the aquifer deformation behavior across the basin.

2.3. The London Basin Test Case

The study area located in the London Basin (Southern England) covers an extension of about 1360 km^2 (Figure 6). This test site was chosen to model land subsidence due to groundwater level change by applying a 1D model, thanks to the large availability of geological, hydrogeological, and geotechnical data. The London Basin is a densely-urbanized area, where, London, the capital and largest city in the United Kingdom, and the most populated in Europe, is located. The basin overlies the Palaeozoic basement, which is bounded to the south by the Variscan Front [41]. The basin is underlain by Paleogene deposits that overlie the Chalk Group [42,43]. These deposits comprise the

Thanet Formation, mainly composed of a fine sand; the Lambeth Group, consisting of vertically- and laterally-variable sequences, mainly of clay with silty and sandy horizons; the Harwich Formation, a silty, sandy clay with gravel beds, and; the London Clay Formation, a dense, fissured clay [44,45]. Quaternary deposits, primarily river terrace deposits, are associated with the River Thames and artificially modified ground. The city of London lies within a graben bounded by the Northern Boundary fault to the north and the Wimbledon-Streatham fault and the Greenwich fault to the south.

The Chalk Group is the principal aquifer in the London Basin supporting both public water and industrial groundwater use supply [46,47]. In the central area of the basin, the chalk aquifer is confined by the overlying Palaeogene formations. The chalk aquifer is recharged on the interfluves—in the Chiltern Hills to the north and North Downs to the south, where the Chalk outcrops and the aquifer becomes unconfined. The unconfined aquifer is characterized by larger seasonal water table variations, where fracture lineaments represent the main river drainage systems [48]. The lithological variations and the faulting on the chalk aquifer have a key control on the variability of the aquifer properties across the Thames Basin [49]. Historic overexploitation of the Chalk aquifer up to the 1960s led to widespread lowering of groundwater levels, reaching a depth of up to 90 m below sea-level in the central London Basin [46]. Thus, the Chalk aquifer became unconfined, leading to under-drainage and desaturation of the London Clay [41,46]. In the mid-1960s, Chalk groundwater level rapidly recovered at a rate of up to 3 m/year [46]. Therefore, an action plan was developed by the London Underground, Thames Water and the Environment Agency (EA) (i.e., the GARDIT, General Aquifer Research Development and Investigation Team) strategy to control groundwater recovery and re-saturation of the London Clay, that could potentially have negative impacts on the foundations of structures and infrastructure in the basin. Thus, the EA implemented an observation borehole network within the basin in order to monitor and manage changes in groundwater levels. Furthermore, an artificial recharge scheme was licensed in North London, NLARS (North London Artificial Recharge Scheme; [46]) to support the groundwater levels control [49,50].

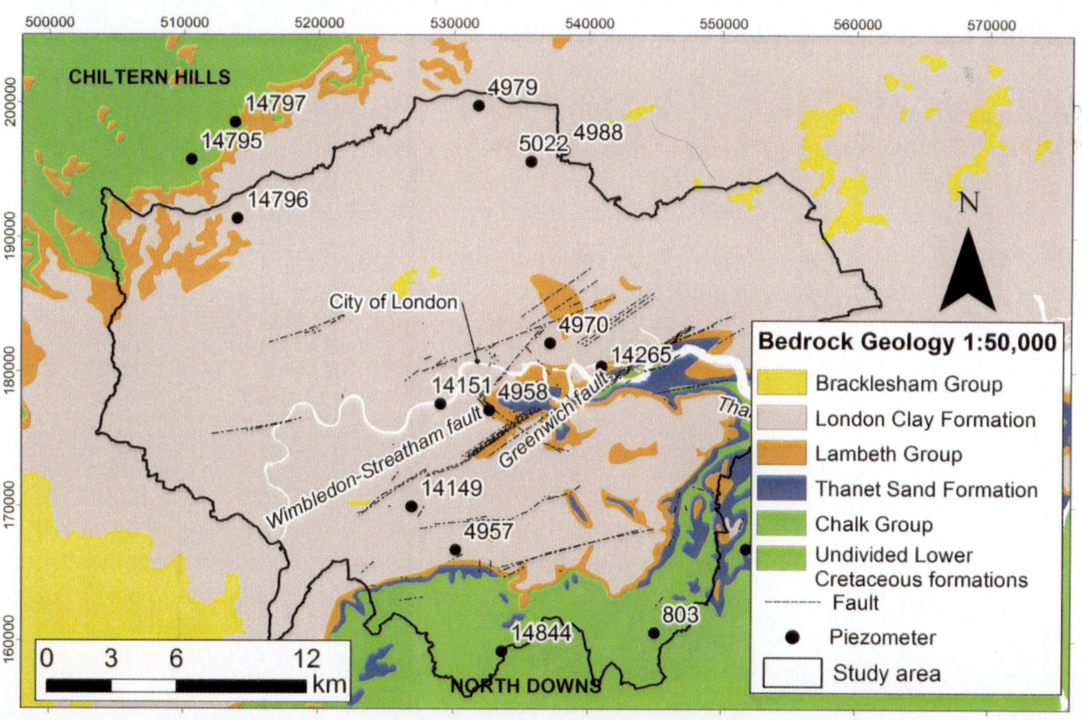

Figure 6. Geological setting of London Basin, modified from [51]. Geological materials © NERC 2016. All rights reserved. British National Grid. Projection: Transverse Mercator. Datum: OSGB 1936. The localization of the piezometers is also reported.

The A-DInSAR datasets available for the London Basin are ERS-1 and ERS-2 SAR scenes acquired in ascending and ENVISAT scenes acquired in descending mode. The first dataset covers the time interval from 19 June 1992 to 31 July 2000 and the second one from 13 December 2002 to 17 September 2010. The available scenes were processed using the GAMMA SAR and Interferometry software (Gamma Remote Sensing, Bern, Switzerland) and, in particular, the Interferometric Point Target Analysis (IPTA) algorithm [52]. Considering the horizontal movements negligible in the study area, LOS displacements and velocities were projected for both datasets along the vertical direction (Figure 7). Ninety-five percent of the PS targets within the study area show an uncertainty of the average LOS velocity in the range between 0.09 and 1.09 mm/year in the ERS-1/2 dataset, and between 0.17 and 1.13 mm/year in the ENVISAT dataset [53].

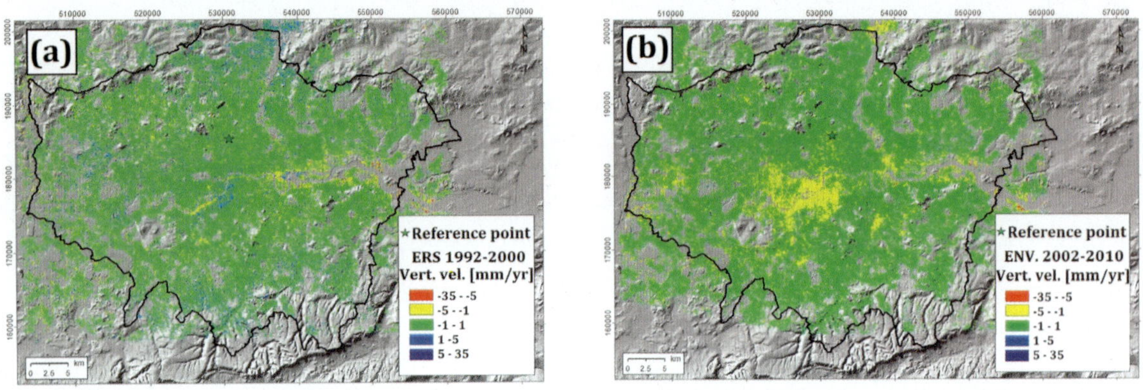

Figure 7. (**a**) Vertical velocity (Vert. vel.) measured by the use of ERS (time interval from 1992 to 2000) and (**b**) ENVISAT (ENV.) data (time interval from 2002 to 2010), overlapped onto shaded relief of NEXTMap® DTM at 50 m resolution; modified from [53]. Reference points for each dataset are also reported. ERS-1/2 and ENVISAT PSI data © CGG NPA Satellite Mapping. NEXTMap® Britain © 2003, Intermap Technologies Inc., All rights reserved.

Principal component analysis was also performed across the London Basin, using ERS-1/2 and ENVISAT data, to identify spatio-temporal deformation patterns.

Furthermore, A-DInSAR time series were also exploited to simulate the ground motion response due to the changes in the groundwater level. Many authors have exploited A-DInSAR data for the calibration and validation of models in which the land subsidence is due to groundwater level variations [54–57]. In this study, A-DInSAR time series were integrated in a one-dimensional model (1-D) to simulate the ground motion in response to changes in hydraulic head in the Chalk aquifer. The 1-D model is based on the inversion of the equation introduced by [58] to calculate the storage coefficient (S):

$$S = \Delta d / \Delta h \tag{1}$$

where Δd is the vertical displacement as estimated by A-DInSAR time series Δh. The assumption of one-dimensional consolidation is motivated by the hypothesis that the ground deformation is only vertical for this area, justified by the fact that the horizontal displacements are believed to not be significant [51]. The 1-D model assumes that the aquifer pore pressure instantaneously equilibrates with piezometric level changes in the aquifer and any time-lag between the piezometer level variations and the compaction of the geological layers is not accounted for. This assumption is justified by the inspection of the ground motion and groundwater records across the basin [51]. The simulation of the ground motion was conducted by using the following equation [59,60]:

$$\Delta d = S \times \Delta h \tag{2}$$

The simulation was performed at nine piezometers representative of semi-confined and confined aquifer condition using the average displacement time series within the buffer areas with a radius of 500 m from each piezometer. In the first step, the geological sequence was compared with the groundwater level variation to classify the aquifer state according to its confined condition (i.e., confined, semi-confined, unconfined) and the geological interval over which the piezometric head varied. Then, the aquifer storage coefficient (S) was calculated by using Equation (1) in the time interval where a good fit between piezometer data and average displacement time series was evident.

Finally, A-DInSAR time series were compared with the simulated ground motion in order to estimate the absolute average error of the model. Furthermore, the capability to use the A-DInSAR time series as the input of the model was assessed in order to simulate the hydraulic head changes. The inverted model was performed at piezometer 4988 (see the location in Figure 6) using the following equation:

$$\Delta h = \Delta d / S \quad (3)$$

3. Results

3.1. Time Series Support for Detection of Land Subsidence: The Oltrepo Pavese Test Case

The outcomes of the A-DInSAR time series analysis acquired by ERS-1/2 and RADARSAT-1 sensors show that the Oltrepo Pavese is affected by three deformational behaviors with linear, non-linear, and seasonal trends. The cumulated displacements observed in the 1992–2000 period is higher than those detected in the period 2003–2010, and a decrease of deformation was observed.

The town of Voghera and the railway of Voghera–Pavia (Figure 8) are located in the most affected area by land subsidence with linear movements (principal component 1, PC1), with average LOS velocity in the range from −2 to −3.7 mm/year, in the period 1992–2000, and in the range of −2 to −4.8 mm/year in range of −2 to −4.8 mm/year, in the period 2003–2010. Areas of moderate uplift were observed via principal component 2 (PC2) of the ERS-1/2 and RADARSAT-1 data, in the sector from Stradella to Monteacuto (Figure 8), where geomorphic evidence of an active emergent thrust was previously observed [61]. Ground motion areas characterized by seasonal movements (principal component 3, PC3) are mainly located in the southern sector of Voghera and in proximity to Lungavilla, Codevilla, Casei Gerola, and Broni. The seasonal deformation behaviors in the Oltrepo Pavese are mainly due to seasonal fluctuations of the groundwater level and swelling-shrinkage of clayey soils [31]. Ground motion areas delineated using the PC3 of ERS-1/2 and RADARSAT-1 data in proximity of Codevilla (see location in Figure 8) match with an area where a high density of damaged buildings was recorded [62]. This site is located over Engineering Geological Unit 8, where shallow alluvial deposits of the first 8 m of depth exhibit swelling potentials from medium to very high [62]. The seasonal component of motion detected in the TS is directly correlated with the effective rainfall detected at the nearest pluviometric station, located in Voghera (Figure 8).

Changes in trends in the displacement time series were observed at the end of 1999 and 2008 in the areas of Broni, Voghera, Lungavilla, and Casei Gerola, and, in most of the cases, the non-linear trend are superimposed on seasonal ground motion areas identified using PC3.

Even though the average velocity may be useful to detect physical processes characterized by linear trends, the same parameter seems not to be efficient in detecting ground motion areas affected by non-linear and seasonal movements. The challenge of exploitation of the time series to detect the ground motion areas is to take into account different deformational behaviors. Hence, the reliability of the results clearly depends on the quality of the time series. Consequently, post-processing checks of the time series are fundamental to remove problems due to phase unwrapping, regional unreal trends, and anomalous displacement detected on certain dates [31]. Anomalous displacements were recorded on 9 March 1997 and 16 July 2000 by the ERS-1/2 descending datasets, and on 10 December 2008, for the RADARSAT-1 ascending dataset in the Oltrepo Pavese. The anomalous displacement identified in the RADARSAT-1 dataset might be related to the snowfall that occurred on the day of the SAR acquisition. Thus, these SAR scenes were not included in the following analyses.

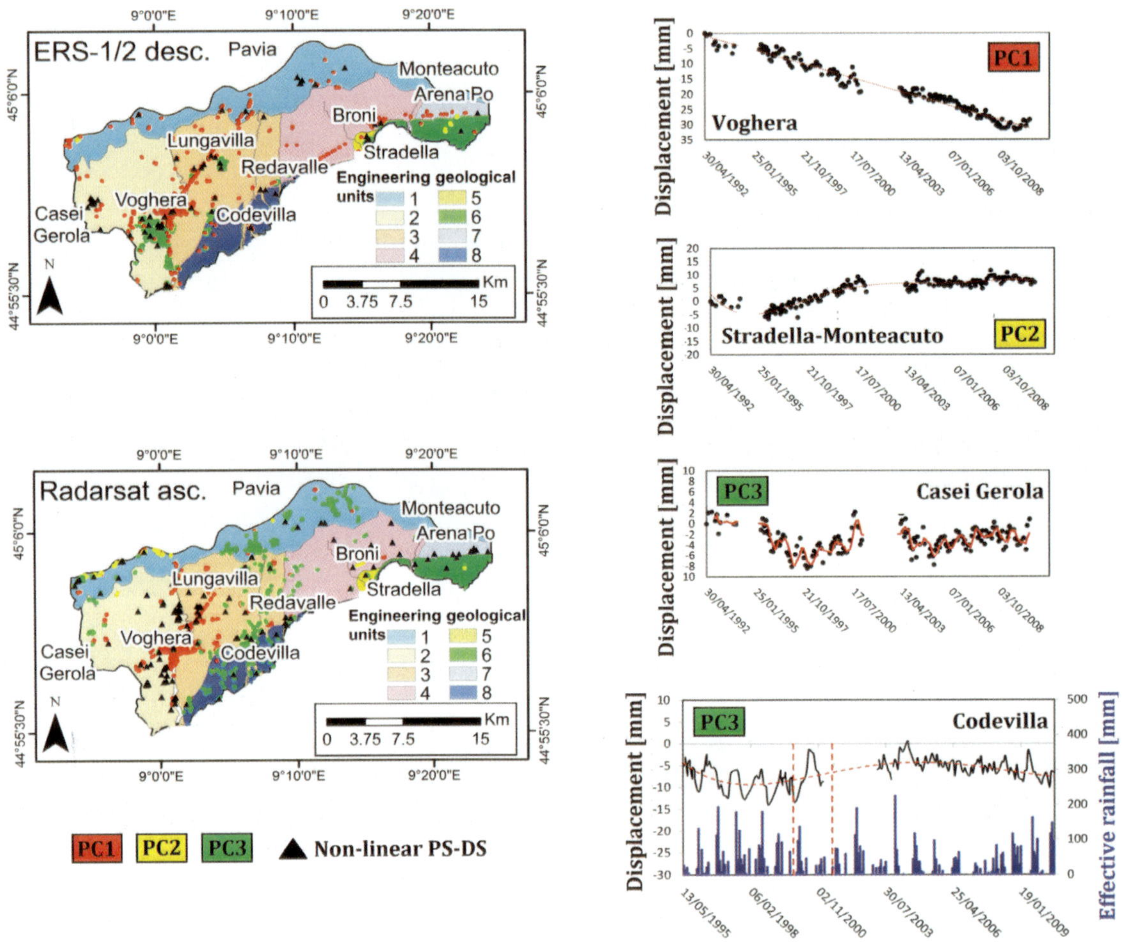

Figure 8. Ground motion areas detected using PC1, PC2, and PC3, and non-linear PS-DS of the ERS-1/2 descending and Radarsat ascending data overlapped on the engineering geological units. Time series (TS) of the deformational behavior detected at Voghera, Stradella–Monteacuto, Casei Gerola, and Codevilla are also reported. For Codevilla, TS were compared with effective rainfall measured at the Voghera weather station.

Furthermore, the same A-DInSAR datasets were analyzed using the methodology based on the average velocity [25] in order to validate the areas detected using the PC1 and to assess the results obtained using these different approaches. The distribution of the so-called "anomalous areas" is represented in Figure 9. Eighty-four percent of the anomalous areas detected in the period from 1992 to 2000 and from 2003 to 2010 coincide with ground motion areas detected using the PC1 of the ERS-1/2 and Radarsat data, respectively. Three and seven percent, respectively, of the anomalous areas detected in the period from 1992 to 2000 and from 2003 to 2010 coincide with ground motion areas detected via PC2. Three percent of the anomalous areas corresponds to ground motion areas detected via PC3 using ERS-1/2 and Rasarsat data. Therefore, 10% and 6% of the anomalous areas are not detected using PCA-based methodology. However, the extension of these areas reaches values of 0.2–0.3 km^2, resulting in localized movements. Otherwise, the approach based on the average velocity allows the detection of 35% and 69% of the ground motion areas detected via PC1 of the ERS-1/2 and Radarsat data. Seventy-two and sixty-seven percent, respectively, of the ground motion areas detected via PC2 can be identified using the average velocity of ERS and Radarsat data. Finally, 51% and 11% of the ground motion areas detected via PC3 of the ERS and Radarsat data, respectively, can be mapped using the average velocity. Overall, the results demonstrate that, when using the PCA-based methodology, most of anomalous areas based on the average velocity can be detected and, as expected, ground motion areas mainly due to seasonal movements can be detected using the time series information.

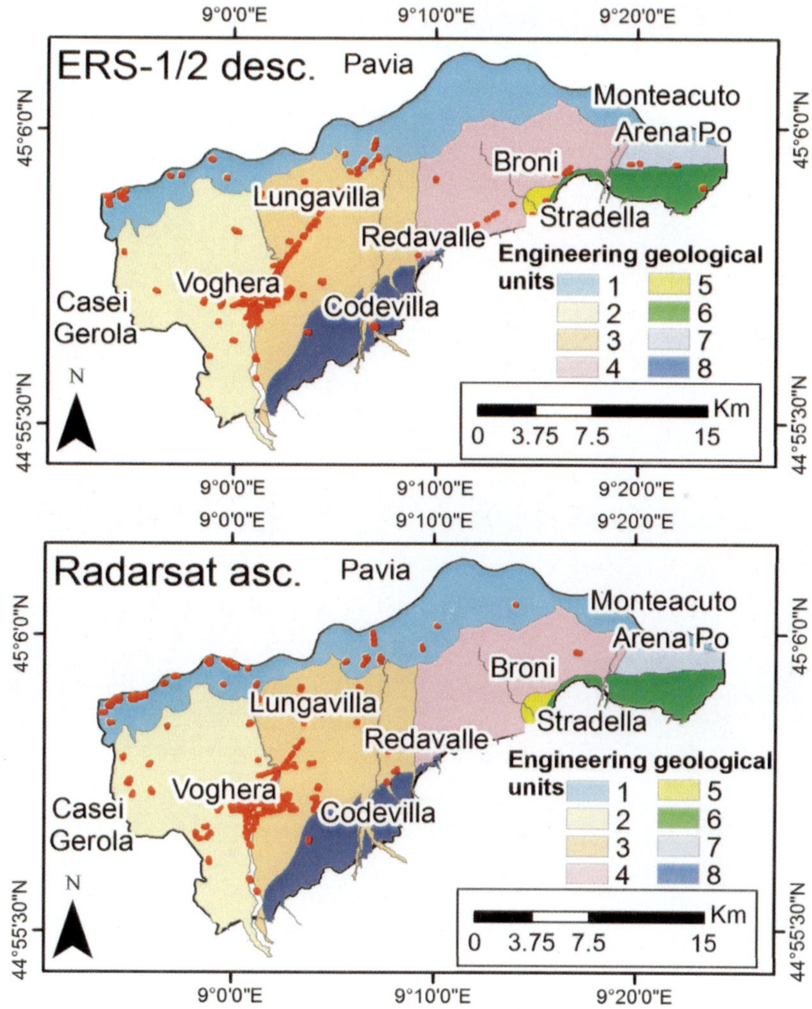

Figure 9. Anomalous areas detected using the average velocity [25] overlapped on the engineering geological units.

3.2. Time Series Support for the Characterization of Land Subsidence Mechanisms: The Alto Guadalentín Test Case

To better understand the land subsidence in the Alto Guadalentín Basin, the time series of multi-temporal A-DInSAR data were analyzed. First, principal component analysis (PCA) was performed for each dataset. Figure 10 shows that the different datasets are characterized by only one principal component (PC1) that explains the percentage of variance higher than 99%. The spatial distribution of principal component (PC) scores, which gives insight into the correlation between the measuring point and the trend of PC1, shows that the area affected by this kind of trend remains the same over the whole monitored period. The PC1 trend is well represented by the eigenvector value, and the land subsidence trend is evident for each dataset.

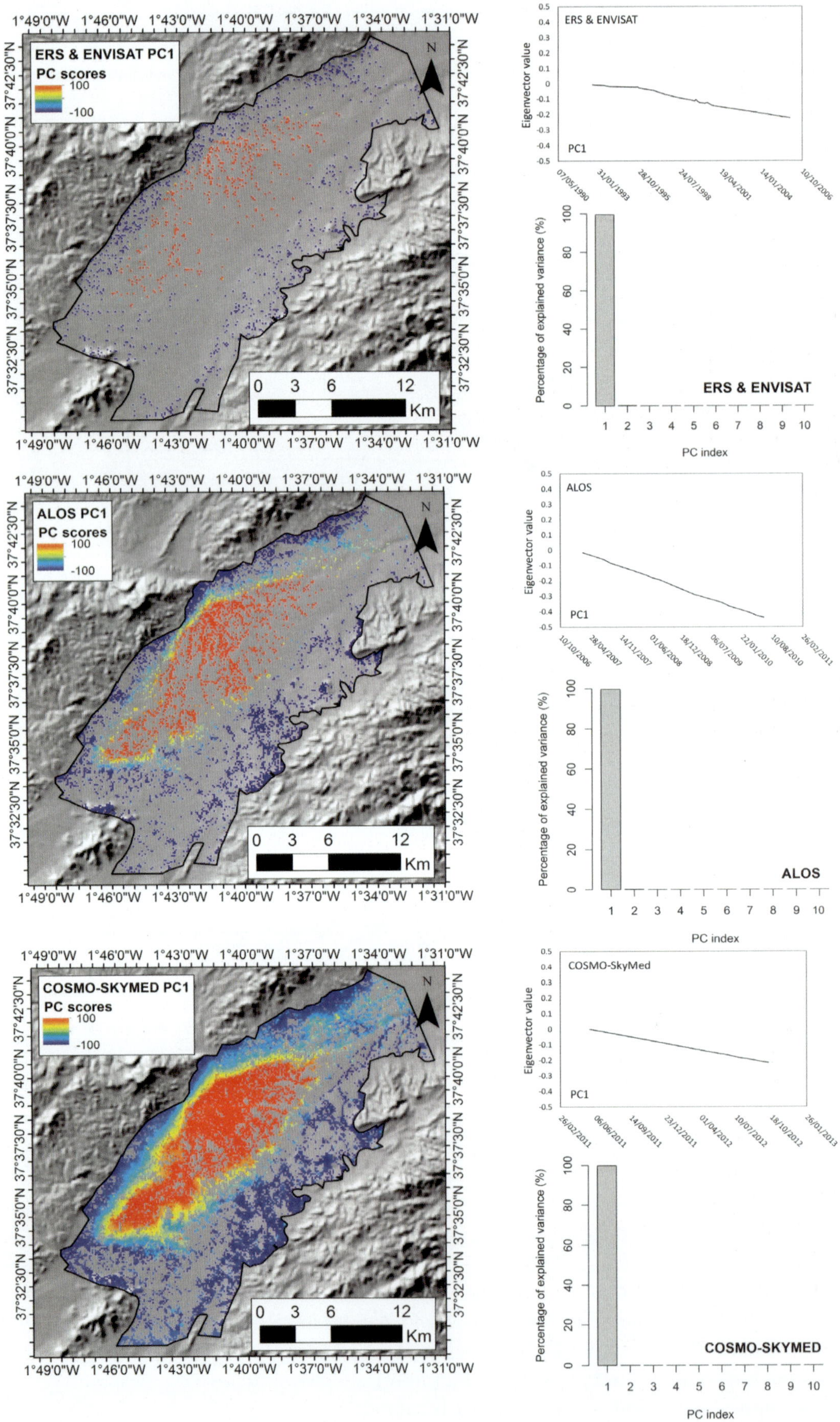

Figure 10. Principal component (PC) score maps of ERS and ENVISAT, ALOS, and COSMO-SkyMed data. For each dataset, the eigenvector value and the percentage of explained variance are also reported.

Furthermore, the cumulative displacement obtained by the use of the ERS-ENVISAT, ALOS PALSAR, and COSMO-SkyMed sensors in the period from 1992 to 2012 were computed (Figure 11). The maximum cumulative land subsidence reached up to 250 cm, with an average subsidence of 180 cm over an extension of 14.8 km^2, located in the central sector of the Alto Guadalentín Basin (Figure 11). The areal extent of the subsidence exhibits a SW–NE elliptical shape parallel to the valley direction, showing an apparent deceleration trend over time. Cross-comparisons were performed between the groundwater level variations, the lithological data, and the displacement time series for five representative sites. Close to the piezometer P1, located in the southern zone, in which the highest groundwater drawdown (162 m) from 1992 to 2007 was measured, the cumulated land subsidence reached 13 cm in the period 1992–2012 (Figure 11). This evidence is linked to the absence of soft soils in the southern border of the basin [39]. In the western part of the basin, where the piezometers P2, P4a, and P4b are located, the soft soil thickness is moderate (from 18 to 21 m). Groundwater drawdowns between 31 and 107 m in the period 1992–2007 correspond to cumulative subsidence in the range of 38–48 cm during the period 1992–2012. Piezometer P2 gives insight into a steady lowering of the piezometric level from 1989 to 2013 (Figure 11) that is correlated with the steady subsidence rate. Otherwise, piezometers P4a–b, which are jointly analyzed due to their proximity and characteristics, exhibit a high groundwater level variability that is not in agreement with the estimated subsidence linear rate. In the eastern part of the basin, piezometers P3a–b display the piezometric level variations are in agreement with subsidence accelerations and decelerations. Thus, cross-comparisons between A-DInSAR time series and piezometric measurements have allowed the recognition of transient aquifer compaction due to piezometric lowering in the period 1992–2007 (Figure 11), which was followed by decelerations of the subsidence when piezometric recovery was observed until 2012. In the central part of the basin where the highest thicknesses of soft soils are found (approximately 190 m), the time series of piezometer P5 reveals a groundwater level recovery from 2007 to 2013 that does not correspond with measured subsidence steady rates (Figure 11). An inelastic, unrecoverable, and permanent deformation of this aquifer was observed, as previously reported [39]. This land subsidence mechanism was triggered by the groundwater exploitation beginning in the early 1970s.

It is worth noting that the A-DInSAR time series reported in Figure 11 were generated considering a linear displacement velocity in the temporal gaps (11 months) between the ALOS PALSAR and COSMO-SkyMed datasets.

3.3. Time Series Support for Modelling of Land Subsidence: The London Basin Test Case

Principal component analysis (PCA) was performed using ERS-1/2 and ENVISAT data, in order to find the spatio-temporal deformation pattern across the London Basin. Figure 12 shows that both datasets are characterized by two principal components of ground motion. The first component represents land subsidence and the second one is an uplift trend. ERS-1/2 dataset shows that 91.6% of the variance is explained by the first component of motion and 4.4% by the second one. The variance is explained be the first and second components of motion (82.5% and 5.4%, respectively) using the ENVISAT data. Such comparisons were performed between the distribution of PC1 and PC2 with the areas of observed natural and anthropogenic geohazards mapped using the PanGeo procedure [26]. These areas were identified combining the same A-DInSAR datasets with geological data [53]. The distribution of PC1 detected using ERS-1/2 and ENVISAT data matches with natural ground movements, such as those due to compressible ground, and anthropogenic ground instability, such as that due to underground construction. The distribution of the PC2 detected using ERS-1/2 and ENVISAT data corresponds to areas where groundwater rise was previously observed [53] and areas in which causes of uplift are unknown [53], whereas geohazard areas due to tectonic uplift movements characterized by low confidence levels [53] do not show evident correlation with the PC2 trend.

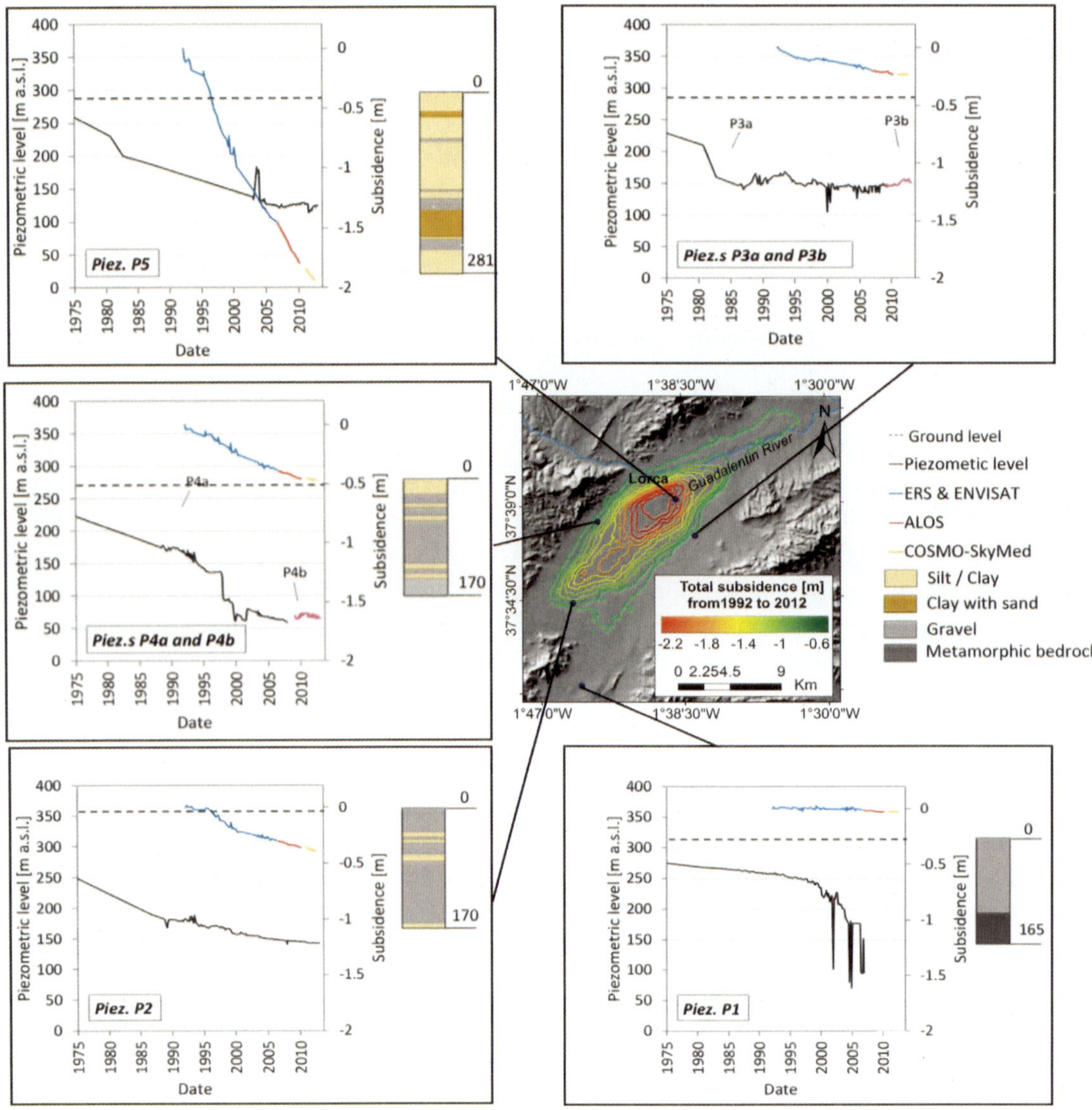

Figure 11. Cumulated displacement from 1992 to 2012 and cross-comparison of the piezometric level variations from 1975 to 2012 (black and pink lines) at the different piezometers (P1, P2, P3a, P3b, P4a, P4b, and P5) with the displacement time series detected by the satellite sensors (blue, red, and yellow lines). For some piezometers the lithological column is reported. The dotted line is the ground level. Piez. stands for piezometer.

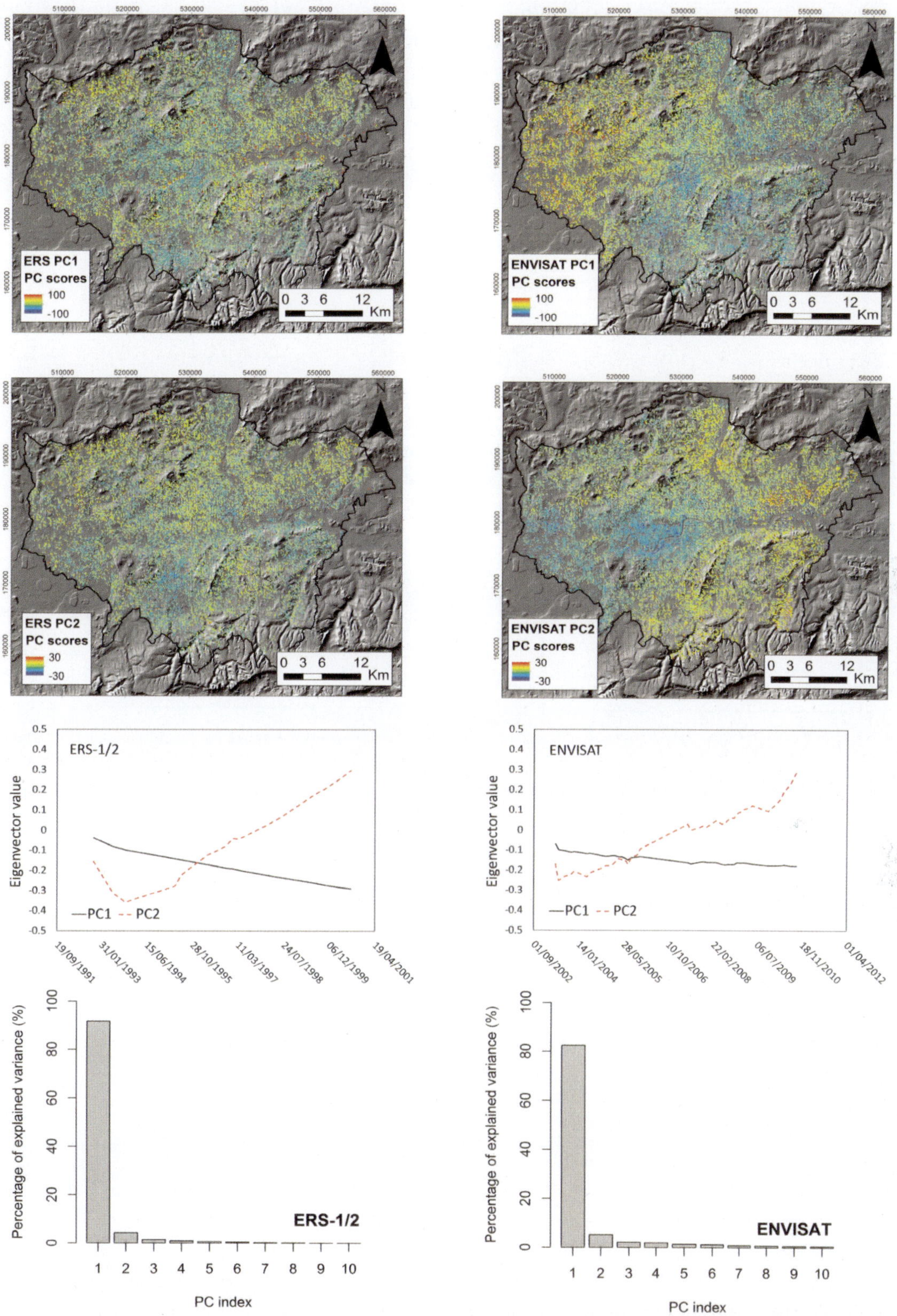

Figure 12. Principal component (PC) score maps of the ERS-1/2 and ENVISAT data using the first and the second component of the movements (PC1, PC2). For each dataset, the eigenvector value and the percentage of the explained variance are also reported.

A-DInSAR time series were used to simulate the acceleration and deceleration of the displacement in response to the changes in the piezometric level across the London Basin. Figure 13 shows the simulated displacements at four piezometers in order to represent the main responses to hydraulic head changes. These are obtained by using the groundwater level data and the estimated storage coefficients as inputs to Equation (2).

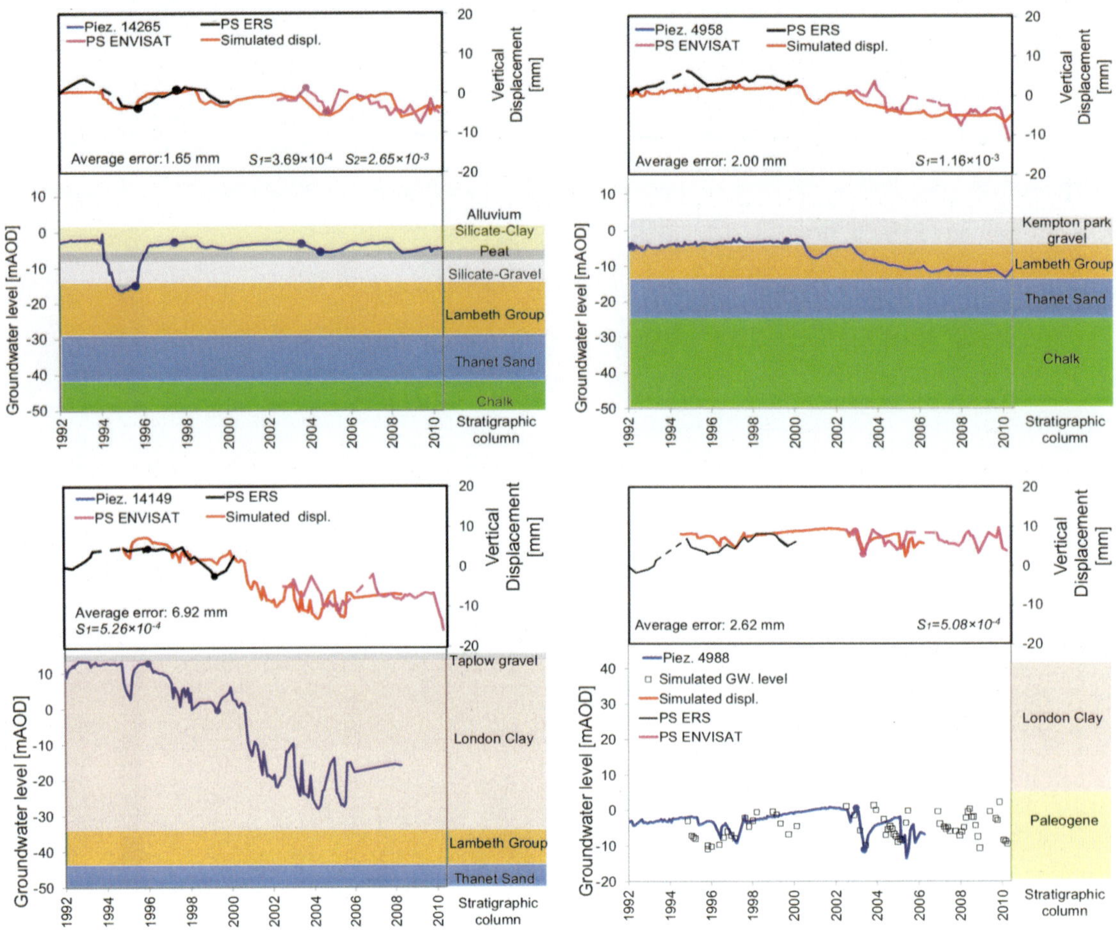

Figure 13. Comparison of A-DInSAR time series for 1992–2000 (PS ERS) and 2002–2010 (PS ENVISAT) and simulated displacements (mm) with the groundwater level variations (m) (modified from [51]). In addition, the stratigraphic column is represented. The dots represent the start and the end of the calibration period for the simulation. Piezometer localization is shown in Figure 6. For piezometer 4988, the simulated groundwater (GW) level is also reported.

In the North London Basin, when the Chalk is confined by the London Clay, the hydraulic head change of 11.50 m coincides with approximately 5.85 mm of surface displacement (Figure 13; piezometer 4988). In the Central London Basin, temporary transition of the water table across the units was detected (Figure 13; piezometer 14265). In this regard, two different storage coefficients were evaluated for the different aquifer conditions in the periods 01 February 1996 to 12 December 1997 and 14 January 2004 to 13 December 2004 for the confined and semi-confined conditions, respectively. The semi-confined conditions in the Lambeth group were observed at piezometer 4958. The hydraulic head change of 2.45 m coincides with approximately 2.85 mm of surface displacement. In the South London Basin, the aquifer is confined by thicker deposits of the London Clay and lower values of the storage coefficient were observed. A hydraulic head change of 13.16 m produces around 6.92 mm of surface displacement (Figure 13; piezometer 14149). The absolute average difference error between the simulated and A-DInSAR time series was also estimated.

The use of A-DInSAR time series for the modelling has allowed finding a correlation between the derived storage coefficient and the aquifer condition. Indeed, where the Chalk is confined by the London Clay the storage coefficient is of the order of 1×10^{-4}, whilst where overlain by the Lambeth Group and semi-confined conditions are expected to exist, the storage is higher, typically 1×10^{-3}, revealing the additional storage provided by sand-rich horizons in the Lambeth Group.

It is noteworthy how another benefit provided by the use of the simulated displacements is the filling of the temporal gap (approximately two years) between the ERS and ENVISAT data. The ENVISAT ground motion time series in Figure 13 were indeed adjusted for the position of the ground on 13 December 2002 to match with the modelled ground motion time series based on the ERS data. A vertical displacement of −5 mm was estimated for piezometer 14149 and −2 mm for piezometer 14265.

Furthermore, Equation (3) was applied in order to simulate the hydraulic head changes based on the observed ground motion at piezometer 4988 (Figure 13). The results show that the A-DInSAR time series could be used to simulate groundwater level changes following this innovative approach. It is worth noting that the relative error between the simulated hydraulic head changes and the measured groundwater level variations using time series acquired by the ERS and ENVISAT sensors is 25% of the hydraulic head change. The frequency of measurements between the displacement and piezometric time series is about 4–6 measurements per year versus 12 measurements per year. Therefore, the difference in the temporal sampling of these measurements could be influencing the error of the model.

4. Conclusions

This work aimed at exploiting the support of A-DInSAR time series for different facets of land subsidence mechanism assessment, starting from detection, up to the characterization and 1-D modelling. The main advantages and limitations of the A-DInSAR time series from an end-user point of view are discussed through three well-studied test sites: the Oltrepo Pavese (Po Plain, in Italy), the Alto Guadalentín (Spain), and the London Basin (UK). In particular, time series analysis was carried out in the Oltrepo Pavese for the detection of the ground motion areas. The time series support for the characterization and the modelling of land subsidence mechanisms was assessed, respectively, in the Alto Guadalentín Basin and in the London Basin.

Although various methodologies were implemented to map the ground motion areas using the average velocity detected by A-DInSAR data [25–30], the use of the time series for the detection of ground motion areas is still not a common practice in the scientific community. The main advantage of A-DInSAR time series for the detection of the ground motion areas is the unprecedented mapping of different deformational behaviors, such as in the case of the Oltrepo Pavese. By exploiting the A-DInSAR time series, the distribution of the areas affected by linear, non-linear, and seasonal movements was assessed. Furthermore, two approaches were applied to detect ground motion areas in order to give insight into the obtained results using the time series and the average velocity. Long-term movements detected using principal component analysis (PCA) are consistent with the results obtained using the average velocity. Otherwise, additional seasonal trends are observed using only PCA.

For the test site of the Alto Guadalentín Basin, PCA analysis was performed using the ERS and ENVISAT, ALOS, and COSMO-SkyMed data. The different datasets show that the Alto Guadalentín Basin is characterized by one component of motion that displays a land subsidence trend. Furthermore, the integrated analysis of A-DInSAR time series acquired by multi-temporal sensors allowed the characterization of the long-term subsidence. The collection of A-DInSAR time series acquired by multi-temporal sensors provided valuable information about the typology of aquifer deformation. Indeed, the use in conjunction of multi-sensor A-DInSAR data prove to be fundamental for the reconstruction of long-term evolution of land subsidence due to groundwater management, as in the case of the history of the Alto Guadalentín aquifer exploitation, confirming the potentials of the back-monitoring approach. Furthermore, the analysis of time series carried out for the Alto Guadalentín Basin has allowed the detection of transient aquifer deformation and inelastic aquifer deformation, not otherwise detectable by exploiting the average velocity. Further consideration to be taken into account for time series acquired by different sensors is how to fill the temporal gap between the datasets. With regard to this issue, the linear velocity was assumed in the temporal gaps (11 months) of the ALOS PALSAR and COSMO-SkyMed datasets. The results are consistent with the measurements obtained by two GPS stations.

Finally, PCA analysis was also performed for the ERS-1/2 and ENVISAT data available for the London Basin. Two components of ground motion were recognized, being related to land subsidence and uplift trends. PCA results are consistent with the areas delineated using the PanGeo procedure. Only geohazard areas due to tectonic uplift movements with low confidence levels show no evident correlation with PC2.

Furthermore, changes, such as acceleration and deceleration in the displacement rates in the London Basin, were simulated by calibrating a 1-D model using A-DInSAR time series. The modelling also has the benefit of filling the temporal gap between ERS and ENVISAT time series (about two years). A-DInSAR time series were also used as input for an inverse model in order to simulate the hydraulic head. This approach could be used to exploit A-DInSAR time series for groundwater level change monitoring in areas where no observation boreholes are available. Thus, we believe that new opportunities and perspectives will emerge for the detection and interpretation of subsiding phenomena using A-DInSAR time series provided by improved acquisition frequency, achievable through new high-resolution SAR satellite sensors, such as COSMO-SkyMed and Sentinel-1 constellations, thanks to the shorter revisiting time.

More widely, the findings of this work confirmed that time series represent an added value for a variety of activities in the framework of land subsidence mitigation and management. These observations have significant implications for similar studies in other geological contexts.

Acknowledgments: Groundwater level data for the London Basin were provided by the Environment Agency (Licence No. A2719; non-commercial use), and the ERS and ENVISAT PS data via the EC FP7 PanGeo project. Francesca Cigna and Stephanie Bricker publish with the permission of the Executive Director of BGS. Part of this work was supported by the Spanish Ministry of Economy and Competitiveness and EU FEDER funds under projects TIN2014-55413- C2-2-P and ESP2013-47780-C2-2-R.

Author Contributions: Roberta Bonì carried out this study and wrote this paper in the framework of her Ph.D. project at the University of Pavia, and during her research visits to the Geological Survey of Spain (IGME) in January–March 2015, and the British Geological Survey (BGS) in June–September 2015. Claudia Meisina provided guidance and support throughout the research process for the three case histories. Francesca Cigna supervised Roberta Bonì during her visit at BGS and revised the manuscript. Davide Notti supported the analysis of the A-DInSAR time series for the Oltrepo Pavese and Alto Guadalentín Basin. Stephanie Bricker provided support for the characterization of hydrogeological properties of the Chalk aquifer. Harry McCormack processed ERS-1/2 and ENVISAT data for the London Basin. Gerardo Herrera supervised Roberta Bonì during her visit to IGME. Roberto Tomás provided support to understand the land subsidence mechanism in the Alto Guadalentín Basin. Marta Béjar-Pizarro contributed analyses of the SAR data for the Alto Guadalentín Basin. Joaquin Mulas supported the analysis of the geological data about the Alto Guadalentín Basin. Pablo Ezquerro supported the analysis of the land subsidence in the Alto Guadalentín Basin. All authors co-wrote and reviewed the manuscript.

Conflicts of Interest: The authors declare no conflict of interest.

References

1. Holzer, T.L.; Galloway, D.L. Impacts of land subsidence caused by withdrawal of underground fluids in the United States. *Rev. Eng. Geol.* **2005**, *16*, 87–99. [CrossRef]
2. Abidin, H.Z.; Andreas, H.; Gumilar, I.; Sidiq, T.P.; Gamal, M. Environmental Impacts of Land Subsidence in Urban Areas of Indonesia. In Proceedings of the FIG Working Week 2015, TS 3—Positioning and Measurement, Sofia, Bulgaria, 17–21 May 2015.
3. Syvitski, J.; Higgins, S. Going under: The world's sinking deltas. *New Sci.* **2012**, *216*, 40–43. [CrossRef]

4. Galloway, D.; Jones, D.R.; Ingebritsen, S.E. *Land Subsidence in the United States*; US Geological Survey: Reston, VA, USA, 1999; p. 177.
5. Cigna, F.; Osmanoğlu, B.; Cabral-Cano, E.; Dixon, T.H.; Ávila-Olivera, J.A.; Garduño-Monroy, V.H.; DeMets, C.; Wdowinski, S. Monitoring land subsidence and its induced geological hazard with Synthetic Aperture Radar Interferometry: A case study in Morelia, Mexico. *Remote Sens. Environ.* **2012**, *117*, 146–161. [CrossRef]
6. Bozzano, F.; Esposito, C.; Franchi, S.; Mazzanti, P.; Perissin, D.; Rocca, A.; Romano, E. Understanding the subsidence process of a quaternary plain by combining geological and hydrogeological modelling with satellite InSAR data: The Acque Albule Plain case study. *Remote Sens. Environ.* **2015**, *168*, 219–238. [CrossRef]
7. Jones, C.E.; An, K.; Blom, R.G.; Kent, J.D.; Ivins, E.R.; Bekaert, D. Anthropogenic and geologic influences on subsidence in the vicinity of New Orleans, Louisiana. *J. Geophys. Res. Solid Earth* **2016**, *121*, 3867–3887. [CrossRef]
8. Herrera, G.; Fernández, J.A.; Tomás, R.; Cooksley, G.; Mulas, J. Advanced interpretation of subsidence in Murcia (SE Spain) using A-DInSAR data-modelling and validation. *Nat. Hazards Earth Syst. Sci.* **2009**, *9*, 647. [CrossRef]
9. Crosetto, M.; Monserrat, O.; Cuevas-González, M.; Devanthéry, N.; Crippa, B. Persistent scatterer interferometry: A review. *ISPRS J. Photogramm. Remote Sens.* **2016**, *115*, 78–89. [CrossRef]
10. Ferretti, A.; Fumagalli, A.; Novali, F.; Prati, C.; Rocca, F.; Rucci, A. A new algorithm for processing interferometric data-stacks: SqueeSAR. *IEEE Trans. Geosci. Remote Sens.* **2011**, *49*, 3460–3470. [CrossRef]
11. Kampes, B.M.; Hanssen, R.F.; Perski, Z. Radar interferometry with public domain tools. In Proceedings of the FRINGE 2003 Workshop, Frascati, Italy, 1–5 December 2003.
12. Arnaud, A.; Adam, N.; Hanssen, R.; Inglada, J.; Duro, J.; Closa, J.; Eineder, M. ASAR ERS Interferometric phase continuity. In Proceedings of the International Geoscience and Remote Sensing Symposium, Toulouse, France, 21–25 July 2003.
13. Berardino, P.; Fornaro, G.; Lanari, R.; Sansosti, E. A new algorithm for surface deformation monitoring based on small baseline differential SAR interferograms. *IEEE Trans. Geosci. Remote Sens.* **2002**, *40*, 2375–2383. [CrossRef]
14. Duro, J.; Closa, J.; Biescas, E.; Crosetto, M.; Arnaud, A. High Resolution Differential Interferometry using time series of ERS and ENVISAT SAR data. In Proceedings of the 6th Geomatic Week Conference, Barcelona, Spain, February 2005; Available online: https://pdfs.semanticscholar.org/d4bc/b3461ddb06da0704815bb40d815d780c8eb8.pdf (accessed on 10 April 2017).
15. Hooper, A. A multi-temporal InSAR method incorporating both persistent scatterer and small baseline approaches. *Geophys. Res. Lett.* **2008**, *35*, L16302. [CrossRef]
16. US National Research Council. *Mitigating Losses from Land Subsidence in the United States*; National Academy Press: Washington, DC, USA, 1991.
17. González, P.J.; Fernández, J. Drought-driven transient aquifer compaction imaged using multitemporal satellite radar interferometry. *Geology* **2011**, *39*, 551–554. [CrossRef]
18. Meisina, C.; Zucca, F.; Fossati, D.; Ceriani, M.; Allievi, J. Ground deformation monitoring by using the permanent scatterers technique: The example of the Oltrepo Pavese (Lombardia, Italy). *Eng. Geol.* **2006**, *88*, 240–259. [CrossRef]
19. Brambilla, G. Prime considerazioni cronologico-ambientali sulle filliti del Miocene superiore di Portalbera (Pavia-Italia settentrionale). In *Nuove Ricerche Archeologiche in Provincia di Pavia, Proceedings of the Convegno di Casteggio, Casteggio, Italy, 14 October 1990*; Civico Museo Archeologico di Casteggio e dell'Oltrepò Pavese: Casteggio, Italy, 1992; pp. 109–113. (In Italian)
20. Pellegrini, L.; Vercesi, P.L. Considerazioni morfotettoniche sulla zona a sud del Po tra Voghera (PV) e Sarmato (PC). *Atti Tic. Sci. Terra* **1995**, *38*, 95–118. (In Italian)
21. Cavanna, F.; Marchetti, G.; Vercesi, P.L. *Idrogeomorfologia e Insediamenti a Rischio Ambientale. Il Caso Della Pianura Dell'Oltrepò Pavese e del Relativo Margine Collinare*; Ricerche & Risultati, Valorizzazione dei progetti di ricerca 1994/97; Fondazione Lombardia Ambiente; Isabel Litografia: Gessate, Italy, 1998; pp. 14–72. (In Italian)
22. Pilla, G.; Sacchi, E.; Ciancetti, G. Studio idrogeologico, idrochimico ed isotopico delle acque sotterranee del settore di pianura dell'Oltrepò Pavese (Pianura lombarda meridionale). *G. Geol. Appl.* **2007**, *5*, 59–74. (In Italian)

23. Boni, A.; Boni, P.; Peloso, G.F.; Gervasoni, S. Dati sulla neotettonica del foglio di Pavia (59) e di parte dei fogli voghera (71) ed alessandria (70). *CNRPF Geodin. Pubbl.* **1980**, *356*, 1199–1223. (In Italian)
24. Meisina, C. Engineering geological mapping for urban areas of the Oltrepo Pavese plain (Northern Italy). In Proceedings of the 10th Congress of the International Association for Engineering Geology and the Environment (IAEG), Nottingham, UK, 6–10 September 2006.
25. Meisina, C.; Zucca, F.; Notti, D.; Colombo, A.; Cucchi, A.; Savio, G.; Giannico, C.; Bianchi, M. Geological interpretation of PSInSAR data at regional scale. *Sensors* **2008**, *8*, 7469–7492. [CrossRef] [PubMed]
26. Bateson, L.; Cuevas, M.; Crosetto, M.; Cigna, F.; Schijf, M.; Evans, H. PANGEO: Enabling Access to Geological Information in Support of GMES: Deliverable 3.5 Production Manual, version 1; 2012. Available online: http://nora.nerc.ac.uk/19289/ (accessed on 10 April 2017).
27. Bianchini, S.; Cigna, F.; Righini, G.; Proietti, C.; Casagli, N. Landslide hotspot mapping by means of persistent scatterer interferometry. *Environ. Earth Sci.* **2012**, *67*, 1155–1172. [CrossRef]
28. Lu, P.; Casagli, N.; Catani, F.; Tofani, V. Persistent Scatterers Interferometry Hotspot and Cluster Analysis (PSI-HCA) for detection of extremely slow-moving landslides. *Int. J. Remote Sens.* **2012**, *33*, 466–489. [CrossRef]
29. Peduto, D.; Cascini, L.; Arena, L.; Ferlisi, S.; Fornaro, G.; Reale, D. A general framework and related procedures for multiscale analyses of DInSAR data in subsiding urban areas. *ISPRS J. Photogramm. Remote Sens.* **2015**, *105*, 186–210. [CrossRef]
30. Di Martire, D.; Novellino, A.; Ramondini, M.; Calcaterra, D. A-differential synthetic aperture radar interferometry analysis of a Deep Seated Gravitational Slope Deformation occurring at Bisaccia (Italy). *Sci. Total Environ.* **2016**, *550*, 556–573. [CrossRef] [PubMed]
31. Bonì, R.; Pilla, G.; Meisina, C. Methodology for Detection and Interpretation of Land subsidence Areas with the A-DInSAR Time Series Analysis. *Remote Sens.* **2016**, *8*, 686. [CrossRef]
32. Bourgois, J.; Mauffret, A.; Ammar, A.; Demnati, A. Multichannel seismic data imaging of inversion tectonics of the Alboran Ridge (Western Mediterranean Sea). *Geo-Mar. Lett.* **1992**, *12*, 117–122. [CrossRef]
33. Martınez-Dıaz, J.J. Stress field variation related to fault interaction in a reverse oblique-slip fault: The Alhama deMurcia fault, Betic Cordillera, Spain. *Tectonophysics* **2002**, *356*, 291–305. [CrossRef]
34. Instituto Geológico y Minero de España (IGME). *Mapa Geologico de España, 1:50.000, Sheet Lorca (953)*; Servicio de Publicaciones Ministerio de Industria: Madrid, Spain, 1981. (In Spanish)
35. Cerón, J.C.; Pulido-Bosch, A. Groundwater problems resulting from CO_2 pollution and overexploitation in Alto Guadalentín aquifer (Murcia, Spain). *Environ. Geol.* **1996**, *28*, 223–228.
36. Confederación Hidrográfica del Segura (CHS). Plan especial ante situaciones de alerta y eventual sequia en la cuenca del Segura: Confederacion hidrografica del Segura. Technical Report. 1996. (In Spanish). Avaliable online: https://www.chsegura.es/chs/cuenca/sequias/pes/eeapes.html (accessed on 11 April 2017).
37. Martín, V.J.M.; Espinosa, G.J.S.; Pérez, R.A. *Mapa geológico de España: E. 1:50,000*; Servicio de Publicaciones, Ministerio de Industria y Energía, Instituto geológico y minero de España (IGME): Madrid, Spain, 1973.
38. González, P.J.; Fernández, J. Error estimation in multitemporal InSAR deformation time series, with application to Lanzarote, Canary Islands. *J. Geophys. Res.* **2011**, *116*, B10404. [CrossRef]
39. Bonì, R.; Herrera, G.; Meisina, C.; Notti, D.; Béjar-Pizarro, M.; Zucca, F.; González, P.J.; Palano, M.; Tomás, R.; Fernández, J.; et al. Twenty-year advanced DInSAR analysis of severe land subsidence: The Alto Guadalentín Basin (Spain) case study. *Eng. Geol.* **2015**, *198*, 40–52. [CrossRef]
40. Zhang, Y.; Xue, Y.-Q.; Wu, J.-C.; Ye, S.-J.; Wei, Z.-W.; Li, Q.-F.; Yu, J. Characteristics of aquifer system deformation in the Southern Yangtse Delta, China. *Eng. Geol.* **2007**, *90*, 160–173. [CrossRef]
41. Royse, K.R.; de Freitas, M.; Burgess, W.G.; Cosgrove, J.; Ghail, R.C.; Gibbard, P.; King, C.; Lawrence, U.; Mortimore, R.N.; Owen, H.G.; et al. Geology of London, UK. *Proc. Geol. Assoc.* **2012**, *123*, 22–45. [CrossRef]
42. Ford, J.R.; Mathers, S.J.; Royse, K.R.; Aldiss, D.T.; Morgan, D.J.R. Geological 3D modelling: Scientific discovery and enhanced understanding of the subsurface, with examples from the UK. *Z. Dtsch. Ges. Geowiss.* **2010**, *161*, 205–218. [CrossRef]
43. Mathers, S.J.; Burke, H.F.; Terrington, R.L.; Thorpe, S.; Dearden, R.A.; Williamson, J.P.; Ford, J.R. A geological model of London and the Thames Valley, southeast England. *Proc. Geol. Assoc.* **2014**, *125*, 373–382. [CrossRef]
44. Ellison, R.A.; Woods, M.A.; Allen, D.J.; Forster, A.; Pharaoh, T.C.; King, C. *Geology of London: Special Memoir for 1:50,000 Geological Sheets 256 (North London), 257 (Romford), 270 (South London), and 271 (Dartford) (England and Wales)*; British Geological Survey: Nottingham, UK, 2004.

45. Sumbler, M.G. *British Regional Geology: London and the Thames Valley*, 4th ed.; HMSO for the British Geological Survey: London, UK, 1996.
46. Jones, M.A.; Hughes, A.G.; Jackson, C.R.; Van Wonderen, J.J. Groundwater resource modelling for public water supply management in London. *Geol. Soc. Lond. Spec. Publ.* **2012**, *364*, 99–111. [CrossRef]
47. Fry, V.A. Lessons from London: Regulation of open-loop ground source heat pumps in central London. *Q. J. Eng. Geol. Hydrogeol.* **2009**, *42*, 325–334. [CrossRef]
48. Bloomfield, J.P.; Bricker, S.H.; Newell, A.J. Some relationships between lithology, basin form and hydrology: A case study from the Thames basin, UK. *Hydrol. Process.* **2011**, *25*, 2518–2530. [CrossRef]
49. O'Shea, M.J.; Sage, R. Aquifer recharge: An operational drought-management strategy in North London. *Water Environ. J.* **1999**, *13*, 400–405. [CrossRef]
50. O'Shea, M.J.; Baxter, K.M.; Charalambous, A.N. The hydrogeology of the Enfield-Haringey artificial recharge scheme, north London. *Q. J. Eng. Geol. Hydrogeol.* **1995**, *28* (Suppl. 2), S115–S129. [CrossRef]
51. Bonì, R.; Cigna, F.; Bricker, S.; Meisina, C.; McCormack, H. Characterisation of hydraulic head changes and aquifer properties in the London Basin using Persistent Scatterer Interferometry land subsidence data. *J. Hydrol.* **2016**, *540*, 835–849. [CrossRef]
52. Werner, C.; Wegmüller, U.; Wiesmann, A.; Strozzi, T. Interferometric point target analysis with JERS-1 L-band SAR data. In Proceedings of the IEEE International Geoscience and Remote Sensing Symposium, IGARSS 2003, Toulouse, France, 21–25 July 2003; Volume 7, pp. 4359–4361.
53. Cigna, F.; Jordan, H.; Bateson, L.; McCormack, H.; Roberts, C. Natural and anthropogenic geohazards in greater London observed from geological and ERS-1/2 and ENVISAT persistent scatterers land subsidence data: Results from the EC FP7-SPACE PanGeo Project. *Pure Appl. Geophys.* **2015**, *172*, 2965–2995. [CrossRef]
54. Chaussard, E.; Bürgmann, R.; Shirzaei, M.; Fielding, E.J.; Baker, B. Predictability of hydraulic head changes and characterization of aquifer-system and fault properties from InSAR-derived ground deformation. *J. Geophys. Res. Solid Earth* **2014**, *119*, 6572–6590. [CrossRef]
55. Hoffmann, J.; Galloway, D.L.; Zebker, H.A. Inverse modelling of interbed storage parameters using land subsidence observations, Antelope Valley, California. *Water Resour. Res.* **2003**, *39*. [CrossRef]
56. Calderhead, A.I.; Therrien, R.; Rivera, A.; Martel, R.; Garfias, J. Simulating pumping-induced regional land subsidence with the use of InSAR and field data in the Toluca Valley, Mexico. *Adv. Water Resour.* **2011**, *34*, 83–97. [CrossRef]
57. Teatini, P.; Castelletto, N.; Ferronato, M.; Gambolati, G.; Janna, C.; Cairo, E.; Marzorati, D.; Colombo, D.; Ferretti, A.; Bagliani, A.; et al. Geomechanical response to seasonal gas storage in depleted reservoirs: A case study in the Po River basin, Italy. *J. Geophys. Res. Earth Surf.* **2011**, *116*. [CrossRef]
58. Hoffmann, J.; Zebker, H.A.; Galloway, D.L.; Amelung, F. Seasonal subsidence and rebound in Las Vegas Valley, Nevada, observed by Synthetic Aperture Radar Interferometry. *Water Resour. Res.* **2001**, *37*, 1551–1566. [CrossRef]
59. Tomás, R.; Herrera, G.; Delgado, J.; Lopez-Sanchez, J.M.; Mallorquí, J.J.; Mulas, J. A ground subsidence study based on DInSAR data: Calibration of soil parameters and subsidence prediction in Murcia City (Spain). *Eng. Geol.* **2010**, *111*, 19–30. [CrossRef]
60. Ezquerro, P.; Herrera, G.; Marchamalo, M.; Tomás, R.; Béjar-Pizarro, M.; Martínez, R. A quasi-elastic aquifer deformational behavior: Madrid aquifer case study. *J. Hydrol.* **2014**, *519*, 1192–1204. [CrossRef]
61. Benedetti, L.C.; Tapponnier, P.; Gaudemer, Y.; Manighetti, I.; van der Woerd, J. Geomorphic evidence for an emergent active thrust along the edge of the Po Plain: The Broni-Stradella fault. *J. Geophys. Res. Solid Earth* **2003**, *108*. [CrossRef]
62. Meisina, C. Light buildings on swelling/shrinking soils: Case histories from Oltrepo Pavese (north-western Italy). *Int. Conf. Probl. Soils* **2003**, *2*, 28–30.

11

Folded Basinal Compartments of the Southern Mongolian Borderland: A Structural Archive of the Final Consolidation of the Central Asian Orogenic Belt

Dickson Cunningham

Department of Environmental Earth Science, Eastern Connecticut State University, Willimantic, CT 06226, USA; cunninghamw@easternct.edu

Academic Editors: Kevin Tansey, Stephen Grebby and Jesus Martinez-Frias

Abstract: The Central Asian Orogenic Belt (CAOB) records multiple Phanerozoic tectonic events involving consolidation of disparate terranes and cratonic blocks and subsequent reactivation of Eurasia's continental interior. The final amalgamation of the CAOB terrane collage involved diachronous closure of the Permian-Triassic Solonker suture in northernmost China and the Jurassic Mongol-Okhotsk suture in northeast Mongolia and eastern Siberia. The distribution, style, and kinematics of deformation associated with these two terminal collision events is poorly documented in southern Mongolia and northernmost China because these regions were later tectonically overprinted by widespread Cretaceous basin and range-style crustal extension and Miocene-recent sinistral transpressional mountain building. These younger events structurally compartmentalized the crust into uplifted crystalline basement blocks and intermontane basins. Consequently, widespread Cretaceous and Late Cenozoic clastic sedimentary deposits overlie older Permian-Jurassic sedimentary rocks in most basinal areas and obscure the deformation record associated with Permian-Triassic Solonker and Jurassic Mongol-Okhotsk collisional suturing. In this report, satellite image mapping of basinal compartments that expose folded Permian-Jurassic sedimentary successions that are unconformably overlapped by Cretaceous-Quaternary clastic sediments is presented for remote and poorly studied regions of southern Mongolia and two areas of the Beishan. The largest folds are tens of kilometers in strike length, east-west trending, and reveal north-south Late Jurassic shortening (present coordinates). Late Jurassic fold vergence is dominantly northerly in the southern Gobi Altai within a regional-scale fold-and-thrust belt. Local refolding of older Permian north-south trending folds is also evident in some areas. The folds identified and mapped in this study provide new evidence for the regional distribution and kinematics of Jurassic and Permian-Triassic contractional tectonism in the southern Mongolia-northern China borderland region. The newly mapped folds are also important potential targets for hydrocarbon exploration and vertebrate paleontological discoveries.

Keywords: Mongolia; Gobi Altai; Fold-and-Thrust-Belt; Permian-Jurassic; Beishan; China; CAOB; intracontinental deformation; Mongol-Okhotsk; orogeny

1. Introduction

The vast region of Central Asia north of Tibet and south of Siberia in Mongolia and China records multiple Phanerozoic tectonic events that have led to the amalgamation of eastern Asia's continental crust ([1–3], and references therein). Modern syntheses of the terrane amalgamation history of southern and Central Mongolia in the Gobi Altai region and in adjacent regions of the Beishan in northern China document the multi-stage history of terrane accretion, seaway closure, and terminal collision along the

Late Permian Solonker-Ugger Us-Hongshishan Suture (Figure 1; [4–9]). The final seaway to close in eastern Asia north of Tibet was the Mongol-Okhotsk Ocean which closed diachronously from west to east during the Latest-Jurassic-Cretaceous [10–12]. Mongolia and western China have since occupied an intraplate, continental interior setting. Distant plate boundary effects have continued to drive intracontinental deformation in Central Asia until the present day ([13–15], and references therein). It is widely perceived that Central Asia is the world's foremost region for studying processes of continental interior crustal reactivation [16].

In this paper, satellite image mapping of folded basinal compartments in isolated and poorly studied regions of southern Mongolia is presented. For comparison, two areas south of the Chinese-Mongolian border in the Beishan region that contain folded basinal compartments are also discussed. Most of the mapped folds are previously undocumented and their recognition and tectonic significance are important for evaluating existing models of the crustal evolution of the southern Central Asian Orogenic Belt (CAOB).

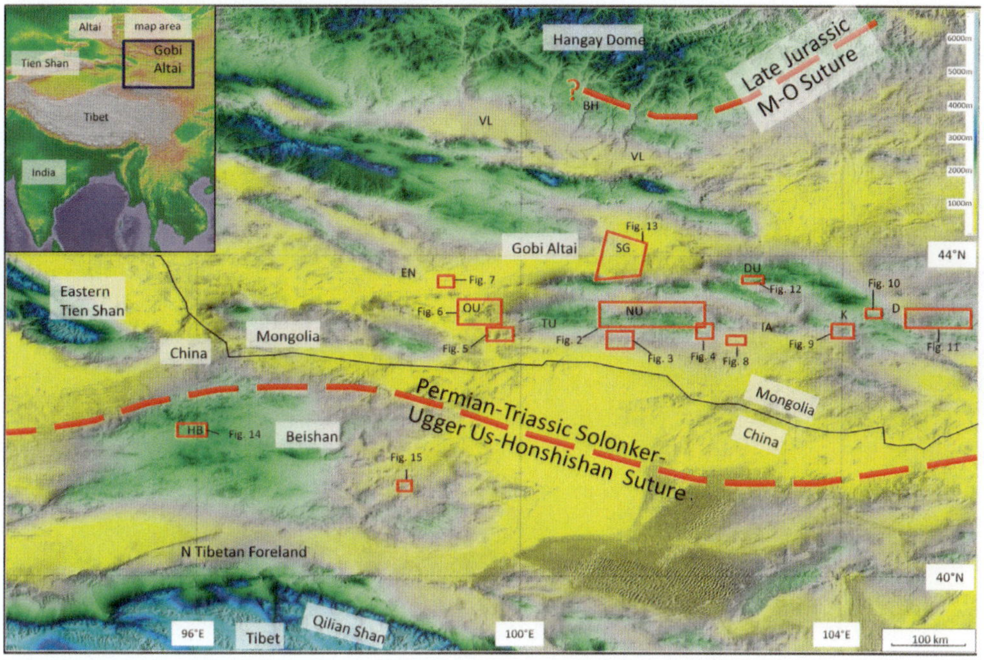

Figure 1. DEM of Gobi Corridor region of southern Mongolia and northern China showing locations of folded basinal compartments (red boxed areas) and subsequent figures (SRTM-90 topographic data). Each location contains contractionally deformed Permian-Jurassic siliciclastic sedimentary rocks unconformably overlain by Cretaceous-Tertiary/Quaternary alluvial sediments that are typically undeformed or less deformed. Locations of Permian Triassic Solonker-Ugger Us-Honshishan suture and Late Jurassic Mongol-Okhotsk suture shown [12,17,18], VL: Valley of Lakes; SG: Sherigyn Gashoon Basin; DU: Dzolen Uul; OU: Ongon Ulaan; EN: Edrengiyn Nuruu; NU: Noyon Uul; TU: Tost Uul; IA: Ikh Argalant; K: Khurmen; D: Dalandzagad; BH: Bayankhongor; HB: Hongyanjing Basin.

1.1. Previous Work and Geological Setting

1.1.1. Paleozoic Terrane Assembly of the Southern CAOB in the Gobi Altai and Beishan Regions

The continental crust of southern Mongolia and the Beishan region consists of a Paleoproterozoic basement complex in southern Mongolia surrounded by various lower through upper Paleozoic arc, accretionary prism/melange, and ophiolitic complexes [4,9,19,20]. Terrane maps of Mongolia with accompanying stratigraphic columns and geochronological data provide a regional lithotectonic framework for understanding the region's crustal evolution [4,21]. More recently, new maps, structural and geochronological data, and gravity and magnetic models from different areas in southern Mongolia further refine the polyphase metamorphic, deformational, and magmatic evolution of the different terranes and call into question the locations of some previously mapped terrane boundaries [8,20,22].

In the Beishan region, Xiao et al. [7] documented the terrane collage which lies directly north of the eastern part of the Archean/Proterozoic Tarim cratonic block and which underlies most of the Beishan region. Xiao et al. [7] recognized multiple arc terranes separated by ophiolitic melanges which amalgamated throughout the Paleozoic until final closure of the Paleo-Asian Ocean in the late Permian along a suture zone that links cryptically eastward with the Permian Solonker suture in Inner Mongolia. The Permian collision marked the end of marine sedimentation in the Mongolian borderland region and the onset of terrestrial sedimentation. Permian sediments from the Tien Shan through the Beishan to southern and eastern Mongolia and adjacent regions of China record the marine-terrestrial transition [23,24] and were affected by crustal shortening due to terminal closure of the final Paleo-Asian ocean seaway [5,7,18,25]. Collision-related crustal shortening likely persisted into the Triassic [5,8,18]. It is proposed that CAOB terrane amalgamation in southern Mongolia throughout the Paleozoic occurred along a north-south oriented active margin and that the accreted terrane assemblage was subsequently oroclinally bent in a counterclockwise sense during the Permo-Triassic into an east-west orientation and cut by sinistral strike-slip faults in the Triassic-mid Jurassic [8,9,24,26,27]. The terrane amalgamation history of southern Mongolia and the Beishan region imparted a dominant east-west structural grain throughout the region (modern coordinates) consisting of terrane boundaries (including suture belts), other faults, metamorphic foliation and sedimentary strike belt trends, and magmatic arc trends [4,8,9,28].

Subsequent to Late Permian consolidation of the CAOB along the Solonker-Ugger Us-Hongshishan suture and collision-related shortening that persisted until the Early Triassic, the Mongolia-China borderland region was affected by at least three other major tectonic events that were intraplate and intracontinental in nature. These consist of: (1) a mid-late Jurassic shortening event which is now recognized over a wide region including the southern Gobi Altai and Beishan [29]; (2) a regional Latest Jurassic-Early-Mid Cretaceous crustal extension event recognized in the Gobi Altai, southeast Mongolia, and in northern China east of the Beishan [30]; (3) a Late Cenozoic sinistral transpression event affecting all of the Gobi Altai, parts of the Beishan and a huge region outside of these areas to the north, west, south, and east [15,31]. In the following section, these tectonic events are reviewed individually.

1.1.2. Mid-Jurassic Contractional Tectonism

During the mid-late Jurassic, southern Mongolia was affected by significant crustal contraction, uplift, and erosional exhumation. This event is recorded by apatite fission-track data that reveal a pulse of mid-Jurassic erosion and cooling [32], regional-scale folding and thrusting along the Mongolia-Chinese border and in the eastern Beishan region [29], and deposition of thick siliciclastic sedimentary successions (up to 2500 m-thick) in the Gobi Altai region as far north as the southern Hangay Dome margin [33–35]. In addition, the Beishan and eastern Tien Shan regions also record a major Jurassic cooling/exhumation event [30,36–38]. The surface expression of this event in Mongolia is obscured in many areas by subsequent tectonic events that have either overprinted Jurassic structures, or by younger sedimentary rocks that unconformably overlie and conceal rocks affected by Jurassic contraction.

Crustal deformation during the mid-late Jurassic increasingly appears to be related to a relatively short period (10 my?) of intense north-south contraction associated with west-to-east, diachronous, terminal closure of the Mongol Okhotsk Ocean [11]. Tomographic images of the relict Mongol-Okhotsk slab show that it is elongated north-south in the lower 1000 km of the mantle beneath Siberia [12], suggesting that subduction was formerly towards the west beneath a north-south trending ribbon-like continent of accreted Paleozoic terranes and Precambrian slivers in western Mongolia [8,27]. The active margin subsequently buckled due to north-south shortening and closed progressively eastwards so that the westernmost Mongol-Okhotsk suture is in the Bayankhongor zone on the southern slopes of the Hangay Dome (Figure 1; [39,40]). Much of southern and eastern Mongolia is suspected to have been affected by this event, because of the wide distribution of Jurassic siliciclastic deposits, a regional Late Jurassic unconformity, and the fission-track evidence for a major exhumation event in the southernmost Gobi Altai and Beishan [32,35,38,39,41,42].

1.1.3. Latest Jurassic-Mid-Cretaceous Regional Extension

Southern and Eastern Mongolia, areas of northern China east of the Beishan, and Transbaikalia (Siberia) experienced diffuse northwest-southeast to north-south crustal extension during the latest Jurassic–mid-Cretaceous [30,34,43–45]. Diffuse crustal extension affected a huge area of central and east Asia estimated to be 1.5–2 times larger than North America's Basin and Range province [45,46]. Crustal extension during the early-mid Cretaceous is interpreted to be due to slab foundering and gravitational collapse of tectonically overthickened crust following terminal closure of the Mongol-Okhotsk Ocean which caused a change in the regional stress field from north-south and northwest-southeast convergence to north-south and northwest-southeast extension [30,42].

In the central and southern Gobi Altai and eastern Mongolia, this event is characterized by metamorphic core complex formation, rift basin development, and up to 1700 m of fluvial, lacustrine and aeolian sedimentation [33,34,47–50]. Organic shales deposited in Cretaceous lakes within individual rifts and linked graben systems comprise important oil source rocks in southern and eastern Mongolia and adjacent regions of China [51,52]. The Cretaceous successions are also world famous for their dinosaur fossil content ([53], and references therein). In addition, significant late Jurassic-mid-Cretaceous volcanism is widely reported in the northern Gobi Altai and in Chinese basins directly southeast and east of Mongolia [30,54]. Syn-rift and post-rift Cretaceous sedimentation overlapped and buried previously folded Permian-Jurassic successions and crystalline basement blocks. From the Late Cretaceous-Miocene, the entire region was tectonically quiescent and characterized by erosion, landscape lowering, and peneplanation [31,34,41].

This latest Jurassic- mid-Cretaceous regional crustal extension event is not documented in the Beishan region, nor is it suggested by any thermochronological data [38,55]. A review of published 1:200,000 scale Chinese geological maps for the entire region indicates that only several small areas of Cretaceous rocks with a total thickness estimated to be around 50 m are documented in the eastern and southeastern Beishan. Instead, the Beishan region typically contains a major upper Jurassic unconformity overlain by Neogene clastic sediments without documented evidence for significant Cretaceous sedimentation or extensional faulting.

1.1.4. Late Cenozoic Tanspressional Reactivation

Late Miocene-Recent transpressional mountain building is well documented in the Gobi Altai region [15,31]. The region is tectonically active today as indicated by widespread historical seismicity including the 1957 M = 8.1 Gobi Altai earthquake [56–59]. Geodetic data indicate 1–10 mm/year northeast and eastward crustal motions relative to a fixed Siberia [60–62]. A wide array of Quaternary northwest striking thrust faults and east-west striking sinistral strike-slip faults is documented throughout the Gobi Altai region [15,31]. The modern mountains of the Gobi Altai are typically thrust ridges, restraining bends, and other transpressional uplifts [15,31,63]. Miocene-Recent alluvial, lacustrine, and aeolian deposits flank the uplifted blocks which typically are basement-cored and fault-bound. The modern basin and range topography of the Gobi Altai is largely a product of transpressional reactivation of the CAOB terrane collage south of the more rigid Archean basement that underlies the Hangay Dome region [15,63].

The Beishan region shows less evidence for Late Cenozoic reactivation, however, several Quaternary deforming belts cut the region and geodetic data suggest northeast-directed crustal displacements on the order of 5 mm/year relative to a fixed Siberia [15,62]. In addition, the tectonic geomorphology of the Beishan region reveals a number of upthrust basement blocks with low mountain front sinuosity and Quaternary fan complexes along their fronts suggesting localized crustal rejuvenation [15].

Late Cenozoic reactivation of the Gobi Altai and parts of the Beishan represents part of the huge deformation field north of Tibet attributed to the Indo-Eurasia collision 2000 km to the south [13,64]. Continued northeast-directed indentation of India into south Asia generates a northeast-directed maximum compressive stress (SHmax) throughout the Beishan and Gobi Altai region [65]. The angular relationship between northeast-directed SHmax and the prevailing east-west basement "grain" in the Beishan and Gobi Altai is a first-order kinematic control on the modern sinistral transpressional deformation in the region [15].

2. Satellite Image Mapping of Folded Basinal Compartments in the Gobi Altai and Beishan

The application of satellite image analysis for documenting the geology and structural evolution of regions in Central Asia is standard practice for tectonicists since the seminal work of Tapponnier and Molnar [13]. The tectonic signal in the landscape is strongly expressed throughout southern Mongolia and adjacent regions of China because of the hyper-arid climate, absence of outlet rivers, and youthful tectonic geomorphology (cf., [15]). Southern Mongolia has few towns or roads and fieldwork in the region requires preliminary remote sensing analysis to efficiently locate areas to investigate in the field and to plan fieldwork logistics. The most detailed Mongolian geological survey quadrangle maps for southern Mongolia are at 1:200,000 scale (1cm on the map = 2 km on the ground). Similarly, in the Beishan region of China, published quadrangle maps are also at 1:200,000 scale and highly generalized in terms of the structural geology. Because Mongolian and Chinese geological quadrangle maps are at such a small scale, large regions of deformed basin strata that are 10–100 km^2 in area typically contain only a few strike and dip measurements, fold hinges are rarely shown, and faults are typically mapped without any indication of their kinematic nature. Consequently, in areas of complex deformation, structural detail is omitted or highly generalized on existing maps and the deformation history of much of southern Mongolia and the Beishan region is thus very poorly documented and incompletely understood.

In contrast, publicly available satellite images can provide much greater detail of the surface geology in the region. Processed Landsat band data allow enhanced discrimination of lithological contacts. SRTM (Shuttle Radar Topography Mission) digital topographic data (90 m and 30 m resolution) commonly reveal important geological information, especially when used as a basemap DEM for visual image drapes with variable transparency and light illumination directions and angles (e.g., [15,66]).

Free public access to Google Earth imagery is especially beneficial for geologists investigating remote regions such as southern Mongolia because of the availability of multiple historical images of the same area and the increasingly high-resolution scenes that are uploaded for viewing. For most areas of Central Asia, including the region discussed in this report, Google Earth provides the following image products: Digital Globe imagery (sub-2 m panchromatic and multispectral resolution), SPOT imagery (2.5–20 m resolution), and Landsat imagery (15–30 m resolution). For this study, Google Earth was dominantly used because the imagery it provides was the best available in terms of public access, resolution, and coverage, and for discriminating folded basinal sequences in unstudied regions in the southern Mongolia borderlands and Beishan region. The wide availability of Google Earth imagery also allows global readers the opportunity to independently check the validity of published geological interpretations, as in this report. In addition to Google Earth, downloaded Landsat TM (Thematic Mapper) imagery was imported into the Global Mapper application (2015, Blue Marble Geographics) and draped over SRTM digital topographic imagery to generate additional 3D models with variable vertical exaggeration capabilities and color blend options to better visualize and confirm fold interpretations in some areas.

3. Folded Basinal Compartments of Southern Mongolia—Mapping Results

Figures 2–15 are satellite photos of individual areas identified in southern Mongolia and the Beishan that contain visible folded sedimentary successions of Permian-Cretaceous age within basinal compartments between uplifted crystalline basement blocks. Although a few other folded basinal areas in the southern Gobi Altai were identified in the region, they were more limited in area and amount of exposure and were therefore not included in this study. In Figures 2–15, traces of bedding are shown and fold hinge-lines are indicated. Interpretation of the folds as either anticlines or synclines is based on interpretation of dip directions, which is based on identification of scarp slopes versus dip slopes on inclined beds. Scarp slopes and dip slopes of sedimentary strata typically generate asymmetric topography and different erosion characteristics making them easily identifiable, unless the topography is eroded flat. Dip slopes are typically smoother and ramp-like when seen from above, whereas scarp slopes are less planar and usually more gullied. In addition, for many of the mapped folds, the steepness of the fold limbs can be qualitatively determined on the imagery, especially when the folds are viewed obliquely down-axis, which then allows the fold vergence to be ascertained. In a few cases, overturned fold limbs were also identified which indicated the vergence direction. The ages of the folded strata are taken from the 1:1,000,000 Mongolian geological map [67]. In the following section, general characteristics of the interpreted fold structures for 14 different map areas are described. Locations of individual map areas are provided in Figure 1.

3.1. Noyon Uul Region

Mapped folds of basinal strata on the north, southwest, and southeast flanks of the Noyon Uul range are presented in Figures 2–4. The north sides of the Noyon and Tost Uul ranges contain the thickest successions of Permian-Jurassic strata exposed in the southern Gobi Altai region with more than 5 km of documented fluvial and lacustrine strata [35]. The stratigraphy is interpreted to be a foreland succession associated with a middle Jurassic north-directed fold-and-thrust belt documented further south [29]. Hendrix et al. [35] also documented a regional unconformity between middle and upper Jurassic rocks and noted that the upper Jurassic-Cretaceous succession is relatively undeformed compared to the folded rocks below.

The dominant structure in the Noyon-Tost Uul region is the 90 km-long Noyon Uul syncline (Figure 2). The eastern end of this regional fold was mapped by Guy et al. [9] who published three stratigraphic and structural cross-sections of the eastern end of the fold. They noted that Carboniferous and Permian strata were initially folded along north-south hinge lines and then the entire Carboniferous-Jurassic succession was folded along east-west hinge lines. Both fold generations produced up-right folds with horizontal fold axes. In addition, Guy et al. [9] identified a north-directed thrust duplex within Permian sandstones in the Noyon succession.

Mapped folds along the northern margins of Noyon and Tost Uul are dominantly east-west trending and multi-km in length and the regional map pattern does not suggest significant refolding, except for an area of refolds around Noyon town in the southeast and a singular refolded anticline southeast of Gurvantes town in the far west of the map area (Figure 2). Fold vergence is not obvious on the imagery, although both [9,35] suggest the east-west folds are part of a north-directed fold-and-thrust belt.

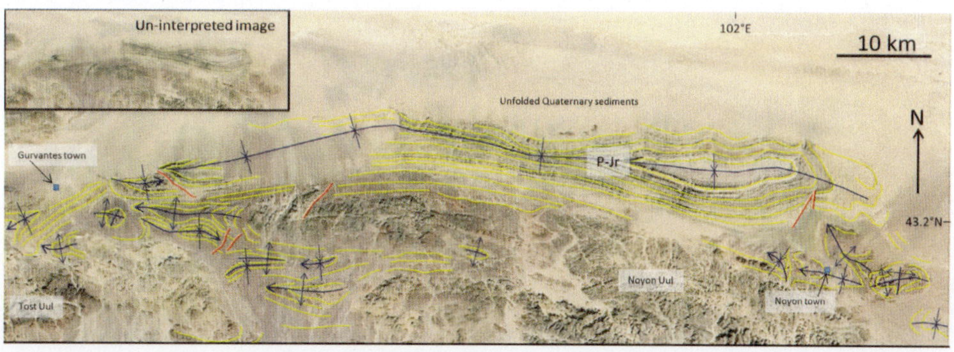

Figure 2. Google Earth (© 2016 DigitalGlobe) image interpretation showing bedding traces (yellow lines) and major fold hinges (dark blue lines) in Permian-Jurassic (P-Jr) rocks in northern Noyon Uul region, southern Mongolia. Red lines are faults. Inset shows un-interpreted scene. See Figure 1 for location.

The southern margin of Noyon Uul contains north- and northwest-vergent folds of Triassic and Jurassic sedimentary rocks in an area of relatively subdued relief (Figure 3). This area also contains active mining operations of Jurassic coal. The folds are exposed on the south side of a major fault that has uplifted the south side relative to the north side. The north side of this fault and the area south of the exposed fold zone are overlapped by unfolded Cretaceous sediments. This area of folding can be traced discontinuously from east to west as a 90 km-long belt of contractional deformation, however, the most extensive area of folding within this belt is the approximately 30-km-wide region shown in Figure 3. Refolds are not clearly identifiable on the imagery, but there is significant fold axis trend variation from dominantly east-west to northeast and north-northeast.

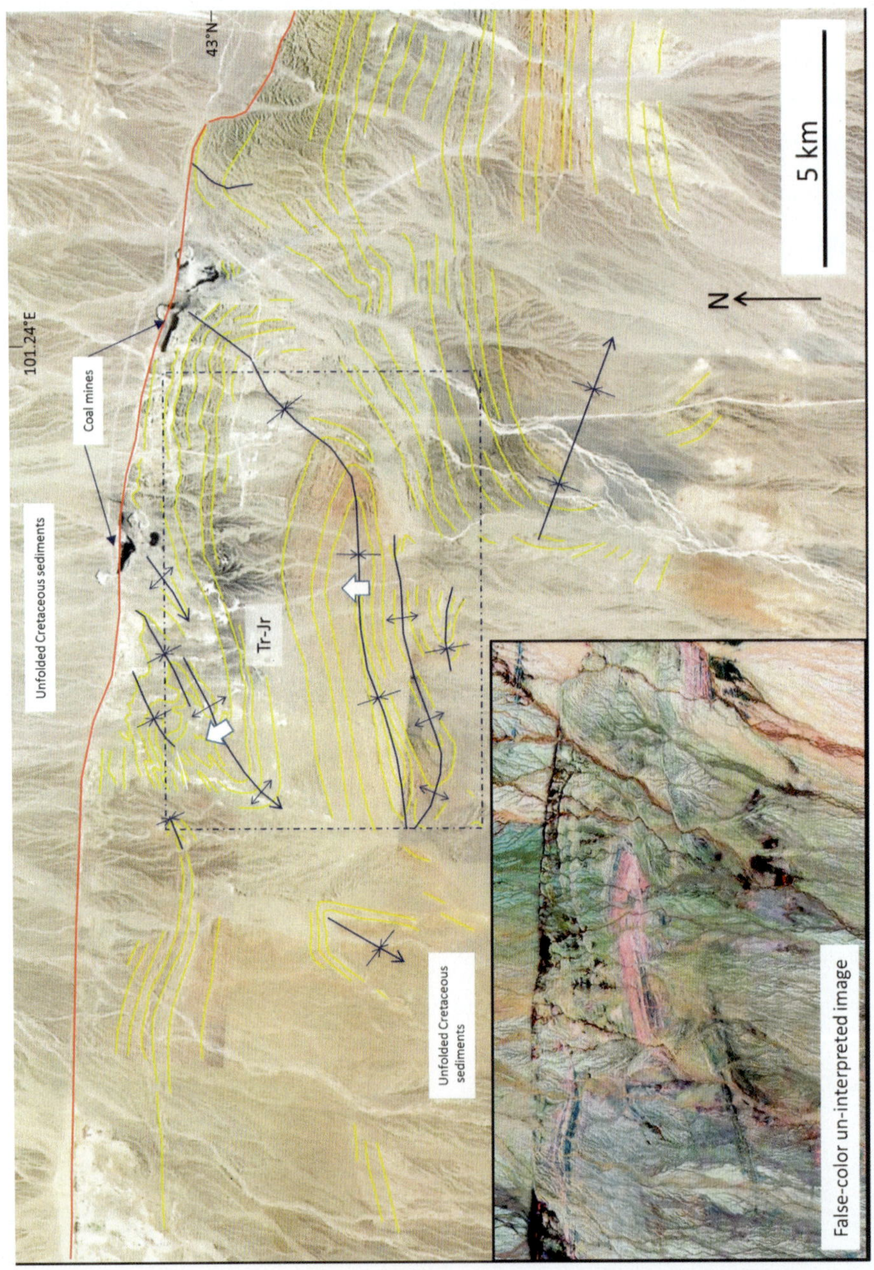

Figure 3. Google Earth (© 2016 DigitalGlobe) image interpretation showing bedding traces (yellow lines) and major fold hinges in mid-upper Triassic-Jurassic (Tr-Jr) succession in southern Noyon Uul region, southern Mongolia. Red line is major fault. Inset shows Landsat TM MrSID (N-47-40_2000) un-interpreted scene that was linearly contrast stretched, filtered (3 × 3 median), and color enhanced using the Global Mapper Difference tool. See Figure 1 for location.

The southeast side of the Noyon Uul range also exposes a folded Permian-Triassic succession (Figure 4). These folds are multi-km in length in east-west and northwest-southeast directions, and are north and northeast-vergent, respectively. The folds are overlapped by undeformed Cretaceous sediments. Thus, for the entire Noyon Uul region, the major folding event must be Late Jurassic in age because the folded middle Jurassic strata are regionally overlain by unfolded Cretaceous clastic sedimentary rocks.

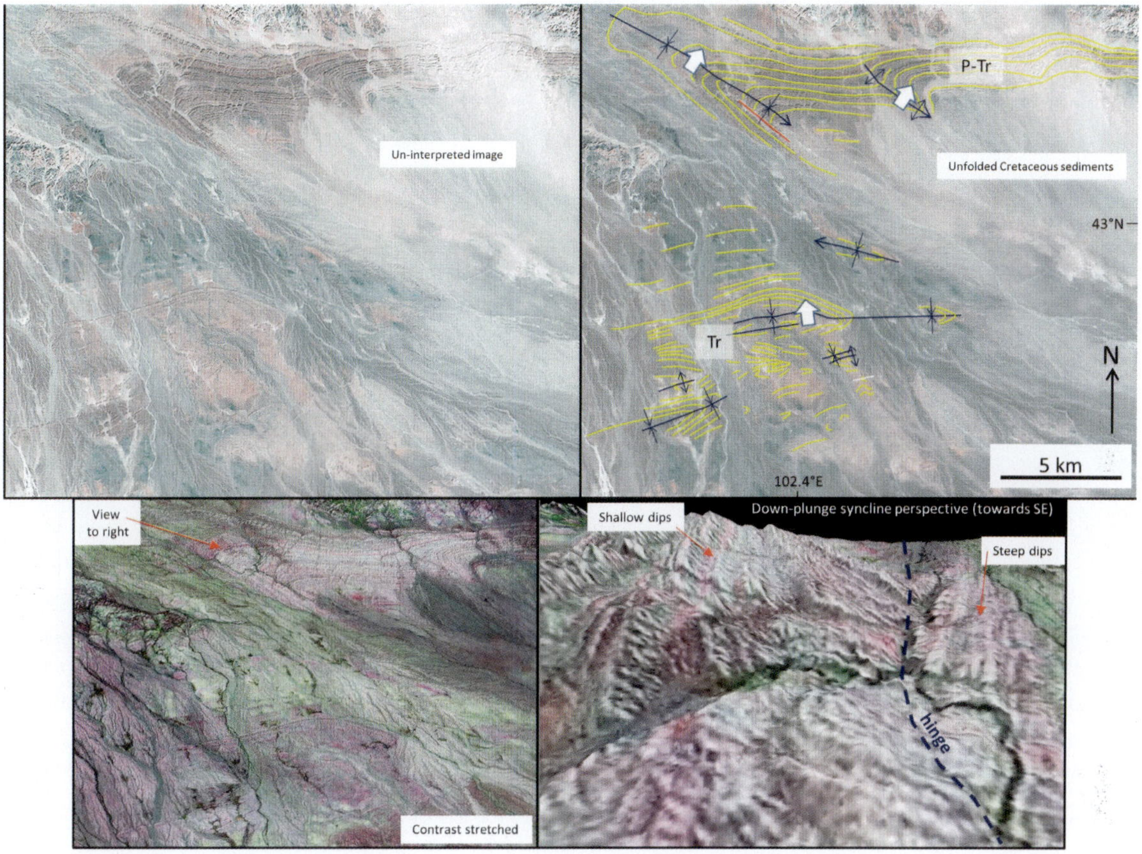

Figure 4. Un-interpreted and interpreted Landsat TM (M48-40_2000 MrSID) image showing bedding traces (yellow lines) and major fold hinges in folded Permian-Triassic rocks (P-Tr), SE Noyon Uul region, southwest Mongolia. Thick white arrows indicate direction of fold vergence. Lower left image shows contrast enhanced image with better discrimination of folded layers (Global Mapper Color Blend Mode-Difference tool). Lower right image shows 3D perspective looking down plunge towards the southeast of asymmetric northeast vergent syncline (the southwest limb is steeper than the northeast limb). Vertical exaggeration visually accentuates differences in limb dip and fold asymmetry. See Figure 1 for location.

3.2. Tost Uul Region

Approximately 50 km southwest of Tost Uul, a basinal area contains refolded Triassic sedimentary rocks (Figure 5). An overturned northeast-trending, northwest-vergent syncline is the dominant structure and can be traced along strike for 16 km. Smaller wavelength northwest-trending folds appear refolded by the larger syncline. Unfolded Cretaceous sediments unconformably overlie the southeast limb of the regional-scale syncline. Thus, the folds are Triassic-Jurassic in age and the large syncline is likely to be a Jurassic structure driven by north-northwest-south-southeast crustal shortening.

3.3. Ongon Ulaan

The Ongon Ulaan inlier preserves folded Triassic-Jurassic rocks that are unconformably overlapped by gently tilted to unfolded Cretaceous sediments (Figure 6). The inlier contains numerous east-west to west-northwest-east-southeast trending open folds that are cut by several faults with similar strike trends. One of the faults shows sinistral strike-slip offsets and appears to cut Quaternary sediments along strike suggesting Late Cenozoic activity (Figure 6; cf., [31], his Figure 4b). The folded sequence records dominantly Jurassic north-south shortening.

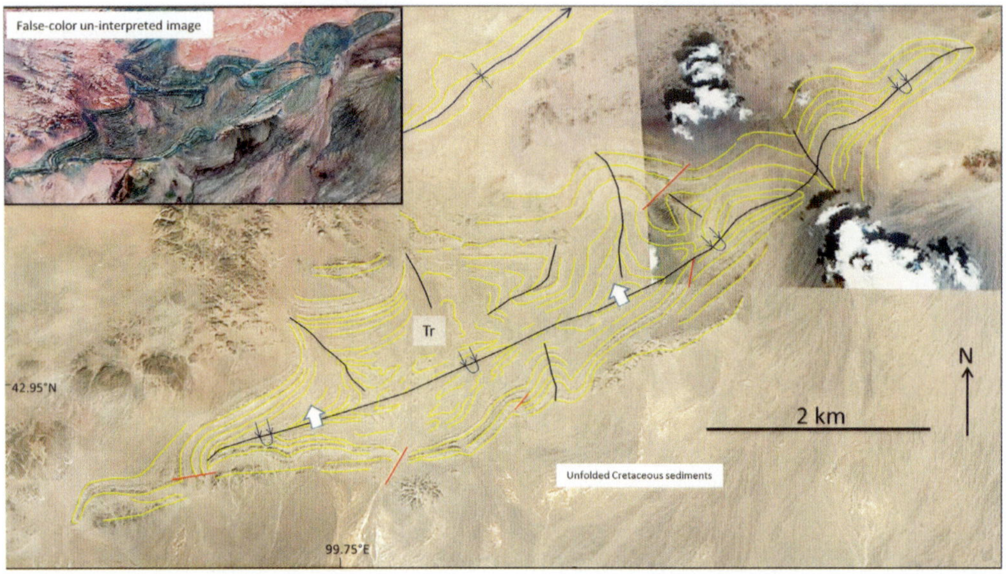

Figure 5. Google Earth (© 2016 CNES/Spot) image interpretation showing bedding traces (yellow lines) and major fold hinges in folded mid-upper Triassic (Tr) rocks in SW Tost Uul region, southern Mongolia. Thick white arrows indicate direction of fold vergence. Inset shows Landsat TM MrSID (N-47-40_2000) un-interpreted scene that was linearly contrast stretched, filtered (3 × 3 median), and color enhanced using the Global Mapper Difference tool. See Figure 1 for location.

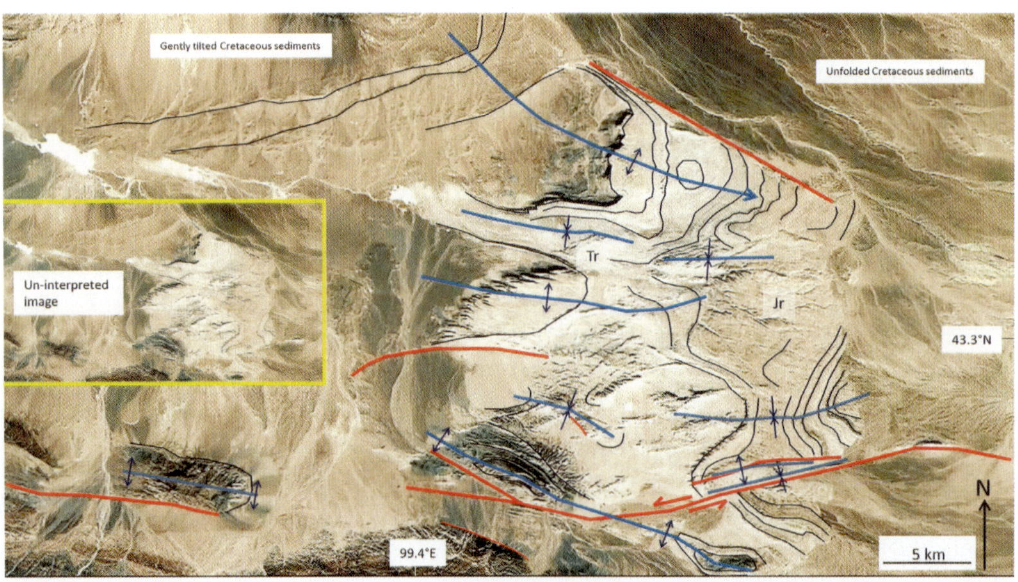

Figure 6. Google Earth (© 2016 DigitalGlobe) image interpretation showing bedding traces (dark blue lines), major fold hinges (light blue lines), and faults (red lines) in folded Triassic-Jurassic (Tr-Jr) succession, Ongon Ulaan, southwest Mongolia. Inset shows un-interpreted scene. See Figure 1 for location.

Another area of folded Triassic sediments is exposed approximately 40 km northwest of Ongon Ulaan in a remote region near the southeast terminus of the Edrengiyn Nuruu (Figures 1 and 7). A northwest trending syncline is exposed adjacent to an east-west fault with north-side-up components of displacement, but unknown kinematics. This fold is unconformably overlapped by gently tilted Cretaceous siliciclastic sediments. A syncline-anticline pair is exposed along strike several kilometers to the northwest. Folding is Triassic-Jurassic age.

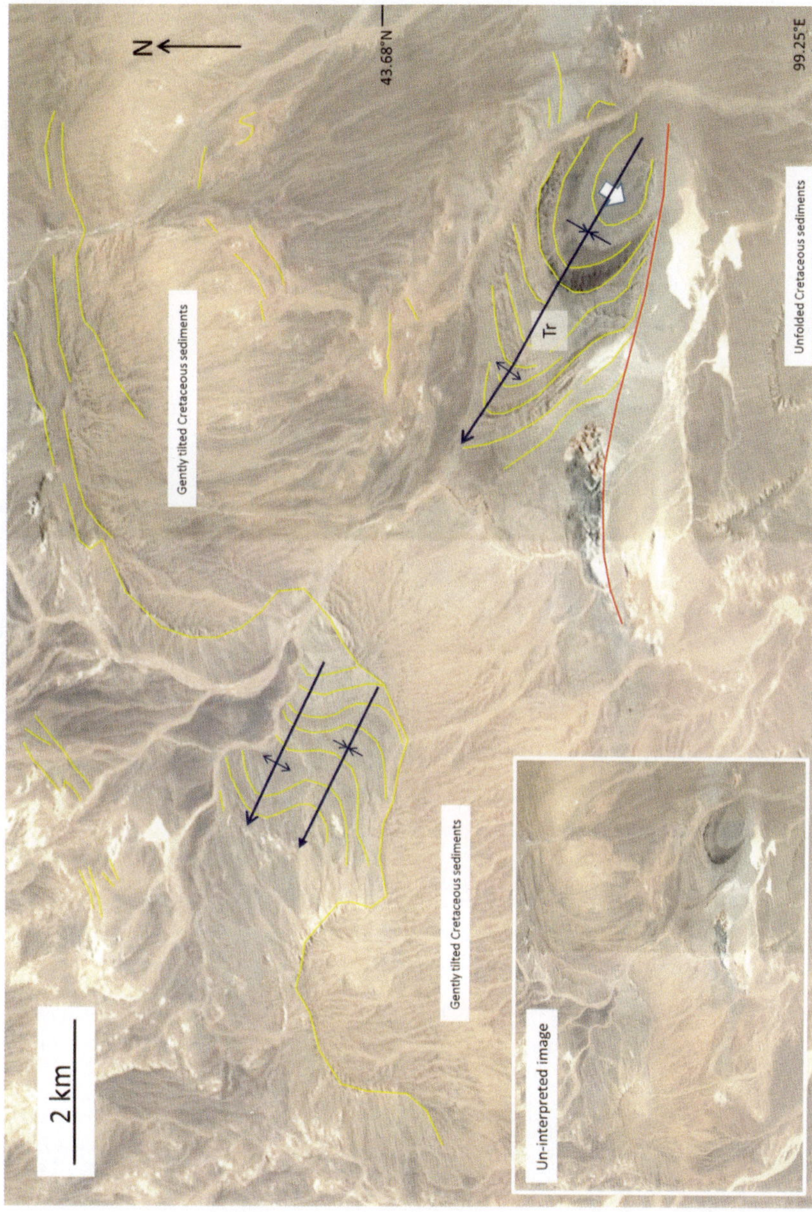

Figure 7. Google Earth (© 2016 DigitalGlobe) image interpretation showing bedding traces (yellow lines) and major fold hinges in folded Triassic (Tr) succession that is unconformably overlain by gently tilted Cretaceous rocks in basinal area northwest of Ongon Ulaan, southwest Mongolia. Thick white arrow indicates direction of fold vergence. Inset shows un-interpreted scene. See Figure 1 for location.

3.4. Ikh Argalant

The large alluvial basin southwest of the Ikh Argalant range contains an area of topographically subdued but exposed Triassic-Jurassic sedimentary rocks that are folded (Figure 8). Two east-west trending anticlines and two synclines are evident on satellite imagery, but their vergence is difficult to discern and the exposure is very incomplete because of overlapping Quaternary alluvial and aeolian deposits. Folding is likely to be Jurassic in age reflecting north-south shortening.

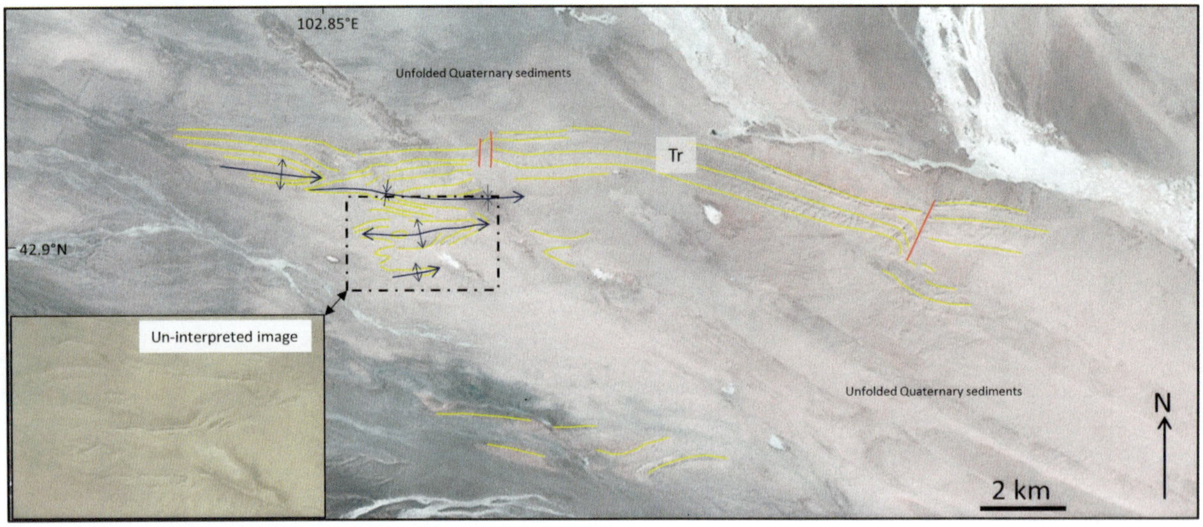

Figure 8. Google Earth (© 2016 DigitalGlobe) image interpretation showing bedding traces (yellow lines) and major fold hinges in Triassic (Tr) sedimentary rocks southwest of Ikh Argalant, southernmost Mongolia. Alluvial and eolian sediments overlap much of the area. Inset shows un-interpreted sub-scene. See Figure 1 for location.

3.5. Khurmen Region

A large inlier (40+ km^2) of folded Permian sedimentary rocks is exposed approximately 25 km south of Khurmen town in the eastern Gobi Altai adjacent to an un-named fault-bound basement block (Figure 9). Fold hinges trend dominantly east-west, but refolding is also evident in one location where a fold hinge curves from north-south to east-northeast trending. The refolded north-south trending hinge line suggests the east-west trending folds are younger. Fold vergence is not obvious on the satellite imagery except for an overturned south-vergent syncline on the north side of the basement block. The southeast margin of the folded sequence is onlapped by unfolded Cretaceous sediments (Figure 9). The fold patterns and Permian age of the sequence suggests early east-west shortening (modern orientation) of perhaps Permian age followed by Jurassic folding due to north-south shortening.

An inlier of folded Permian-Jurassic siliciclastic rocks also occurs approximately 30 km east of Khurmen town and 35 km south of Dalanzadgad in a low basinal area at the eastern end of the Gobi Altai (Figure 10). The largest fold is a 4 km-long west-northwest trending, doubly plunging anticline. A buried east-west trending syncline and northeast trending anticline are also inferred from scattered outcrops visible in the imagery. Smaller north and northwest trending folds are also evident in the northern part of the area (Figure 10) and appear re-oriented by later folding. Unfolded Cretaceous sedimentary rocks unconformably overlie the folded Permian rocks in the north. The folded rocks suggest an east-work shortening event (modern coordinates) of probable Permian age succeeded by north-south shortening of likely Jurassic age.

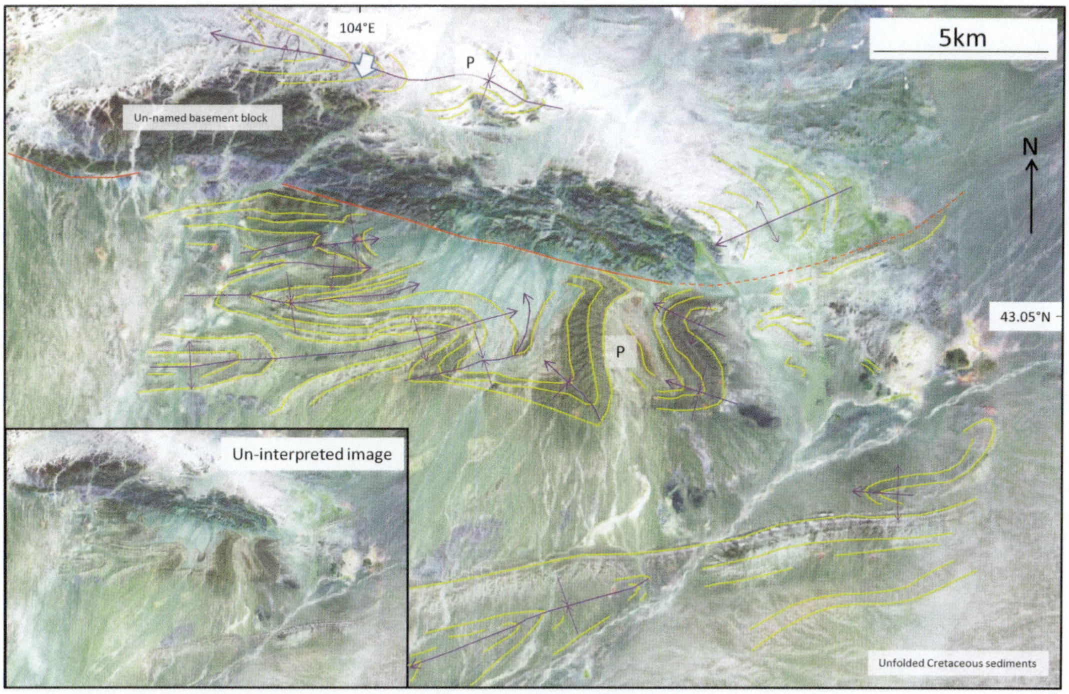

Figure 9. Landsat TM image (4-2-5 false-color scene processed using ArcMap, ESRI, Redlands, CA, USA, 2015) showing bedding traces (yellow lines) and major fold hinges in refolded Permian (P) sedimentary succession south of Khurmen, southernmost Mongolia. Thick white arrow indicates direction of fold vergence. Inset shows un-interpreted scene. See Figure 1 for location.

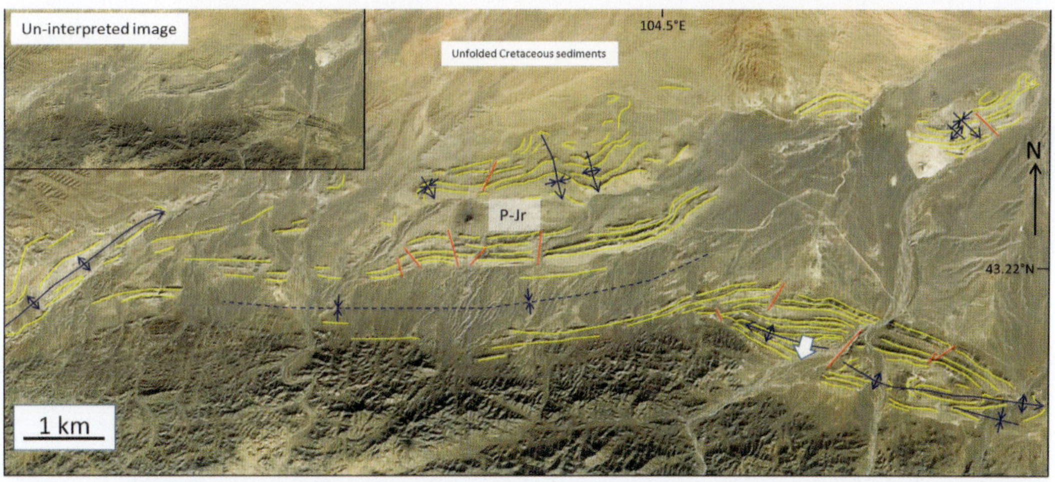

Figure 10. Google Earth image interpretation (2016, CNES Spot Image) showing bedding traces (yellow lines) and major fold hinges in folded Permian-Jurassic (P-Jr) sequence, 30 km east of Khurmen, easternmost Gobi Altai. Thick white arrow indicates direction of fold vergence. Inset shows un-interpreted scene. See Figure 1 for location.

3.6. Folds Southeast of Dalanzadgad

A 70-km long east-west belt of folded Permian-Jurassic rocks is exposed approximately 50 km southeast of the town of Dalanzadgad in the easternmost Gobi Altai region (Figure 11). The folds consist of a series of east-west trending north or northwest-vergent anticlines and synclines that plunge westward. The numerous "Z"-shaped fold forms in plan-view may suggest a dextral component of shear either accompanied folding or was later superimposed onto the fold belt. A north-directed thrust fault is visible in the imagery where there is clear truncation of footwall bedding and can be mapped for more than 30 km in an east-west sense (Figure 11, Inset A). The majority of the exposed folded sequence is on the thrust fault's upper plate. Tear faults also cut the fold belt in many locations and compartmentalize the folded thrust sheet (Figure 11, Inset B). Unfolded Cretaceous sediments that unconformably overlap the sequence constrain the folding to Permian-Jurassic age. The dominant east-west hinge lines and north-direction of fold vergence and overthrusting is consistent with Jurassic age crustal contraction documented elsewhere in the region [29,35].

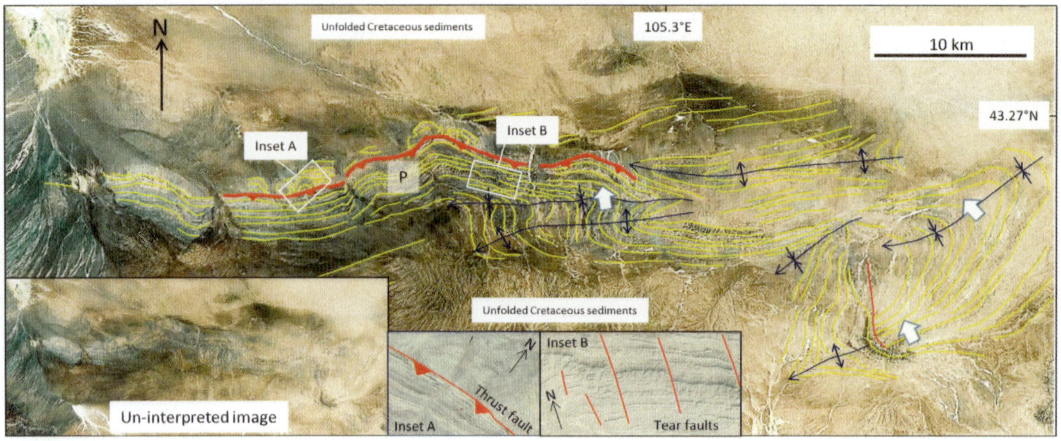

Figure 11. Bing Map image (© 2016 DigitalGlobe) interpretation showing bedding traces (yellow lines) and major fold hinges in folded and thrusted Permian sequence (P), 50 km southeast of Dalanzadgad, easternmost Gobi Altai. Insets show un-interpreted scene and detail of thrust fault and tear faults. Thick white arrows indicate direction of fold vergence. See Figure 1 for location.

3.7. Dzolen Uul

Directly west of the Dzolen Uul range in the eastern Gobi Altai a narrow anticlinal ridge of Permian sedimentary rocks is exposed within the wide alluvial basin (Figures 1 and 12). The fold trends approximately east-west and is onlapped by unfolded Cretaceous alluvial sediments. The steeper north rim suggests northward fold vergence. The age of folding is likely to be Jurassic based on the east-west fold trend and northerly vergence which is documented elsewhere in the region (this study; [9,29,35]).

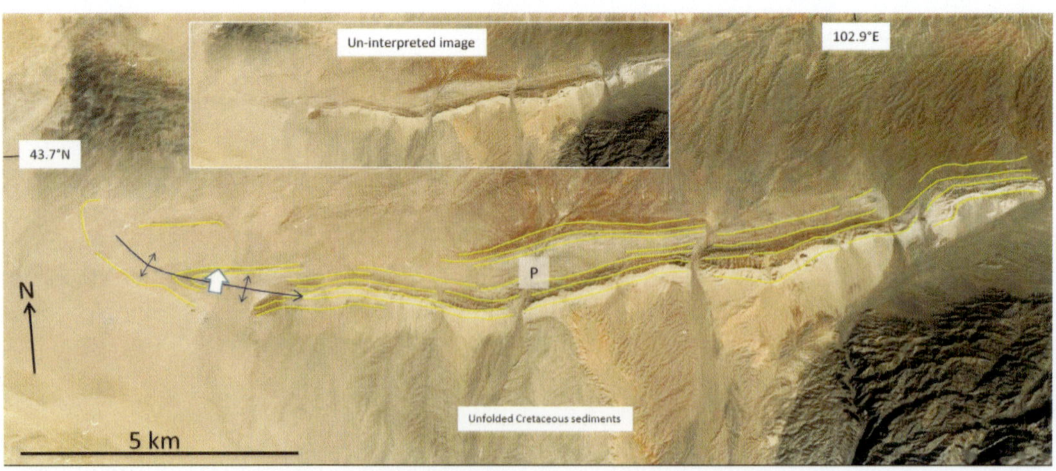

Figure 12. Google Earth image interpretation (2016, CNES Spot Image) showing bedding traces (yellow lines) and major fold hinges in folded Permian (P) sedimentary succession, west end of Dzolen Uul, southeast Gobi Altai. Thick white arrow indicates direction of fold vergence. Inset shows un-interpreted scene. See Figure 1 for location.

3.8. Sherigyn Gashoon Basin

The Sherigyn Gashoon Basin contains a regional north-vergent open anticline that overlies a basement arch that further west is upthrust by Late Cenozoic thrust faults (Figure 13; cf., [15]). Thus, the anticline is probably a passive response to Late Cenozoic blind thrusting at depth. Other multi-km-scale open folds deform the basin fill to the south and southwest. Two sub-km scale folds are also visible in one location and trend more north-northeast-south-southwest (Figure 13). Unlike regions to the south, folds in the Sherigyn Gashoon Basin deform Cretaceous alluvial sedimentary rocks and so are post-Cretaceous in age. The regional, long wavelength, open folds of the Sherigyn Gashoon Basin contrast with the tighter folds to the south of Permian-Triassic and Jurassic age. They are more similar to Late Cenozoic folds documented in the Valley of Lakes region directly north (Figure 1; [15,50]).

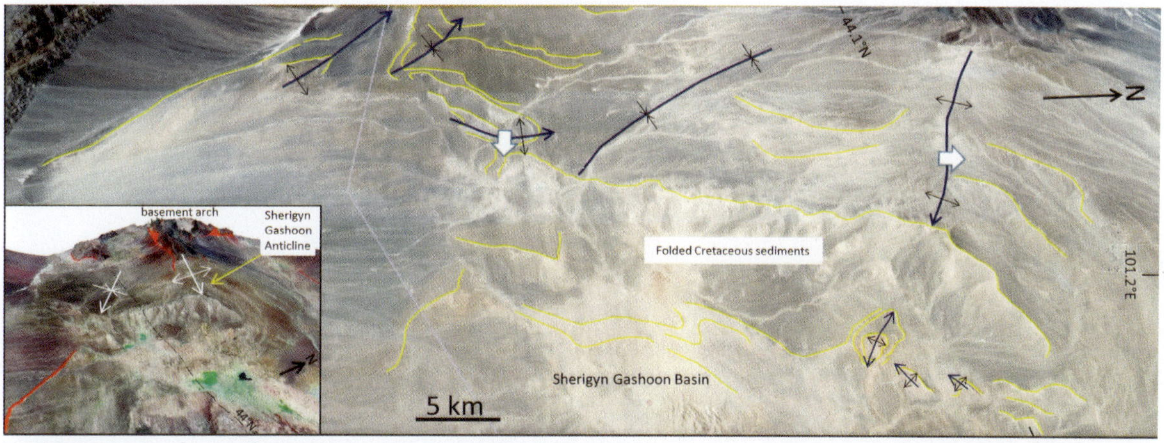

Figure 13. Google Earth (© 2016 DigitalGlobe) image interpretation showing bedding traces (yellow lines) and major fold hinges in Cretaceous basin fill, Sherigyn Gashoon Basin, central Gobi Altai. Thick white arrows indicate direction of fold vergence. Inset shows vertically exaggerated (3×) DEM with Landsat TM image drape of main Sherigyn Gashoon anticline which is passive response to upthrusted basement block along-strike to west. See Figure 1 for location.

3.9. Beishan Region

The Beishan region of northern China contains wide areas of spectacularly exposed folded basement and basinal rocks. Some of the finest examples of regional-scale refolded folds in the world have been documented within basinal compartments containing Permian-Early Jurassic strata [68,69]. Complete documentation of all the folded Permian-Jurassic basinal compartments of the Beishan is beyond the scope of this paper. However, two examples are included here for comparison with the southern Mongolian examples (Figures 14 and 15). Refolded folds of the Hongyanjing Basin in the central Beishan record a Late Permian-Triassic north-south contractional event followed by early-mid Jurassic dextral east-west displacements on east-west basin bounding faults and continued north-south basinal contraction (Figure 14; [55,68]). The Permian rocks were originally deposited in an extensional basin that was compressed and inverted [68]. The fold interference patterns are typically Ramsay Type 2 [70] with kilometer-scale wavelengths.

Further east in the easternmost Beishan, Jurassic sedimentary rocks are folded into northwest- and southwest-vergent folds adjacent to bounding bedrock massifs (Figure 14). The largest fold is a northwest-vergent synform that appears to be slightly refolded, because its hinge line bends from north-northeast to northeast trending (Figure 15). The south-vergent folds are east-west to northwest-southeast trending and overturned to the south and southeast. East and northeast striking tear faults cut the western fold limb and the entire folded area is unconformably overlain by unfolded uppermost Jurassic alluvial sediments. Thus, the deformation in both Beishan areas is Permian-upper

Jurassic in age. The differences in fold trends may reflect buttressing effects of the adjacent bedrock blocks directly to the west during crustal shortening. More work is needed in the central and eastern Beishan region to further refine the different Permian-Jurassic deformation events and their regional kinematics.

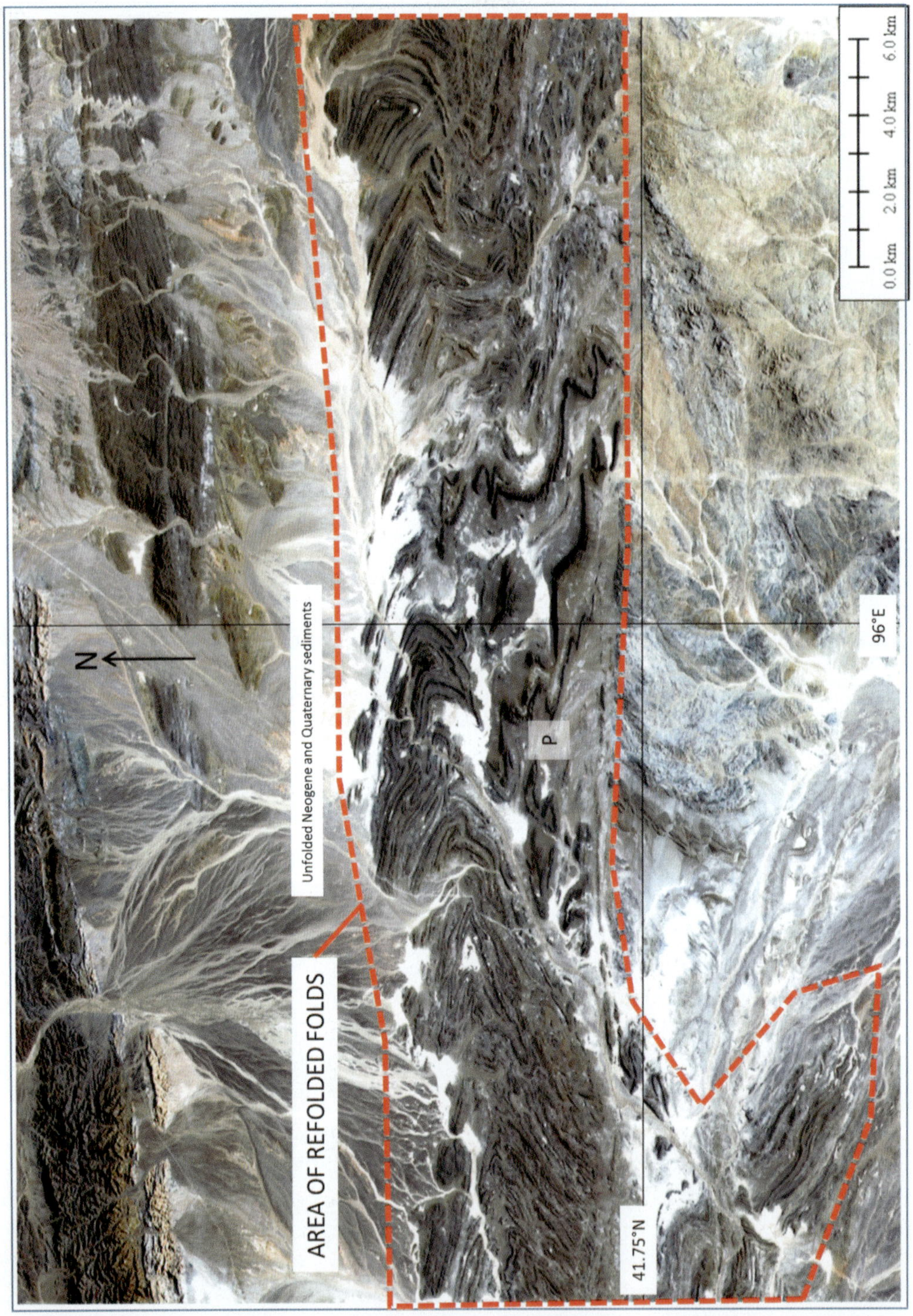

Figure 14. ALOS 10 m resolution multispectral image draped on SRTM-90 digital topography showing refolded Permian (P) succession in Hongyanjing Basin, central Beishan region, northwest China. See [68] for detailed fold interpretation and discussion. See Figure 1 for image location.

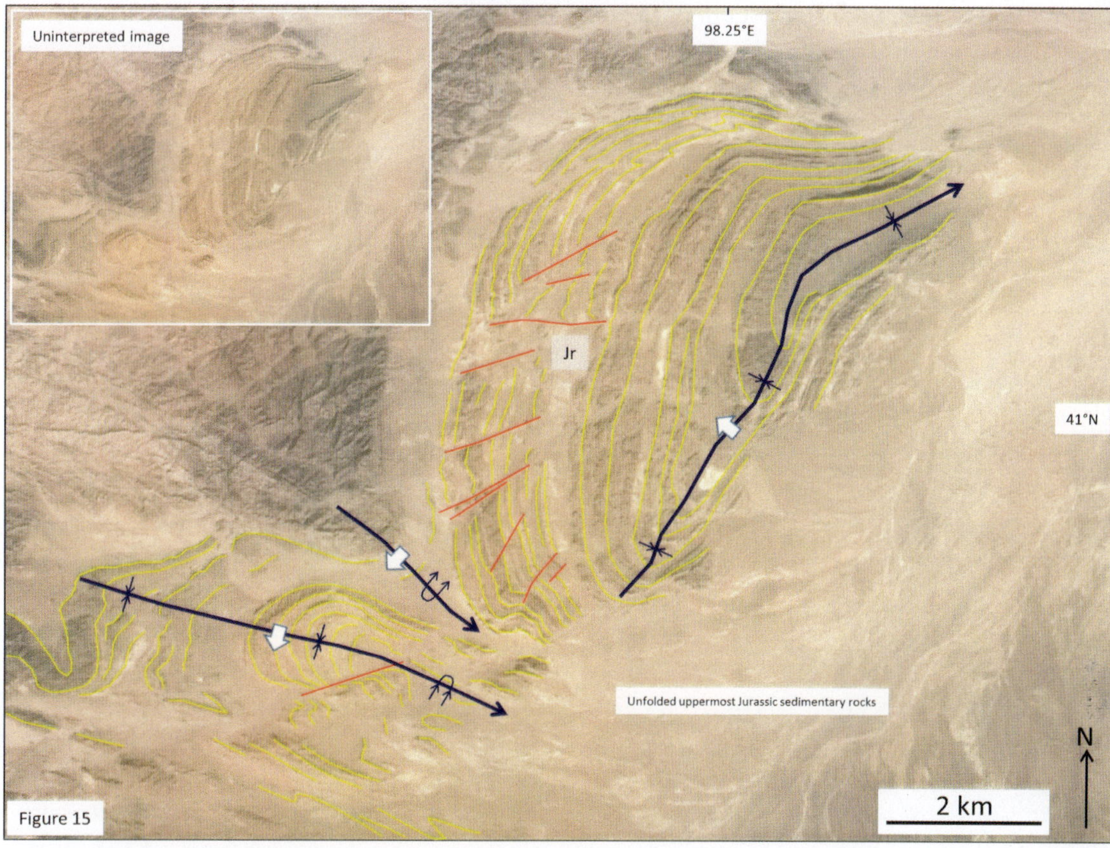

Figure 15. Google Earth (© 2016 DigitalGlobe) image interpretation showing bedding traces (yellow lines) and major fold hinges in folded Jurassic succession (Jr), eastern Beishan, China. Thick white arrows indicate direction of fold vergence. See Figure 1 for location.

4. Discussion

The twelve map areas analyzed in this study in the Gobi Altai region (Figures 2–13) all contain continental siliciclastic basin fill that ranges in age from Permian-Cretaceous. The strata are regionally folded and despite some local strike-trend variations, major folds are east-west striking suggesting north-south shortening. Fold vergence is both towards the north and south, but more commonly northerly. The largest areas of basinal fold exposures around Noyon Uul and southeast of Dalanzadgad (Figures 3–5 and 11) also reveal evidence for north-directed thrusting of the folded strata. Local evidence for an earlier phase of east-west shortening around north-south hinge lines is indicated in at least four areas (Figures 2, 5, 9 and 10) and previously reported in the southern Gobi Altai [9]. Thus, some folds show evidence for refolding and are therefore composite contractional structures. Folded Jurassic strata overlain by unfolded Cretaceous sedimentary rocks in 9 of the 12 areas suggests that the major north-south shortening event is late Jurassic in age. The older and less visibly evident north-south trending folds are interpreted to be Permian in age (cf., [9]). Thus, the southern Gobi Altai locally reveals evidence for a more regionally developed continental interior Late Jurassic fold-and-thrust belt that overprints a Permian contractional event (Figure 16; [29]). Cretaceous extensional basins subsequently compartmentalized the Permian-Jurassic shortened crust and eroding horst blocks provided the sedimentary infill that buried much of the older Permian-Jurassic fold-and thrust belts. Post-extension erosion of the diffuse Cretaceous basin-and-range province beveled the region to a low-relief plain and further buried much of the shortened Permian-Jurassic strata. Late Cenozoic sinistral transpressional uplift reactivated the entire Gobi Altai region generating new thrust blocks and restraining bend mountains leading to local erosional exhumation of Permian-Jurassic folded compartments and further burial of other deformed zones by Quaternary alluvial sediments. Thus, the Gobi Altai region's poly-deformational history and associated syn-tectonic clastic sedimentation history has resulted in a complex compartmentalized crustal architecture with limited exposures of deformed Permian-Jurassic strata (Figure 16).

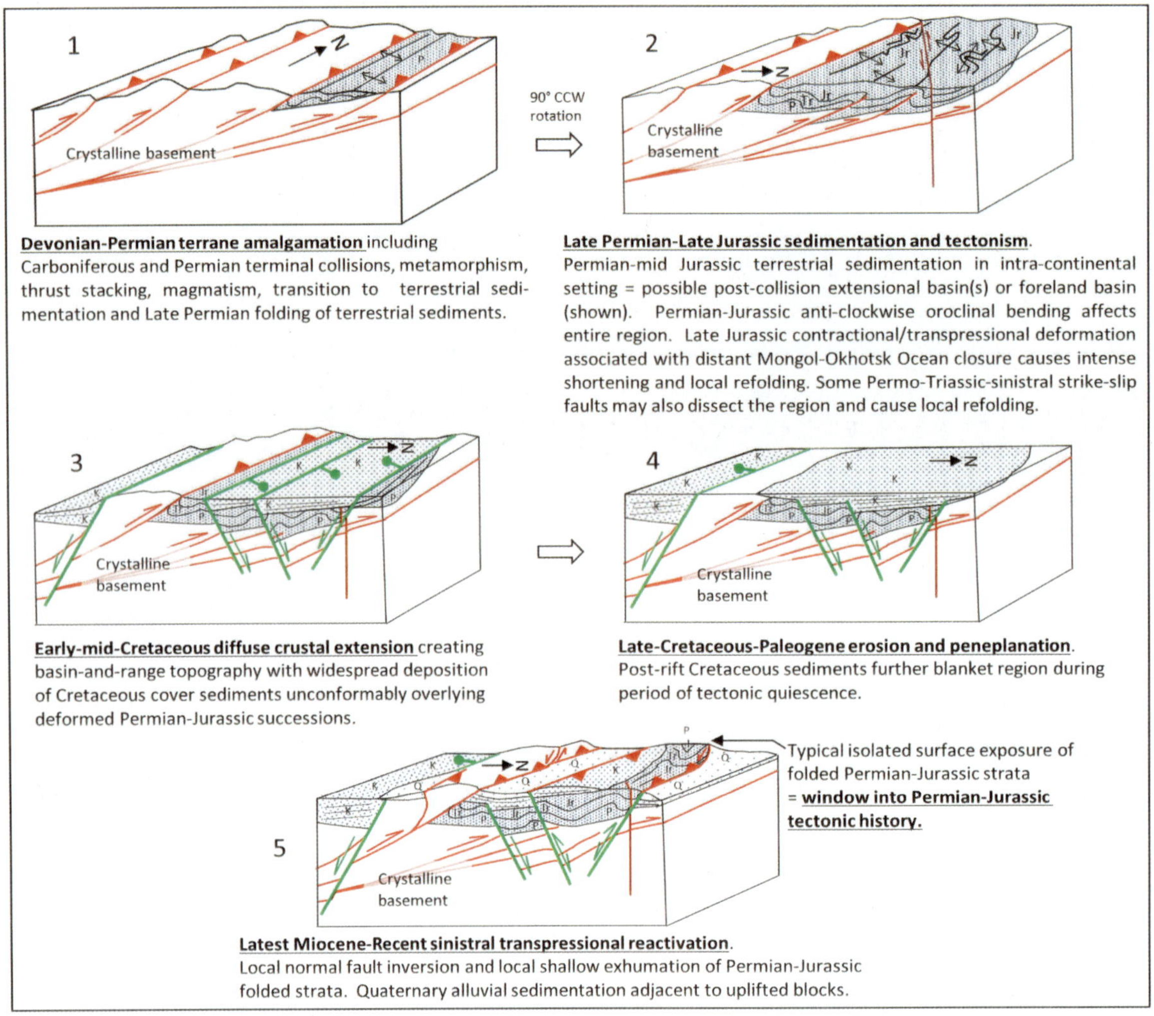

Figure 16. Block model showing five stages of crustal evolution responsible for typical structural/stratigraphic relationships within folded basinal compartments in the southernmost Gobi Altai region, Mongolia. **Stage 1** involved Devonian-Triassic terrane consolidation, progressive crustal shortening, mountain building, metamorphism and magmatic additions to the crust (not shown), progressive erosion and terrestrial sedimentation, and folding of eroded sediments in Permian foreland. **Stage 2** involved Late Permian-Triassic-mid-Jurassic erosion and deposition of thick terrestrial sediments that were subsequently north-south shortened into a Late Jurassic fold-and-thrust belt. Left-lateral strike-slip faults also cut the region during this stage and may cause some refolding of Permian strata [26]. **Stage 3** involved diffuse crustal extension during the early-mid Cretaceous that created basin and range-type compartmentalized rift basins and deposition of widespread alluvial sediments that unconformably overlapped the older deformed Permian-Jurassic strata [30,34]. **Stage 4** involved Late Cretaceous-Miocene tectonic quiescence and erosion, sedimentation, and peneplanation of the landscape ([15,34], and references therein). **Stage 5** involved Miocene-Recent sinistral transpressional reactivation of the Gobi Altai region creating a transpressional basin and range landscape and overprinting to variable degrees all previous deformation events [15,31]. Uplifted and eroded basinal areas in scattered localities in the southern Gobi Altai region reveal folded Permian-Jurassic strata that remain buried elsewhere.

In the Beishan, refolded folds of Permian-Triassic strata are documented in the eastern Hongyanjing Basin and are interpreted to be products of either two or three separate Permian-Triassic or Permian-Jurassic contractional and transpressional events [55,68,71]. The latter Triassic or Jurassic event is interpreted to be characterized by significant north-south shortening. In the eastern Beishan, large areas of folded Permian-Jurassic basinal rocks are unconformably overlain by unfolded late Jurassic strata. Fold axis trends in the eastern Beishan are widely variable as shown in Figures 14 and 15. The variation in fold trend is due to refolding and the varied orientations of rigid basement block boundaries that shortened the sedimentary infill between basement blocks in different directions.

The results of this study provide further evidence that Permian shortening associated with the closure of the Solonker suture caused initial folding of Permian clastic sedimentary rocks in the Beishan and southern Gobi Altai region. The regional distribution of this event is uncertain, but appears to extend from the Tien Shan in the west through the Beishan, southern Gobi Altai and Inner Mongolian regions of China [18,72,73]. Paleomagnetic, structural, and seismic tomographic evidence suggest that during the Permian-Jurassic the north-south CAOB terrane collage was oroclinally bent during progressive subduction of the Mongol-Okhotsk seaway so that the Gobi Altai-Beishan region became east-west oriented [12,27]. The tectonic explanation for the older north-south trending folds identified in this study is unclear because they predate the 90° CCW rotation of the southern arm of the CAOB orocline and thus they would have originally formed in an east-west direction along a former north-south convergent boundary. Perhaps a better explanation is that the folds are related to left-lateral wrench-related deformation in southern Gobi Altai crust that was a result of oblique closure of the Paleo-Asian ocean along the Solonker Suture (cf., [26]).

Terminal closure of the Mongol-Okhotsk seaway in the late Jurassic caused extensive north-south shortening in the southern Gobi Altai and northern China borderland region. Focused Late Jurassic shortening in the Mongolia-China borderland region 250–400 km south of the Mongol-Okhotsk suture suggests that sediment-filled intracontinental weak zones were reactivated. The tectonic setting of Permian-Jurassic sedimentation in the southern Gobi Altai has previously been interpreted to have been a non-marine foreland basin [35]. However, perhaps equally plausibly, Genyao et al. [74] suggested that following Solonker collisional suturing, post-orogenic extensional collapse generated rift trough(s) in the southern Gobi Altai and northern China border region in the Triassic-Late Jurassic that were predisposed to subsequent Late Jurassic rift inversion and folding of sedimentary infill. This interpretation is based on the existence of preserved mid-Jurassic rifts in eastern Mongolia and adjacent regions of China [44]. In addition, the Permian and possibly Triassic-Jurassic basin setting in the Beishan region may also have been extensional or transtensional based on the observation of syn-sedimentary normal faults in the Hongyanjing Basin [68].

The Late Jurassic contractional event in the China-Mongolia borderland region created the dominant east-west striking folds of Permian-Jurassic strata in the southern Gobi Altai that strongly overprint the north-south Permian folds. Interestingly, the northern Gobi Altai region and Valley of Lakes (Figure 1) contains locally significant Triassic to early-mid-Jurassic basaltic lavas and intercalated siliciclastic sedimentary strata accumulations [54]. However, syn-tectonic sedimentary deposits associated with the Late Jurassic Mongol-Okhotsk terminal collision are not documented in the Valley of Lakes basins [52]. This is probably because during the late Jurassic, the northernmost Gobi Altai region was mountainous, uplifted, and eroding near the collisional suture and thus not an area of sediment accumulation [52]. Subsequently, during the early Cretaceous, extensional rifting strongly affected the northern Gobi Altai and Valley of Lakes region as it did throughout the southern Gobi Altai and eastern Mongolia and adjacent areas of China [30]. Folding of Cretaceous rocks in northern Gobi Altai basins is attributed to a pulse of mid-Cretaceous basin inversion and/or Late Neogene-Recent transpressional deformation [15,52,75]. Folding of Cretaceous rocks associated with these later events is also documented in the Sherigyn Gashoon Basin in this study (Figure 13; cf., [15]).

Because much of the Gobi Altai region is now buried by Cretaceous and Quaternary sediments and because late Cenozoic transpressional faulting has created new mountains in the region, the older Permian and Jurassic folds reported in this study are only locally exposed. Thus, these few scattered surface exposures are especially important because they provide critical archival records of Permian-Jurassic tectonic events that were likely regional in scope, but are now largely hidden within basinal substrata. The Permian-Jurassic events are also less likely to be structurally distinguishable in the older, multiply deformed, metamorphic and igneous basement rocks that are exposed in surrounding mountain ranges.

Finally, the Permian-Jurassic folds mapped in this study are not only important because of the deformation events they record, but they also may be important for localizing hydrocarbons and thus should be regarded as potential exploration targets. The Gobi Altai region contains Permian, Jurassic, and Cretaceous oil source rocks and widespread coal deposits [76], and fossil fuel exploration is active in the region. In addition, the identification of large exposures of folded Permian-Jurassic strata reported here in remote unexplored regions may provide new discovery prospects for vertebrate paleontologists, especially those searching for dinosaur fossils that are so widely reported elsewhere in the Gobi Desert region [53].

5. Conclusions

(1) Folds of Permian-Jurassic basinal strata in remote regions of the southern Gobi Altai and Beishan were mapped by analyzing satellite images that are, for the most part, freely available. Most folds are previously undocumented; thus, this paper fills a gap in knowledge concerning the crustal architecture of basinal areas of the southern Gobi Altai region. The folds and cross-cutting relationships reveal Permian, Late Jurassic, and Late Cenozoic crustal shortening.

(2) Most folds analyzed in this study in the Mongolia-China borderland are east-west trending and record north-south shortening and northward vergence. Permian-Jurassic folds of sedimentary rocks are only exposed in a limited number of localities throughout the region, because Cretaceous extensional basin development and Late Cenozoic transpressional mountain building have further compartmentalized the crust, and overlying Cretaceous-Recent syn-tectonic sedimentary rocks have buried the Permian-Jurassic fold structures.

(3) Permian-Triassic shortening is attributed to the terminal Paleo-Asian ocean closure along the originally north-south trending Solonker-Ugger-Uus-Hongshshan suture. Late Jurassic north-south shortening is attributed to the Late Jurassic terminal closure of the Mongol-Okhotsk Ocean 250–400 km to the north which reactivated and further shortened the Permian-Jurassic basins.

(4) Counterclockwise oroclinal bending of the southern CAOB including the southern Gobi Altai and Beishan region during diachronous Mongol-Okhotsk west-to-east closure, rotated north-south trending Permian-Triassic contractional structures into east-west orientations. Mongol-Okhotsk terminal collision in the Late Jurassic reactivated the foldbelt and further contracted it. North-south trending folds of Permian sequences in the Beishan and southern Gobi Altai are locally refolded by Late Jurassic east-west striking folds.

(5) Regional-scale folds of Permian-Jurassic age in remote basinal compartments of the southernmost Gobi Altai and Beishan may provide new target areas for fossil fuel exploration and vertebrate paleontological discoveries.

Acknowledgments: The author appreciates the patient editorial support of Stephen Grebby who encouraged submission of the manuscript. The research presented here builds on prior fieldwork in the Gobi Altai region funded by the Natural Environment Research Council NERC CONNECT B Award NER/D/S2003/00671.

Conflicts of Interest: The author declares no conflict of interest.

References

1. Sengor, A.M.C.; Natal'In, B.A.; Burtman, V.S. Evolution of the Altaid tectonic collage and Palaeozoic crustal growth in Eurasia. *Nature* **1993**, *364*, 299–307. [CrossRef]
2. Windley, B.F.; Alexeiev, D.; Xiao, W.; Kroner, A.; Badarch, G. Tectonic models for accretion of the Central Asian Orogenic Belt. *J. Geol. Soc.* **2007**, *164*, 31–47. [CrossRef]

3. Wilhem, C.; Windley, B.F.; Stampfli, G.M. The Altaids of Central Asia: A Tectonic and Evolutionary Innovative Review. *Earth Sci. Rev.* **2012**, *113*, 303–341. [CrossRef]
4. Badarch, G.; Cunningham, W.D.; Windley, B.F. A new terrane subdivision for Mongolia: Implications for the Phanerozoic crustal growth of Central Asia. *J. Asian Earth Sci.* **2002**, *21*, 87–110. [CrossRef]
5. Xiao, W.; Windley, B.F.; Hao, J.; Zhai, M. Accretion leading to collision and the Permian Solonker suture, Inner Mongolia, China: Termination of the central Asian orogenic belt. *Tectonics* **2003**, *22*. [CrossRef]
6. Xiao, W.; Windley, B.F.; Badarch, G.; Sun, S.; Li, J.; Qin, K.; Wang, Z. Palaeozoic accretionary and convergent tectonics of the southern Altaids: Implications for the growth of Central Asia. *J. Geol. Soc.* **2004**, *161*, 339–342. [CrossRef]
7. Xiao, W.J.; Mao, Q.G.; Windley, B.F.; Han, C.M.; Qu, J.F.; Zhang, J.E.; Ao, S.J.; Guo, Q.Q.; Cleven, N.R.; Lin, S.F.; et al. Paleozoic multiple accretionary and collisional processes of the Beishan orogenic collage. *Am. J. Sci.* **2010**, *310*, 1553–1594. [CrossRef]
8. Lehmann, J.; Schulmann, K.; Lexa, O.; Corsini, M.; Kröner, A.; Štípská, P.; Tomurhuu, D.; Otgonbator, D. Structural constraints on the evolution of the Central Asian Orogenic Belt in SW Mongolia. *Am. J. Sci.* **2010**, *310*, 575–628. [CrossRef]
9. Guy, A.; Schulmann, K.; Clauer, N.; Hasalova, P.; Seltmann, R.; Armstrong, R.; Benedicto, A. Late Paleozoic–Mesozoic tectonic evolution of the Trans-Altai and South Gobi Zones in southern Mongolia based on structural and geochronological data. *Gondwana Res.* **2014**, *25*, 309–337. [CrossRef]
10. Zorin, Y.A. Geodynamics of the western part of the Mongolia-Okhotsk collisional belt, Trans-Baikal region (Russia) and Mongolia. *Tectonophysics* **1999**, *306*, 33–56. [CrossRef]
11. Yang, Y.T.; Guo, Z.X.; Song, C.C.; Li, X.B.; He, S. A short-lived but significant Mongol–Okhotsk collisional orogeny in latest Jurassic–earliest Cretaceous. *Gondwana Res.* **2015**, *28*, 1096–1116. [CrossRef]
12. Van der Voo, R.; van Hinsbergen, D.J.; Domeier, M.; Spakman, W.; Torsvik, T.H. Latest Jurassic–earliest Cretaceous closure of the Mongol-Okhotsk Ocean: A paleomagnetic and seismological-tomographic analysis. *Geol. Soc. Am. Spec. Pap.* **2015**, *513*, 589–606.
13. Tapponnier, P.; Molnar, P. Active faulting and Cenozoic tectonics of the Tien Shan, Mongolia, and Baykal regions. *J. Geophys. Res.* **1979**, *84*, 3425–3459. [CrossRef]
14. De Grave, J.; Buslov, M.M. Distant effects of India–Eurasia convergence and Mesozoic intracontinental deformation in Central Asia: Constraints from apatite fission-track thermochronology. *J. Asian Earth Sci.* **2007**, *29*, 188–204. [CrossRef]
15. Cunningham, D. Mountain building processes in intracontinental oblique deformation belts: Lessons from the Gobi Corridor, Central Asia. *J. Struct. Geol.* **2013**, *46*, 255–282. [CrossRef]
16. Raimondo, T.; Hand, M.; Collins, W.J. Compressional intracontinental orogens: Ancient and modern perspectives. *Earth Sci. Rev.* **2014**, *130*, 128–153. [CrossRef]
17. Gong, Q.S.; Liu, M.Q.; Li, H.L.; Liang, M.H.; Dai, W.J. The type and basic characteristics of Beishan orogenic belt, Gansu. *Northwest. Geol.* **2002**, *35*, 28–34.
18. Xiao, W.J.; Windley, B.F.; Huang, B.C.; Han, C.M.; Yuan, C.; Chen, H.L.; Li, J.L. End-Permian to mid-Triassic termination of the accretionary processes of the southern Altaids: Implications for the geodynamic evolution, Phanerozoic continental growth, and metallogeny of Central Asia. *Int. J. Earth Sci.* **2009**, *98*, 1189–1217. [CrossRef]
19. Helo, C.; Hegner, E.; Kröner, A.; Badarch, G.; Tomurtogoo, O.; Windley, B.F.; Dulski, P. Geochemical signature of Paleozoic accretionary complexes of the Central Asian Orogenic Belt in South Mongolia: Constraints on arc environments and crustal growth. *Chem. Geol.* **2006**, *227*, 236–257. [CrossRef]
20. Guy, A.; Schulmann, K.; Janoušek, V.; Štípská, P.; Armstrong, R.; Belousova, E.; Hanžl, P. Geophysical and geochemical nature of relaminated arc-derived lower crust underneath oceanic domain in southern Mongolia. *Tectonics* **2015**, *34*, 1030–1053. [CrossRef]
21. Tomurtogoo, O. A new tectonic scheme of the Paleozoides in Mongolia. *Mong. Geosci.* **1997**, *3*, 12–19.
22. Guy, A.; Schulmann, K.; Munschy, M.; Miehe, J.M.; Edel, J.B.; Lexa, O.; Fairhead, D. Geophysical constraints for terrane boundaries in southern Mongolia. *J. Geophys. Res. Solid Earth* **2014**, *119*, 7966–7991. [CrossRef]
23. Manankov, I.N.; Shi, G.R.; Shen, S.Z. An overview of Permian marine stratigraphy and biostratigraphy of Mongolia. *J. Asian Earth Sci.* **2006**, *26*, 294–303. [CrossRef]
24. Johnson, C.L.; Amory, J.A.; Zinniker, D.; Lamb, M.A.; Graham, S.A.; Affolter, M.; Badarch, G. Sedimentary response to arc-continent collision, Permian, southern Mongolia. *Geol. Soc. Am. Spec. Pap.* **2008**, *436*, 363–390.

25. Heumann, M.J.; Johnson, C.L.; Webb, L.E.; Taylor, J.P.; Jalbaa, U.; Minjin, C. Paleogeographic reconstruction of a late Paleozoic arc collision zone, southern Mongolia. *Geol. Soci. Am. Bull.* **2012**, *124*, 1514–1534. [CrossRef]
26. Lamb, M.A.; Hanson, A.D.; Graham, S.A.; Badarch, G.; Webb, L.E. Left-lateral sense offset of upper Proterozoic to Paleozoic features across the Gobi Onon, Tost, and Zuunbayan faults in southern Mongolia and implications for other Central Asian faults. *Earth Planet. Sci. Lett.* **1999**, *173*, 183–194. [CrossRef]
27. Edel, J.B.; Schulmann, K.; Hanžl, P.; Lexa, O. Palaeomagnetic and structural constraints on 90 anticlockwise rotation in SW Mongolia during the Permo–Triassic: Implications for Altaid oroclinal bending. Preliminary palaeomagnetic results. *J. Asian Earth Sci.* **2014**, *94*, 157–171. [CrossRef]
28. Lamb, M.A.; Badarch, G. Paleozoic sedimentary basins and volcanic arc systems of southern Mongolia: New geochemical and petrographic constraints. *GSA Memoir* **2001**, *194*, 117–149.
29. Zheng, Y.; Zhang, Q.; Wang, Y.; Liu, R.; Wang, S.G.; Zuo, G.; Wang, S.Z.; Lkaasuren, B.; Badarch, G.; Badamgarav, Z. Great Jurassic thrust sheets in Beishan (North Mountains)—Gobi areas of China and southern Mongolia. *J. Struct. Geol.* **1996**, *18*, 1111–1126. [CrossRef]
30. Meng, Q.R. What drove late Mesozoic extension of the northern China-Mongolia tract? *Tectonophysics* **2003**, *369*, 155–174. [CrossRef]
31. Cunningham, D. Tectonic Setting and Structural Evolution of the Late Cenozoic Gobi Altai Orogen. *Geol. Soc. Lond. Spec. Publ.* **2010**, *338*, 361–387. [CrossRef]
32. Dumitru, T.A.; Hendrix, M.S. Fission-track constraints on Jurassic folding and thrusting in southern Mongolia and their relationship to the Beishan thrust of northern China. *GSA Memoir* **2001**, *194*, 215–229.
33. Jerzykiewicz, T.; Russell, D.A. Late Mesozoic stratigraphy and vertebrates of the Gobi Basin. *Cretaceous Res.* **1991**, *12*, 345–377. [CrossRef]
34. Traynor, J.J.; Sladen, C. Tectonic and stratigraphic evolution of the Mongolian People's Republic and its influence on hydrocarbon geology and potential. *Mar. Pet. Geol.* **1995**, *12*, 35–52. [CrossRef]
35. Hendrix, M.S.; Graham, S.A.; Amory, J.Y.; Badarch, G. Noyon Uul syncline, southern Mongolia: Lower Mesozoic sedimentary record of the tectonic amalgamation of central Asia. *Geol. Soc. Am. Bull.* **1996**, *108*, 1256–1274. [CrossRef]
36. Hendrix, M.S.; Graham, S.A.; Carroll, A.R.; Sobel, E.R.; McKnight, C.L.; Schulein, B.J.; Wang, Z. Sedimentary record and climatic implications of recurrent deformation in the Tian Shan: Evidence from Mesozoic strata of the north Tarim, south Junggar, and Turpan basins, northwest China. *Geol. Soc. Am. Bull.* **1992**, *104*, 53–79. [CrossRef]
37. Cunningham, D.; Owen, L.A.; Snee, L.W.; Li, J. Structural framework of a major intracontinental orogenic termination zone: The easternmost Tien Shan, China. *J. Geol. Soc.* **2003**, *160*, 575–590. [CrossRef]
38. Gillespie, J.; Glorie, S.; Xiao, W.; Zhang, Z.; Collins, A.S.; Evans, N.; De Grave, J. Mesozoic reactivation of the Beishan, southern Central Asian Orogenic Belt: Insights from low-temperature thermochronology. *Gondwana Res.* **2016**, in press. [CrossRef]
39. Tomurtogoo, O.; Windley, B.F.; Kröner, A.; Badarch, G.; Liu, D.Y. Zircon age and occurrence of the Adaatsag ophiolite and Muron shear zone, central Mongolia: Constraints on the evolution of the Mongol–Okhotsk ocean, suture and orogen. *J. Geol. Soc.* **2005**, *162*, 125–134. [CrossRef]
40. Bussien, D.; Gombojav, N.; Winkler, W.; von Quadt, A. The Mongol-Kkhotsk belt in Mongolia—An appraisal of the geodynamic development by the study of sandstone provenance and detrital zircons. *Tectonophysics* **2011**, *510*, 132–150. [CrossRef]
41. Berkey, C.P.; Morris, F.K. The peneplanes of Mongolia. *Am. Mus. Novit.* **1924**, *136*, 1–11.
42. Shuvalov, V.F. Continental red beds of the Upper Jurassic, Mongolia. *Doklady Akademii Nauk SSSR* **1969**, *189*, 1088.
43. Zorin, Y.A.; Sklyarov, E.V.; Mazukabzov, A.M.; Belichenko, V.G. Metamorphic core complexes and Early Cretaceous rifting in Transbaikalia. *Geologiya I Geofizika* **1997**, *38*, 1574–1583.
44. Graham, S.A.; Hendrix, M.S.; Johnson, C.L.; Badamgarav, D.; Badarch, G.; Amory, J.; Porter, M.; Barsbold, R.; Webb, L.E.; Hacker, B.R. Sedimentary record and tectonic implications of Mesozoic rifting in southeast Mongolia. *Bull. Geol. Soc. Am.* **2001**, *113*, 1560–1579. [CrossRef]
45. Graham, S.A.; Cope, T.; Johnson, C.L.; Ritts, B. Sedimentary basins of the late Mesozoic extensional domain of China and Mongolia. In *Phanerozoic Rift Systems and Sedimentary Basins*; Roberts, D.G., Bally, A.W., Eds.; Elsevier: Amsterdam, The Netherlands, 2012; pp. 443–461.

46. Ren, J.; Tamaki, K.; Li, S.; Junxia, Z. Late Mesozoic and Cenozoic rifting and its dynamic setting in Eastern China and adjacent areas. *Tectonophysics* **2002**, *344*, 175–205. [CrossRef]
47. Meng, Q.R.; Hu, J.M.; Jin, J.Q.; Zhang, Y.; Xu, D.F. Tectonics of the late Mesozoic wide extensional basin system in the China-Mongolia border region. *Basin Res.* **2003**, *15*, 397–416. [CrossRef]
48. Webb, L.E.; Graham, S.A.; Johnson, C.L.; Badarch, G.; Hendrix, M.S. Occurrence, age, and implications of the Yagan-Onch Hayrhan metamorphic core complex, Southern Mongolia. *Geology* **1999**, *27*, 143–146. [CrossRef]
49. Cunningham, D.; Davies, S.; McLean, D. Exhumation of a Cretaceous rift complex within a Late Cenozoic restraining bend, southern Mongolia: Implications of the crustal evolution of the Gobi Altai region. *J. Geol. Soc.* **2009**, *166*, 1–13. [CrossRef]
50. Johnson, C.L.; Constenius, K.C.; Graham, S.A.; Mackey, G.; Menotti, T.; Payton, A.; Tully, J. Subsurface evidence for late Mesozoic extension in western Mongolia: Tectonic and petroleum systems implications. *Basin Res.* **2015**, *27*, 272–294. [CrossRef]
51. Sladen, C.; Traynor, J.J. Lakes during the evolution of Mongolia. *Am. Assoc. Pet. Geol. Stud. Geol.* **2000**, *46*, 35–57.
52. Johnson, C.L. Sedimentary basins in transition: Distribution and tectonic settings of Mesozoic strata in Mongolia. *Geol. Soc. Am. Spec. Pap.* **2015**, *513*, 543–560.
53. Benton, M.J.; Shishkin, M.A.; Unwin, D.M. *The Age of Dinosaurs in Russia and Mongolia*; Cambridge University Press: Cambridge, UK, 2003.
54. van Hinsbergen, D.J.; Cunningham, D.; Straathof, G.B.; Ganerød, M.; Hendriks, B.W.; Dijkstra, A.H. Triassic to Cenozoic multi-stage intra-plate deformation focused near the Bogd Fault system, Gobi Altai, Mongolia. *Geosci. Front.* **2015**, *6*, 723–740. [CrossRef]
55. Tian, Z.; Xiao, W.; Zhang, Z.; Lin, X. Fission-track constrains on superposed folding in the Beishan orogenic belt, southernmost Altaids. *Geosci. Front.* **2016**, *7*, 181–196. [CrossRef]
56. Florensov, N.A.; Solonenko, V.P. *The Gobi-Altai Earthquake*; Akademiya Nauk USSR: Moscow, Russia, 1963. (In Russian)
57. Baljinnyam, I.; Bayasgalan, A.; Borisov, B.A.; Cisternas, A.; Dem'yanovich, M.G.; Ganbataar, L.; Kochetkov, V.M.; Kurushin, R.A.; Molnar, P.; Hervé, P.; et al. *Ruptures of Major Earthquakes and Active Deformation in Mongolia and Its Surroundings*; Geological Society of America Memoir: Boulder, CO, USA, 1993; Volume 181, p. 62.
58. Bayarsayhan, C.; Bayasgalan, A.; Enhtuvshin, B.; Hudnut, K.W.; Kurushin, R.A.; Molnar, P.; Olziybat, M. 1957 Gobi-Altay, Mongolia, earthquake as a prototype for southern California's most devastating earthquake. *Geology* **1996**, *24*, 579–582. [CrossRef]
59. Adiya, M.; Ankhtsetseg, D.; Baasanbat, T.; Bayar, G.; Bayarsaikhan, Ch.; Erdenezul, D.; Mungunsuren, D.; Munkhsaikhan, A.; Munkhuu, D.; Narantsetseg, R.; et al. *One Century of Seismicity in Mongolia (1900–2000)*; Research Centre of Astronomy & Geophysics of the Mongolian Academy of Sciences (RCAG), Mongolia and Laboratoire de Teledetection et Risque Sismique, BP12: Bruyeres le Chatel, France, 2003.
60. Wang, Q.; Zhang, P.-Z.; Freymueller, J.T.; Bilham, R.; Larson, K.M.; Lai, X.; You, X.; Niu, Z.; Wu, J.; Li, Y.; et al. Present-day crustal deformation in China constrained by global positioning system measurements. *Science* **2001**, *94*, 574–577. [CrossRef] [PubMed]
61. Calais, E.; Vergnolle, M.; San'kov, V.; Lukhnev, A.; Miroshnitchenko, A.; Amarjargal, S.; Deverchere, J. GPS measurements of crustal deformation in the Baikal-Mongolia area (1994–2002): Implications for current kinematics of Asia. *J. Geophys. Res. B Solid Earth* **2003**, *108*. [CrossRef]
62. Liu, M.; Yang, Y.; Shen, Z.; Wang, S.; Wang, M.; Wan, Y. Active tectonics and intracontinental earthquakes in China: The kinematics and geodynamics. *Spec. Pap. Geol. Soc. Am.* **2007**, *425*, 299–318.
63. Cunningham, D. Structural and Topographic Characteristics of Restraining Bend Mountain Ranges of the Altai, Gobi Altai and easternmost Tien Shan. *Geol. Soc. Lond. Spec. Publ.* **2007**, *290*, 219–237. [CrossRef]
64. Holt, W.E.; Chamot-Rooke, N.; Le Pichon, X.; Haines, A.J.; Shen-Tu, B.; Ren, J. Velocity field in Asia inferred from Quaternary fault slip rates and Global Positioning System observations. *J. Geophys. Res. B Solid Earth* **2000**, *105*, 19185–19209. [CrossRef]
65. Heidbach, O.; Tingay, M.; Barth, A.; Reinecker, J.; Kurfeß, D.; Müller, B. The 2008 release of the World Stress Map. 2008. Available online: http://www.world-stress-map.org/ (accessed on 29 December 2016).

66. Roberts, N.; Cunningham, D. Automated alluvial fan discrimination, Quaternary fault identification, and the distribution of tectonically reactivated crust in the Gobi Altai region, Southern Mongolia. *Int. J. Remote Sens.* **2008**, *29*, 6957–6969. [CrossRef]
67. Tomurtogoo, O. *Geological Map of Mongolia, 1:1,000,000*; Mongolian Academy of Science, Institute of Geology and Mineral Resources: Ulaan Baatar, Mongolia, 1999.
68. Zhang, J.; Cunningham, D. Kilometer-scale refolded folds caused by strike-slip reversal and intraplate shortening in the Beishan region, China. *Tectonics* **2012**, *31*. [CrossRef]
69. Cleven, N.R.; Lin, S.; Xiao, W. The Hongliuhe fold-and-thrust belt: Evidence of terminal collision and suture-reactivation after the Early Permian in the Beishan orogenic collage, Northwest China. *Gondwana Res.* **2015**, *27*, 796–810. [CrossRef]
70. Ramsay, J.G. *Folding and Fracturing of Rocks*; McGraw-Hill: London, UK, 1967.
71. Tian, Z.; Xiao, W.; Shan, Y.; Windley, B.; Han, C.; Zhang, J.E.; Song, D. Mega-fold interference patterns in the Beishan orogen (NW China) created by change in plate configuration during Permo-Triassic termination of the Altaids. *J. Struct. Geol.* **2013**, *52*, 119–135. [CrossRef]
72. Xiao, W.; Han, C.; Liu, W.; Wan, B.; Zhang, J.E.; Ao, S.; Zhang, Z.; Song, D.; Tian, Z.; Luo, J. How many sutures in the southern Central Asian Orogenic Belt: Insights from East Xinjiang–West Gansu (NW China)? *Geosci. Front.* **2014**, *5*, 525–536. [CrossRef]
73. Jian, P.; Liu, D.; Kröner, A.; Windley, B.F.; Shi, Y.; Zhang, W.; Zhang, F.; Miao, L.; Zhang, L.; Tomurhuu, D. Evolution of a Permian intraoceanic arc–trench system in the Solonker suture zone, Central Asian Orogenic Belt, China and Mongolia. *Lithos* **2010**, *118*, 169–190. [CrossRef]
74. Genyao, W.; Yuan, W.; Min, L. Palinspastic reconstruction and geological evolution of Jurassic basins in Mongolia and neighboring China. *J. Palaeogeogr.* **2013**, *2*, 306–317.
75. Constenius, K.N.; Coogan, J.; Erdenebat, B.; Tully, J.; Johnson, C.; Graham, S.; Cunningham, D. Late Jurassic-Early Cretaceous Rifting in the Tugrug and Taatsiin Tsagaan Nuur Basins, Gobi-Altai Region of SW, Mongolia—Implications for Petroleum Exploration. Available online: https://www.researchgate.net/profile/Kurt_Constenius2/publication/277476649_Late_Jurassic-Early_Cretaceous_Rifting_in_the_Tugrug_and_Taatsiin_Tsagaan_Nuur_Basins_Gobi-Altai_Region_of_SW_Mongolia__Implications_for_Petroleum_Exploration/links/556b8de708aeccd7773a1be2.pdf (accessed on 6 January 2017).
76. Erdenetsogt, B.O.; Lee, I.; Bat-Erdene, D.; Jargal, L. Mongolian coal-bearing basins: Geological settings, coal characteristics, distribution, and resources. *Int. J. Coal Geol.* **2009**, *80*, 87–104. [CrossRef]

12

Perennial Lakes as an Environmental Control on Theropod Movement in the Jurassic of the Hartford Basin

Patrick R. Getty [1,*], Christopher Aucoin [2], Nathaniel Fox [3], Aaron Judge [4], Laurel Hardy [5] and Andrew M. Bush [1,6]

1 Center for Integrative Geosciences, University of Connecticut, 354 Mansfield Road, U-1045, Storrs, CT 06269, USA
2 Department of Geology, University of Cincinnati, 500 Geology Physics Building, P.O. Box 210013, Cincinnati, OH 45221, USA; aucoincd@mail.uc.edu
3 Environmental Systems Graduate Group, University of California, 5200 North Lake Road, Merced, CA 95340, USA; n.s.fox01@gmail.com
4 14 Carleton Street, South Hadley, MA 01075, USA; ajudge01@gmail.com
5 1476 Poquonock Avenue, Windsor, CT 06095, USA; leraducha@gmail.com
6 Department of Evolutionary Biology, University of Connecticut, 75 North Eagleville Road, U-3403, Storrs, CT 06269, USA; andrew.bush@uconn.edu
* Correspondence: patrick.getty@uconn.edu

Academic Editors: Neil Donald Lewis Clark and Jesús Martínez Frías

Abstract: *Eubrontes giganteus* is a common ichnospecies of large dinosaur track in the Early Jurassic rocks of the Hartford and Deerfield basins in Connecticut and Massachusetts, USA. It has been proposed that the trackmaker was gregarious based on parallel trackways at a site in Massachusetts known as Dinosaur Footprint Reservation (DFR). The gregariousness hypothesis is not without its problems, however, since parallelism can be caused by barriers that direct animal travel. We tested the gregariousness hypothesis by examining the orientations of trackways at five sites representing permanent and ephemeral lacustrine environments. Parallelism is only prominent in permanent lacustrine rocks at DFR, where trackways show a bimodal orientation distribution that approximates the paleoshoreline. By contrast, parallel trackways are uncommon in ephemeral lacustrine facies, even at sites with large numbers of trackways, and those that do occur exhibit differences in morphology, suggesting that they were made at different times. Overall, the evidence presented herein suggests that parallelism seen in Hartford Basin *Eubrontes giganteus* is better explained as a response to the lake acting as a physical barrier rather than to gregariousness. Consequently, these parallel trackways should not be used as evidence to support the hypothesis that the trackmaker was a basal sauropodomorph unless other evidence can substantiate the gregariousness hypothesis.

Keywords: ichnology; theropod; gregarious; Early Jurassic; Newark Supergroup

---

1. Introduction

Trace fossils, such as dinosaur trackways, are sedimentary structures that are produced in situ by living animals interacting with their environment. As such, they offer insights into the behavior of their makers that are sometimes unavailable from body fossils, which are often transported and disarticulated before preservation. Among the many behaviors that can be evaluated from trackways is gregariousness, or group behavior. Historically, one of the initial lines of evidence that has been used to support gregariousness is the occurrence of parallel trackways [1]. Later researchers, however, have noted that parallelism by itself is not strong enough evidence to state that a particular trackmaker was gregarious because physical barriers, such as shorelines, could dictate the direction of travel as the animals went about their daily activities [2]. Over time, individual animals might produce large numbers of parallel trackways as they each interact with the physical barrier. In one recent example, Razzolini et al. [3] described a suite of Middle Jurassic theropod dinosaur tracks preserved on a tidal flat in which the animals moved parallel and away from the shore. These researchers interpreted the behavior to represent the feeding of individual or very small groups of theropods and the bimodal orientations to represent the animals heading toward and then away from the water during low tide. Additionally, Moratalla and Hernán [4] have reported dinosaur tracks with preferred orientations that they inferred to result from paleogeographic barriers. Thus, care must be taken to rule out physical barriers and other environmental controls before parallelism can be used to infer gregariousness.

Additional lines of evidence are now used to further test whether or not a particular suite of trackways represents gregarious behavior. Among these is constant spacing between trackways, which is inferred to result from animals maintaining enough distance to avoid colliding [5]. Recently, Castanera et al. [6] used the constant spacing of Early Cretaceous sauropod trackways as an indicator that the animals they were studying were gregarious. Constant spacing is considered even stronger evidence for gregariousness when the trackways all turn the same direction [7]. Velocities derived from trackway calculations are also used as evidence for group behavior when they are the constant because individual animals must move at similar speeds in order to prevent the group from scattering [8]. Barco et al. [9], for example, considered the similar orientations and velocities of trackways at a site in Spain as evidence of gregariousness in Early Cretaceous theropods, although the trackways were later reinterpreted as those of ornithischians [10]. Yet another line of evidence used to support gregariousness is the general morphology of the tracks. For example, track depth will vary with time as a sediment surface dries or floods, such that tracks made in wet sediment will be deeper than those made in drier sediment [11,12]. Trackways formed by groups traveling simultaneously should, therefore, exhibit similar morphologies. Finally, Getty et al. [13] proposed that examining the trackway orientations of the same ichnospecies in different environments could be used as a way to test for gregariousness; if the animal were indeed gregarious, then it is likely that trackway parallelism would occur regardless of sedimentary environment.

The purpose of this paper is to follow up on the research presented by Getty et al. [13] and provide a more comprehensive evaluation of whether or not the *Eubrontes giganteus* trackmaker in the Hartford Basin, which is commonly regarded as a large theropod, was gregarious. This is accomplished by summarizing the number and orientation of trackways at five tracksites that had been previously mapped, including one in a permanent lacustrine paleoenvironment and four in an ephemeral lacustrine one. The evidence presented herein shows that trackway parallelism is only strongly developed at the permanent lacustrine site, where the trackways are oriented parallel to the paleoshoreline. The evidence from these five tracksites suggests that the parallelism observed in the permanent lacustrine environment is better explained as a result of the animals avoiding the lake rather than from gregarious behavior.

2. Geological Context

The tracksites examined in this study are located within the Hartford Basin, in western Connecticut and Massachusetts, USA (Figure 1). The Hartford Basin is a part of a larger group of rift basins in eastern North America called the Newark Supergroup, which formed in the Late Triassic and Early Jurassic as Pangea broke up and the Atlantic Ocean formed [14–16]. Deposition is thought to have spanned from either the Landinian or Carnian stages of the Triassic Period to at least the Sinemurian Stage of the Jurassic Period [17,18]. The sedimentary rocks that filled the basin are primarily of lacustrine, fluvial, and alluvial origin, with minor aeolian deposits, and are interspersed with mafic intrusive and extrusive igneous rocks. Although dinosaur tracks are found in rocks representing nearly all sedimentary environments, they are most abundant in the lacustrine rocks [19]. These lake-derived rocks exhibit cyclical alternations of red shale exhibiting abundant desiccation cracks to gray and black laminated shale lacking evidence of subaerial exposure. Olsen [20] proposed that the cyclically alternating lacustrine beds are indicative of fluctuations in lake level resulting from Milankovich Cycle climate forcing. Red, mud-cracked shales are thought to represent deposition under playa conditions, whereas black shales represent deposition during lake highstands.

The dinosaur tracks examined for this study come from five localities, including Dinosaur Footprint Reservation (DFR), William Murray quarry (WMQ), Gary Gaulin tracksite (GGT), Powder Hill Dinosaur Park (PHDP), and Dinosaur State Park (DSP) (Table 1). The rocks at DFR belong to the Portland Formation, whereas those from the other four localities belong to the East Berlin Formation, which underlies the Portland and is separated from it by Hampden Formation lava flows [12,13,21,22]. The beds at these localities, their sedimentary structures, and their inferred depositional environments will be described in more detail below.

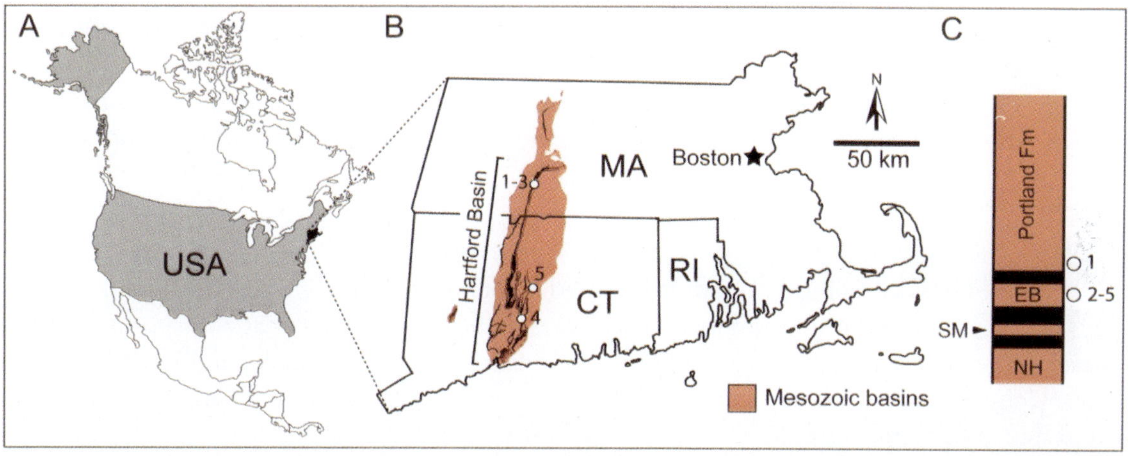

Figure 1. Geographical and geological context of the dinosaur track sites discussed herein. (**A**) Map of North America showing the United States in gray and southern New England in black; (**B**) Inset showing the distribution of Mesozoic basins belonging to the Newark Supergroup in southern New England; (**C**) Simplified stratigraphic column of the Hartford Basin showing the approximate position of the different sites. Numbers for the sites are as follows: 1, Dinosaur Footprint Reservation; 2, William Murray Quarry; 3, Gary Gaulin Tracksite; 4, Powder Hill Dinosaur Park; 5, Dinosaur State Park. Abbreviations in (**B**,**C**) are: CT, Connecticut; EB, East Berlin Formation; MA, Massachusetts; NH, New Haven Formation; RI, Rhode Island; SM, Shuttle Meadow Formation.

Table 1. Summary of dinosaur footprint localities examined for this study. DFR: Dinosaur Footprint Reservation; WMQ: William Murray quarry; GGT: Gary Gaulin tracksite; PHDP: Powder Hill Dinosaur Park; DSP: Dinosaur State Park.

Site	Coordinates	Formation	No. of Tracks	No. of Animals	Environment	Orientations
DFR	42°14′29.99″ N, 72°37′24.83″ W	Portland	236	53	Permanent lake	Bimodal
WMQ	42°12′02.59″ N, 72°38′51.54″ W	East Berlin	30	9	Playa lake	Random
GGT	42°11′51.74″ N, 72°38′41.16″ W	East Berlin	12	1	Playa lake	N/A
PHDP	41°30′11.48″ N, 72°43′48.73″ W	East Berlin	64	32	Playa lake	Random
DSP	41°39′06.82″ N, 72°39′24.66″ W	East Berlin	~2000	Unknown	Playa lake	Random

N/A: Not applicable.

3. Materials and Methods

3.1. *Dinosaur Track Identification, Ichnotaxonomic Assignment, and Track Depth*

The tracks were identified under low-angle light conditions in the early morning and late afternoon. We recorded dimensions, such as length, width, angle of divarication, and digit III projection ratio for each track, and used these parameters to assign the tracks to ichnotaxa following revised diagnoses for *Eubrontes* and *Eubrontes giganteus* provided by Olsen et al. [23]. These workers diagnosed *Eubrontes* as a track of a bipedal, functionally tridactyl dinosaur that is greater than 25 cm long, has a relatively short third digit and broad pes, and an angle of divarication between 25° and 40°. These authors provided the following defining characteristics for *Eubrontes giganteus*: the track is greater than 30 cm long, the digit III projection ratio is about 2.2, the length to width ratio is 1.4 to 1.5, digits II and IV project about equally along the axis of digit II, and the divarication is 30° to 40°. Track depth was measured by placing a ruler over the proximal pad of digit III and measured down from the ruler to the bottom of the track. Depth was considered in conjunction with qualitative characteristics of the track, such as the presence or absence of pad and claw details and mud push-ups or collapse structures to determine the timing if the tracks were made in wet or dry sediment.

3.2. *Tracksite Mapping*

All sites and excavated slabs that we mapped, with the exception of the main (Ostrom) bed at DFR and the Mount Holyoke College slab, were mapped by hand using a 1 m^2 chalk grid applied directly on the track surface. Preliminary maps, with a scale of 39.4 to 1, were constructed on graph paper with grid squares measuring 2.54 cm on a side. The maps were later digitized in the lab. The Mount Holyoke College slab is mounted vertically and a map was constructed by tracing the outline of the slab and tracks from photographs.

Due to its large size, the main track bed of the DFR site was mapped using a Leica Flexline TS/02 total station (Leica Geosystems AG, Heerbrugg, Switzerland) with its northing, easting, and elevation setup on three points along the perimeter of the site. The outline of the perimeter was plotted by moving the prism along the edge of the outcrop at irregular intervals every couple of cm. The position of each footprint was characterized by two points: one proximal (usually the metatarsal-phalangeal pad behind digit four) and one distal (usually the tip of digit three). The northing, easting, and elevation data for the tracks and the perimeter were sorted in excel and then uploaded to Autocad software (Autodesk, San Rafael, CA, USA) to produce a preliminary map that showed the data in plan view. Data were then rotated 13° to account for the current dip of the beds. Lastly, the Autocad map was then exported to Adobe Illustrator (Adobe Systems, San Jose, CA, USA) to produce the final map.

3.3. *Estimating the Number of Individuals*

Although it is possible that a trackmaker crossed the surface more than once, thus leaving more than one trackway, we estimated the maximum number of trackmakers by assuming that a single animal crossed the surface only once, thus leaving only one trackway or isolated track. On Bed 2 at Powder Hill Dinosaur Park, this approach was difficult because the bed is exposed discontinuously throughout the quarry. There, we attempted to link trackway segments based on similarities in size, shape, and orientation.

3.4. *Orientation of Sedimentary Structures and Tracks*

The orientations of sedimentary structures and tracks were measured using a Brunton compass. The crests of oscillation ripples, which form parallel or oblique to modern shores [24], and the long axes of fossilized wood fragments, which typically orient parallel or perpendicular to currents [25], were recorded to estimate the direction of the paleoshoreline. Most dinosaur trackways were relatively straight and their orientations were measured from the bearing of a string run through their midlines.

The direction of movement for dinosaurs that left isolated tracks was estimated from the compass bearing of the track, as measured along the length of digit III.

3.5. Statistical Analysis and Rose Diagram Construction

For each site and bedding plane, we tested the track and trackway orientations for directional preference using the PAST software package, version 3.14 [26]. PAST performs three tests for preferred direction, including Rayleigh's R, Rao's U, and chi-squared. Given the apparent bimodality in some data, and the fact that bimodality is to be expected if geographic boundaries controlled direction of travel, we performed the tests on orientations (e.g., treating travel due north and due south as equivalent). For all localities, the three tests gave congruent results (either significant or not significant at the 0.05 level). A significant result indicates that there is a preferred direction to travel; that is, orientations deviate significantly from random. Rose diagrams for sedimentary structures were initially generated in PAST and then exported to Adobe Illustrator for finishing touches; those for trackways were constructed by hand.

4. Site Descriptions and Trackway Orientations

4.1. Dinosaur Footprint Reservation

DFR is located in Holyoke, Massachusetts along US Highway 5 at 42°14′29.99″ N 72°37′24.83″ W. Approximately 69 m of lacustrine strata are discontinuously exposed there [22]. The strata belong to the basal portion of the Portland Formation [23], which is the uppermost sedimentary formation within the Hartford Basin. It is an historically and scientifically significant site because it is the first location at which dinosaur gregariousness was inferred [1], and because it is the location from which the type specimen of *Eubrontes giganteus* was collected [23]. Beds at the base of the section, including the main track bed discussed by Hitchcock [1] and Ostrom [2], are gray and were deposited as sediment in a shallow, yet permanent, lake, as is evidenced by the lack of subaerially produced sedimentary structures, such as raindrop imprints and desiccation cracks [23,27]. By contrast, beds near the top of the section exhibit mud cracks and some are red, suggesting that the depositional regime shifted to a playa lake with at least intermittent drying [23].

Oscillation ripple marks are prominent features of many DFR beds, with the exception of the main track bed mapped by Ostrom [2], although faint ones occur on that bed as well [27]. Two beds, one approximately 1–2 cm below, and the other about 13 cm above the main track bed, exhibit ripple marks with crest orientations of 200°–277° and 220°–279°, respectively, with averages of 252° and 245° (Figure 2A). Similar ripple crest trends, which range from 220° to 280°, are seen in the beds near the top of the section. The orientations of the plant fossils at this site vary based on size (Figure 2B). Small plant fragments with diameters less than 10 cm are oriented between 300° and 340°, or approximately northwest to southeast, and larger fragments with diameters above 10 cm are oriented between 230° and 260°. Thus, the larger woody fragments are nearly parallel to the ripples, whereas the smaller fragments are oriented approximately perpendicular to the ripples. Considering the orientation of the ripple crests, we infer that the paleoshoreline was oriented approximately northeast to southwest, and that the larger woody fragments aligned parallel to the shore. By contrast, it appears that the smaller woody fragments aligned perpendicular to the shore.

Recent remapping of the site has shown that there are 805 tracks impressed on the main track bed [27]. Of the 805 tracks, 236 of them are *Eubrontes giganteus* (Figure 2C). These tracks are estimated to have been made by 53 different animals. There is a statistically significant bimodal distribution of tracks and trackways on the main track bed ($p < 0.01$ for all three tests; $n = 53$), with forty of them, or 75% of the total, trending to the west, and the remainder oriented to the east (Figure 2D). Restricting the tests to trackways (i.e., removing isolated tracks) did not change these results ($n = 39$). The tracks and trackways are approximately parallel to the ripple crests (Figure 2E). A similar statistically significant, east-to-west bimodal track distribution has been observed on beds near the top of the DFR section ($p < 0.05$ for all three tests; $n = 49$) [27,28], although there is more variability in the orientations (Figure 2F).

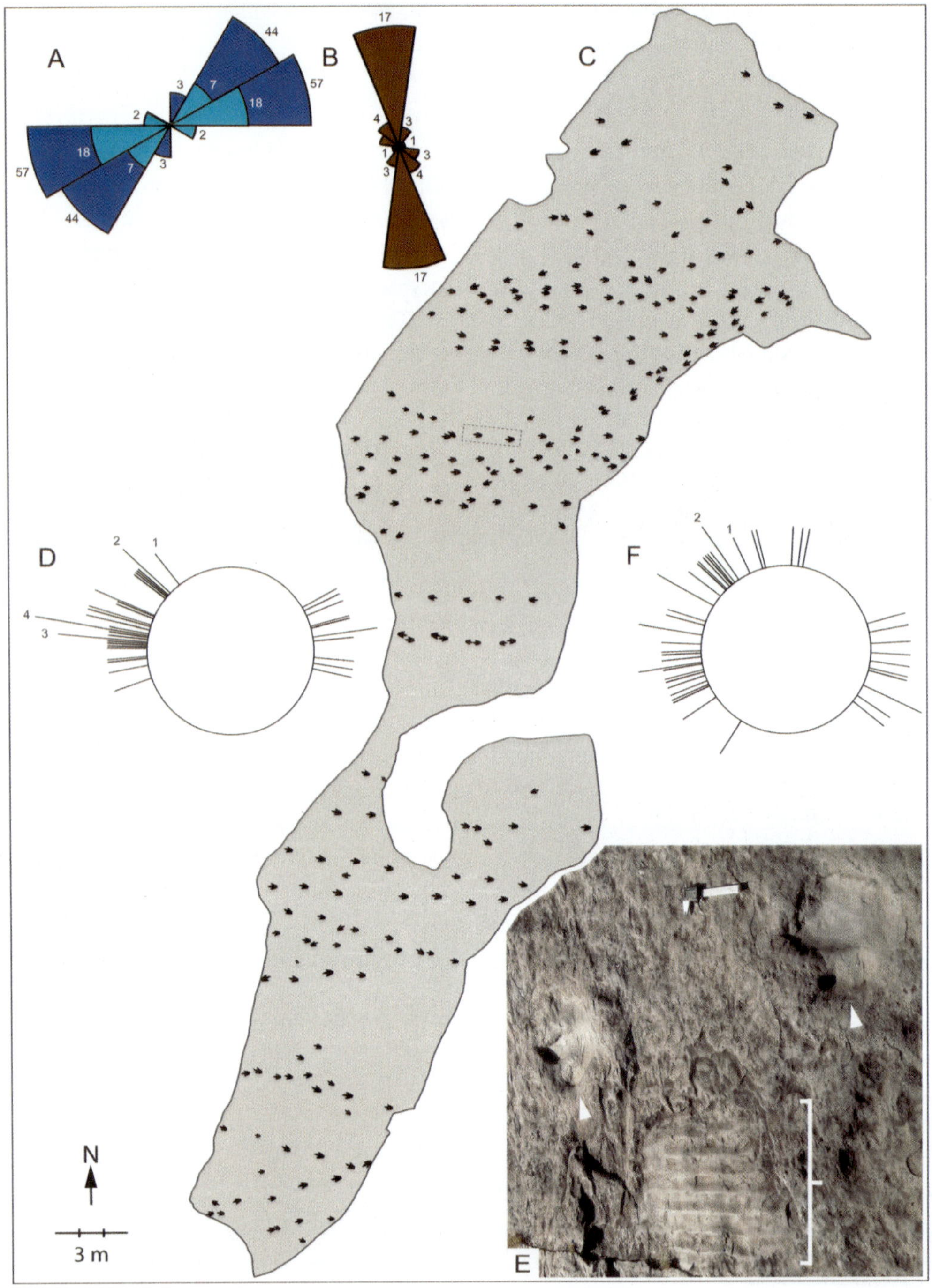

Figure 2. Sedimentary structures, trackways, and orientations from Dinosaur Footprint Reservation, in Holyoke, Massachusetts. (**A**) Rose diagram of the orientation of oscillation ripples above (light blue) and below (dark blue) the main track bed; (**B**) Rose diagram of the long axes of fossilized logs on beds ~69 m above the main track bed; (**C**) Sketch map of the main track bed; (**D**) Rose diagram of track and trackway orientations on the main track bed; (**E**) Photograph of *Eubrontes giganteus* showing the trackway paralleling the ripples (bracketed) underneath the bed. Digital caliper for scale; (**F**) Rose diagram of trackway orientations on 23 different beds ~69 m above the main track bed.

4.2. William Murray Quarry

William Murray operated a small stone quarry in Holyoke, Massachusetts in the 1920s. The rock he excavated is derived from the East Berlin Formation and its red color, along with abundant ripple marks and desiccation cracks, indicate that it was produced in a shallow, ephemeral lake. The rock was primarily used for wall construction, but when Mr. Murray found a bed bearing dinosaur tracks he began selling slabs with the tracks to wealthy locals, such as Ms. Belle Skinner and Mrs. Aaron Bagg, both of Holyoke, Massachusetts. He also sold track-bearing slabs to institutions such as Clark University, in Worcester, Massachusetts; Mount Holyoke College, in South Hadley, Massachusetts, and Forest Park, in Springfield, Massachusetts [12]. Unfortunately, the slabs sold to Mrs. Aaron Bagg and to Clark University have been lost and they are presumed to no longer exist. The other three slabs remain, and are available for study. Belle Skinner's residence is now the Wistariahurst Museum and the tracks are on display in an old driveway [29].

4.2.1. In Situ Tracks

When Murray ceased quarrying operations in the 1930s, due to lack of demand, a portion of the track-bearing surface remained exposed on the quarry floor (Figure 3A). The track bed was reburied to prevent vandalization, but was subsequently uncovered and the tracks on it were described and evaluated [12]. Twenty-three animals are thought to have made the 65 footprints that were found. Of those, 30 tracks in eight trackways, along with an isolated footprint, are large enough to be considered *E. giganteus*. Sample size for the in situ tracks (Figure 3A) is small ($n = 9$), and, visually, there is no clear alignment (Figure 3B). The three statistical tests failed to detect a preferred orientation ($p > 0.05$). Two of the trackways are oriented to the northwest and are nearly parallel (Figure 3B), but they differ in depth and morphology, suggesting that they were made at different times, as shown by Getty [12]. The shallower trackway consists of five tracks with an average depth of only 6 mm, and each track exhibits claw and toe-pad details. By contrast, the deeper trackway, which consists of six steep-walled footprints with an average depth of 23 mm, shows evidence for mud collapse and mud push-ups, and exhibits only the gross outline of the foot.

4.2.2. Tracks on Excavated Slabs

The slab sold to Mount Holyoke College is the smallest of the three remaining slabs from Murray Quarry, and the footprints on it are relatively faint. Only a single track on this slab is large enough to be considered *E. giganteus*. The Forest Park and Wistariahurst Museum slabs are much larger and have numerous tracks and trackways on them, including four *E. giganteus* trackways and two isolated tracks (Figure 3C). These two slabs lay adjacent to each other in the quarry with little or no intervening stone, as is evidenced by three trackways that cross both slabs [12]. The orientation of the slabs when in the quarry is not known, so the orientations of trackways on them cannot be compared with those remaining in situ. They can, however, be compared to each other. Two of the trackways are approximately parallel (Figure 3D), but like the trackways remaining in the quarry, these trackways are of different depths and exhibit morphological variations that suggest that they were made at different times. Figure 3E,F shows a 12-mm deep track, from a five-step trackway with average track depth of 14 mm. This track, along with the others in the trackway, exhibits claw and toe-pad details. By contrast, Figure 3G,H shows a 32-mm deep track that belongs to a 12-step trackway with an average track depth of 22 mm. This track, along with the others in the trackway, shows only the gross outline of the foot along with mud push-ups between the toes and evidence of mud collapse.

4.3. Gary Gaulin Dinosaur Tracksite

Trace fossils at the Gaulin Dinosaur Tracksite in Holyoke, Massachusetts, were initially discovered by property owner Gary Gaulin while constructing a fishpond in March 1996. Since then, continued excavations have exposed approximately 2 m of reddish brown mudstone and fine-grained sandstone of the East Berlin Formation. All of the beds exhibit prominent mud cracks and oscillation ripples indicative of an ephemeral playa-lake environment, and one bed exhibits desiccation cracks that are sinuous in cross section that are indicative of incipient soil formation [30]. Hundreds of small to medium-sized tridactyl vertebrate tracks have been identified, most of which are attributed to the ichnogenera *Anomoepus*, *Anchiauripus*, and *Grallator* [31]. *Treptichnus bifurcus* and *Skolithos* are common at the site [32], and the type and only specimen of an arthropod body imprint *Cheliceratichnus lockleyi,* attributed to a camel-spider-like arthropod, was found there [33].

Most footprints at the Gaulin Site, including the theropod tracks *Anchisauripus* and ornithischian tracks *Anomoepus*, belong to relatively small trackmakers. However, two *Eubrontes giganteus* trackways have been found at the site as well. The first is a trackway composed of 12 tracks that was found on a bed, near the top of the outcrop, which has an exposed surface area of ~63 m^2 (Figure 3I–K). This bed is larger than others, such as the Murray Quarry, in the Hartford Basin on which larger numbers of *E. giganteus* trackways have been found. The second occurrence was found recently on a bed approximately 12 cm below the other *E. giganteus* trackway. The track is still partly obscured by overlying layers on the edge of a ledge, and too little of the bed is exposed to determine if other trackways are preserved on it. The two *E. giganteus* trackways were undoubtedly produced at different times, as evidenced by their being on different beds.

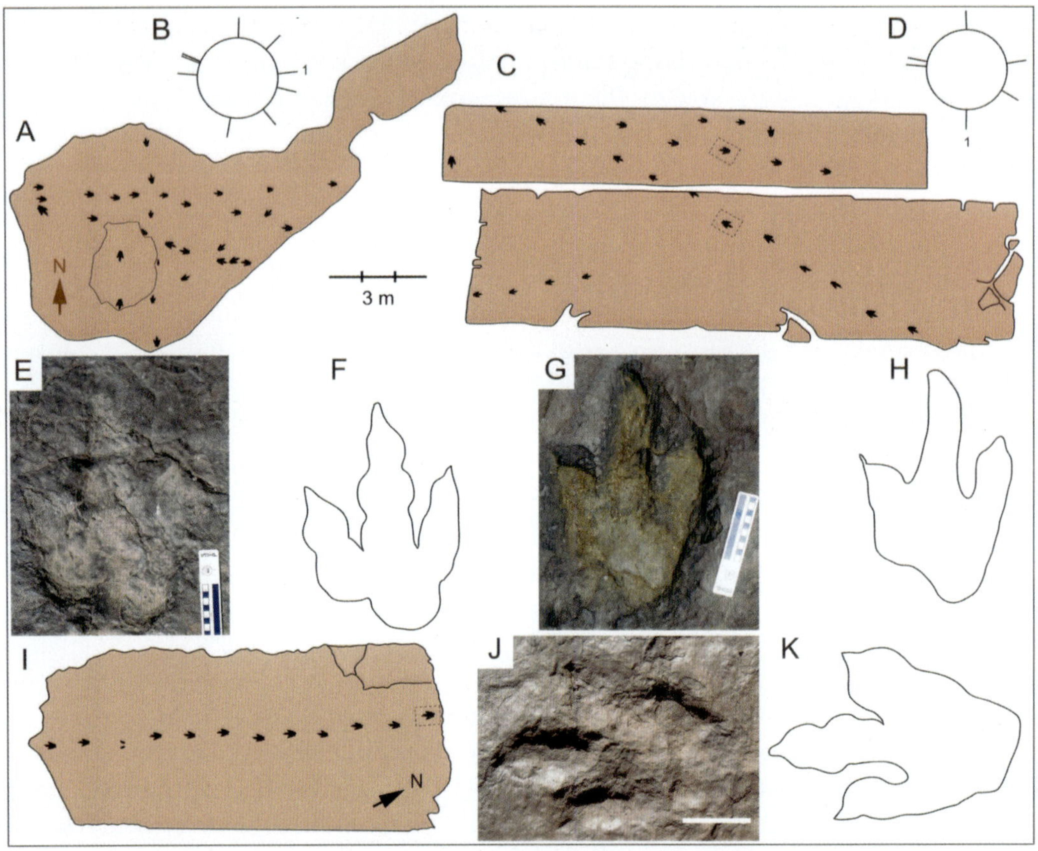

Figure 3. *Eubrontes giganteus* tracks, trackways, and orientations from the William Murray quarry, in Holyoke, Massachusetts. (**A**) Sketch map of the in situ surface; (**B**) Rose diagram of track and trackway orientations on the in situ surface; (**C**) Sketch map of the Forest Park (bottom) and Wistariahurst Museum (top) slabs; (**D**) Rose diagram of the orientations of the tracks and trackways on the slabs; (**E,F**) photograph and line drawing, respectively, of a 12 mm deep track boxed on the map of the Wistariahurst Museum slab; (**G,H**) photograph and line drawing, respectively, of a 30 mm deep track boxed on the map of the Forest Park slab; (**I**) Sketch map of a portion of the Gary Gaulin Dinosaur Tracksite showing the isolated trackway on that bed; (**J,K**) photograph and line drawing, respectively, of the track boxed in (**I**). Sketch maps are to scale. Scale bars in (**E,G**) are in cm; that in (**J**) is 10 cm.

4.4. *Powder Hill Dinosaur Park*

Powder Hill Dinosaur Park, in Middlefield, Connecticut is the site of a small brownstone quarry that was active in 1848–1849 [34]. Due to the presence of dinosaur tracks, the site was donated to Yale University in 1929, and ownership was ultimately transferred to the town of Middlefield in 1976 [13]. Footprints are exposed on four beds (numbered one to four from the quarry bottom to top) of the East Berlin Formation that are composed of muddy, fine-grained sand. Oscillation ripple marks crosscut by desiccation cracks indicate shallow-water deposition followed by drying. Beds of coarse-grained, channeled sands exposed in cross section within the quarry walls indicate a fluvial influence.

Getty et al. [13] identified the tracks and trackways of 71 dinosaurs on the four beds at PHDP, 32 of which were large enough to be considered *Eubrontes giganteus* (Figure 4A). Bed 1 had four trackways on it, with two trackways diverging in a northerly direction and two diverging in a westerly direction (Figure 4B). Bed 2 had 17 trackways, with a slight clustering of five trackways toward the east (Figure 4C), but statistical analysis of trackway orientations showed this not to be significant ($p > 0.05$; $n = 17$). Furthermore, the various tracks are a combination of true tracks and undertracks (Figure 4D–G), indicating that the animals crossed the surface at different times. Bed 3 had four *E. giganteus* impressed on it, with three trackways diverging to the north (Figure 4H). The seven trackways on Bed 4 are clustered into three groups (Figure 4I), with one group oriented to the northeast, a second group oriented to the south, and a third group oriented to the northwest. Analysis of the trackway orientations on Bed 4 were not statistically significant, however, although sample size was small ($p > 0.05$; $n = 7$).

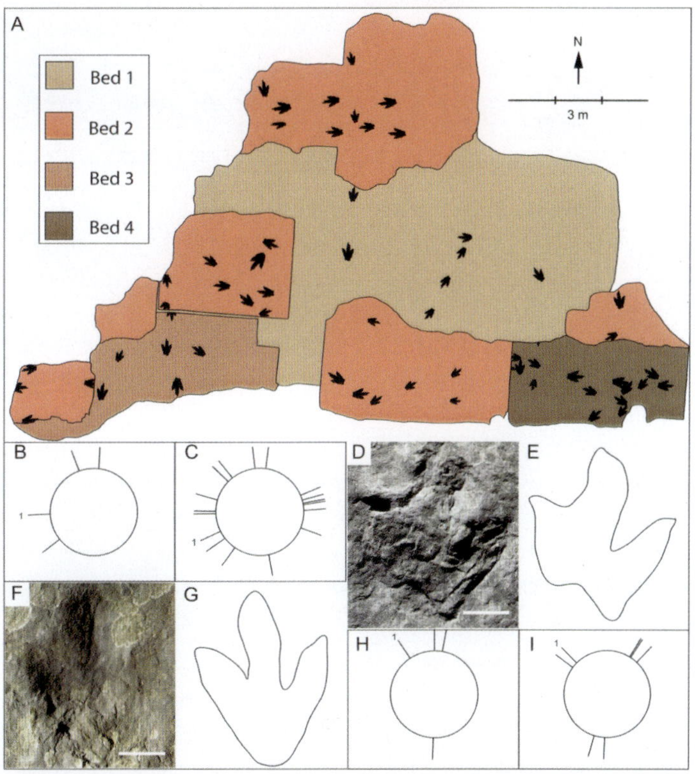

Figure 4. *Eubrontes giganteus* tracks, trackways, and orientations from Powder Hill Dinosaur Park in Middlefield, Connecticut. (**A**) Sketch map of the quarry floor showing all track-bearing beds; (**B**) Rose diagram of track and trackway orientations on Bed 1; (**C**) Rose diagram track and trackway orientations on Bed 2; (**D**,**E**) photograph and interpretive drawing of *E. giganteus* inferred to be a true track; (**F**,**G**) photograph and line drawing, respectively, of a *E. giganteus* undertrack; (**H**) Rose diagram of track and trackway orientations on Bed 3; (**I**) Rose diagram of track and trackway orientations on Bed 4. Scale bars in (**D**,**F**) are 10 cm.

4.5. Dinosaur State Park

The DSP site, which is located in Rocky Hill, Connecticut, was discovered in 1967 during construction of a new state building [30]. The locality is considered to be one of the largest dinosaur tracksites in the world, with an estimated 2000 tracks preserved on two beds [35–38]. Of those, approximately 600 are on display under a geodesic dome; the remainder were reburied for protection [39]. The majority of the tracks are referred to *Eubrontes giganteus*, although tracks of smaller theropods, including *Anchisauripus* and *Grallator*, and those of protosuchian crocodilians called *Batrachopus*, have been found [22].

The rock at the site belongs to the East Berlin Formation and is composed of gray arkosic siltstone and mudstone. The presence of raindrop imprints, ripple marks, and desiccation cracks suggests deposition in shallow water followed by subaerial exposure [22]. Unusual tracks consisting primarily of scratch marks have been interpreted as theropod swim tracks [40], which would suggest much deeper water at times, but they also could have been produced by animals walking on firm sediment [37].

Despite the large number of trackways, no preferred orientation has been observed [2,37,38] (Figure 5A). Furthermore, the trackways show variations in morphology, such as depth and sharpness of outline, that suggest that they were made over a period of time during which the sediment dried (Figure 5B). Some of the trackways on the lower bed are known to be undertracks because they occur more distinctly in other areas of the tracksite where the upper bed has not been peeled away.

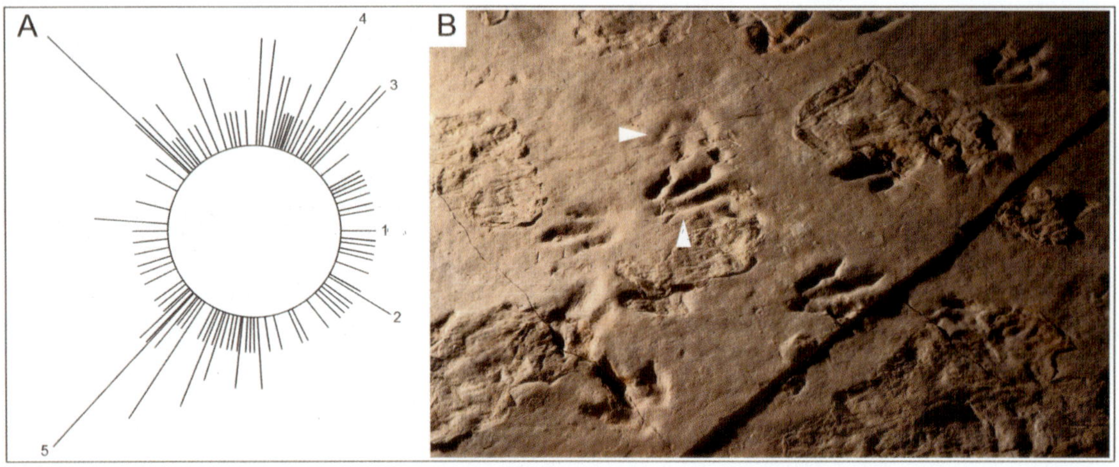

Figure 5. *Eubrontes giganteus* orientations and trackways from Dinosaur State Park in Rocky Hill, Connecticut. (**A**) Rose diagram of track and trackway orientations modified from [2] and used with permission of Elsevier; (**B**) Photograph of part of the in situ track surface showing random orientation and tracks of different depths. The horizontal arrowhead points to a shallow track with a diffuse outline, whereas the vertical arrowhead points to an adjacent deep track with very distinct margins.

5. Discussion

5.1. Trackways and Sedimentary Structures at Dinosaur Footprint Reservation

Hitchcock [1] first cited the trackways at DFR as evidence of gregarious behavior in 1836, although he thought that the trackmakers were birds. Over a century later, Ostrom [2] mapped the site and wrote a seminal paper on dinosaur gregariousness in which the evidence from DFR figured prominently. Ostrom argued that, in the absence of evidence for a physical barrier at the site, the large number of trackways was best explained by the animals traveling together as a herd. Since the publication of his paper, the trackways at DFR have been cited by numerous other authors discussing the evidence for dinosaur social behavior [41–43]. Not all researchers, however, have been convinced that trackway parallelism at the site indicates gregariousness [22,40,44]. These other authors have pointed to the presence of ripple marks at the site as an indication that the shoreline was nearby and that it might have dictated the animals' direction of movement. Following up on the work of these other researchers, Getty et al. [27] measured the compass bearing of oscillation ripple marks, which form approximately parallel to the shoreline [24], on beds immediately above and below the main track bed and showed that orientations of the ripple marks and the trackways were strongly coincident. They also reported faint ripple marks with the same general trend on the main bed along with the trackways. These authors also pointed out that the main track bed lacks evidence of subaerial exposure, which means that the lake was stable over a period of time during which more than one individual could have walked alone along the shore, leaving parallel trackways. Thus, these researchers concluded that the animals were following the shoreline. The same trend in trackways and ripples occurs 69 m above the main bed at this site [27,28], indicating that the shoreline dictated the direction of movement for the *Eubrontes giganteus* trackmaker over a long period of time. Finally, Getty et al. [27] pointed out that beds near the top of the section on which desiccation cracks occur exhibit more divergence in trackway orientations than those on the main bed, which suggests that as the lake dried the animals were free to move in directions that they could not when the lake was full. Overall, the evidence from DFR indicates that the animals followed a shoreline, although it does not falsify the hypothesis that the animals were traveling as a group.

5.2. Trackways and Sedimentary Structures at Other Study Sites

The beds at all of the other sites examined for this study are red in color and exhibit abundant desiccation cracks, indicating that they were deposited in shallow water that later dried up. One locality, the Gary Gaulin Dinosaur Tracksite, exhibits only a single complete *E. giganteus* trackway on the main bedding plane and thus inferences about gregariousness based on the orientations of multiple trackways cannot be made. Nonetheless, Getty and Fox [21] pointed out that the trackway at the site is at least suggestive of individual, rather than group, behavior because the surface area exposed is as large as that of the Murray Quarry, where multiple trackways have been found. All of the sites other than the Gary Gaulin Dinosaur Tracksite exhibit multiple *E. giganteus* trackways and therefore gregariousness can be tested for by examining for trackway parallelism. The smaller sites that we examined directly, including the Murray Quarry and Powder Hill Dinosaur Park, exhibit mostly non-parallel suites of trackways. Of those parallel trackways that do occur, those from the Murray Quarry and from Bed 2 at Powder Hill Dinosaur Park were clearly made at different times because they vary in depth and morphology. Furthermore, none of the statistical analyses of trackway orientations suggested non-randomness at the 0.05 level. The evidence at Dinosaur State Park is similar to that from the Murray Quarry and PHDP in that no preferred orientation has been observed, and in that the trackways show signs of having been formed over long periods of time. Thus, the tracksites from the ephemeral lake environments of the East Berlin Formation show no convincing evidence for gregariousness. Considering the stratigraphic difference between the playa lake tracksites and Dinosaur Footprint Reservation, one might be more likely to argue that a non-gregarious trackmaker made the tracks in the underlying East Berlin Formation and a gregarious one made the tracks in the Portland Formation. We deem this hypothesis to be less likely than the one that the Portland Formation perennial lake dictated the direction of movement because the bimodal distribution of trackways in the Portland Formation is not as strong in the uppermost beds at DFR, where desiccation cracks and red mudstones indicate a return to playa lake conditions.

5.3. Eubrontes giganteus *Parallelism and the Hypothesis that the Trackmaker Was a Sauropodomorph*

Based on comparisons of skeletal material with footprint proportions, most researchers [23,44,45] consider the *Eubrontes giganteus* trackmaker to have been a theropod. A minority opinion, however, is that the trackmaker was in fact a basal sauropodomorph [46–49]. Among the evidence that has been presented to support the latter hypothesis is that the parallel trackways at Dinosaur Footprint Reservation indicate that the trackmaker was gregarious. As outlined above, however,

DFR is unique in the Hartford Basin in exhibiting numerous parallel trackways, and the parallelism observed there is directly associated with a permanent lacustrine sedimentary facies. The tracksites stratigraphically below DFR in playa lake facies show no statistically significant parallelism, and the beds above the main track bed at DFR show more scattered orientations with the return to playa lake conditions. Based on these facts, we have argued above that the parallelism seen at DFR resulted from the animals' interaction with the lake rather than from group interaction. Consequently, we argue also that the parallel trackways at DFR should not be presented as evidence in favor of a sauropodomorph trackmaker.

6. Conclusions

In the Hartford Basin of Connecticut and Massachusetts, USA, parallel *Eubrontes giganteus* trackways are most common in beds at Dinosaur Footprint Reservation that were deposited as sediment in a shallow, yet permanent, lacustrine setting. The rocks upon which the parallel trackways occur are gray in color and lack sedimentary structures, such as mud cracks and raindrop imprints, which indicate subaerial exposure. The orientation of the paleoshoreline at this site was determined using the crests of oscillation ripples and fossilized wood. The ripples and large woody fragments were oriented northeast to southwest, which we hypothesize represents the approximate trend of the shoreline; smaller woody fragments were oriented approximately perpendicular to the inferred shoreline. The parallel trackways at this site exhibit a bimodal distribution that approximates the paleoshoreline. By contrast, trackways in ephemeral lacustrine facies at other sites, which is typified by reddish colored rocks with abundant desiccation cracks, are primarily oriented randomly, even when found in large numbers. The few parallel trackways that do occur in ephemeral lacustrine rocks exhibit morphological variations indicative of changes in sediment saturation, and therefore suggest that the animals crossed the surface at different times. The evidence presented herein suggests that the parallelism seen in *Eubrontes giganteus* in the Hartford Basin is better explained as a response to a physical barrier (i.e., a large lake) than to gregariousness. Consequently, the parallel trackways from DFR should not be used as evidence to support the hypothesis that the trackmaker was a basal sauropodomorph unless further evidence can substantiate the gregariousness hypothesis.

Acknowledgments: P.R.G. wishes to thank Margery Coombs for initially bringing this topic to his attention. Thanks are due to Carol Constant, Gary Gaulin, Laurie Gaulin, Josh Knox, Mark McMenamin, William Murray, Patrick Sullivan, and the town of Middlefield, Connecticut for granting access to study the dinosaur tracks in their care. We are indebted to Margery Coombs, Walter Coombs, Sebastian Dalman, Eric Dewar, Gary Gaulin, Kevin Getty, Benjamin Hamilton, Taylor Jacobson, Dariusz Mokos, Davien Pares, and Xavier Pares for invaluable field assistance. Portions of P.R.G.'s work were funded by grants from the Commonwealth College (Honors Program) at the University of Massachusetts and from the Center for Environmental Sciences and Engineering at the University of Connecticut.

Author Contributions: P.R.G. and C.A. proposed the mapping projects. All authors participated in mapping and analyzing data from one or more of the dinosaur track sites. P.R.G., C.A., A.J., and A.M.B. wrote the paper.

Conflicts of Interest: The authors declare no conflict of interest. The funding sponsors had no role in the design of the study; in the collection, analyses, or interpretation of data; in the writing of the manuscript, and in the decision to publish the results.

Abbreviations

The following abbreviations are used in this manuscript:

DFR	Dinosaur Footprint Reservation
DSP	Dinosaur State Park
WMQ	William Murray Quarry
GGT	Gary Gaulin Tracksite
PHDP	Powder Hill Dinosaur Park

References

1. Hitchcock, E. Ornithichnology—Description of the foot marks of birds (Ornithichnites) on New Red Sandstone in Massachusetts. *Am. J. Sci.* **1836**, *29*, 308–340.
2. Ostrom, J.H. Were some dinosaurs gregarious? *Palaeogeogr. Palaeoclimatol. Palaeoecol.* **1972**, *11*, 287–301. [CrossRef]
3. Razzolini, N.L.; Oms, O.; Castanera, D.; Vila, B.; dos Santos, V.F.; Galobart, À. Ichnological evidence of megalosaurid dinosaurs crossing Middle Jurassic tidal flats. *Sci. Rep.* **2016**, 31494. [CrossRef] [PubMed]
4. Moratalla, J.J.; Hernán, J. Probable palaeogeographic influences of the Lower Cretaceous Iberian rifting phase in the Eastern Cameros Basin (Spain) on dinosaur trackway orientations. *Palaeogeogr. Palaeoclimatol. Palaeoecol.* **2010**, *295*, 116–130. [CrossRef]
5. Lockley, M.G.; Houck, K.J.; Prince, N.K. North America's largest dinosaur trackway site: Implications for Morrison Formation paleoecology. *Geol. Soc. Am. Bull.* **1986**, *97*, 1163–1176. [CrossRef]
6. Castanera, D.; Barco, J.L.; Díaz-Martínez, I.; Gascón, J.H.; Pérez-Lorente, F.; Canudo, J.I. New evidence of a herd of titanosauriform sauropods from the lower Berriasian of the Iberian range (Spain). *Palaeogeogr. Palaeoclimatol. Palaeoecol.* **2011**, *310*, 227–237. [CrossRef]
7. Currie, P.J. Hadrosaur trackways from the lower Cretaceous of Canada. *Acta Palaeontol. Pol.* **1983**, *28*, 63–73.
8. Cotton, W.D.; Cotton, J.E.; Hunt, A.P. Evidence for social behavior in ornithopod dinosaurs from the Dakota Group of northeastern New Mexico, USA. *Ichnos* **1998**, *6*, 141–149. [CrossRef]
9. Barco, J.L.; Canudo, J.I.; Ruiz-Omeñaca, J.I. New data on *Therangospodus oncalensis* from the Berriasian Fuentesalvo tracksite (Villar del Río, Soria, Spain): An example of gregarious behaviour in theropod dinosaurs. *Ichnos* **2006**, *13*, 237–248. [CrossRef]
10. Castanera, D.; Pascual, C.; Razzolini, N.L.; Vila, B.; Barco, J.L.; Canudo, J.I. Discriminating between medium-sized tridactyl trackmakers: Tracking ornithopod tracks in the base of the Cretaceous (Berriasian, Spain). *PLoS ONE* **2013**, *8*, e81830. [CrossRef] [PubMed]
11. Lockley, M.G. *Tracking Dinosaurs: A New Look at an Ancient World*; Cambridge University Press: New York, NY, USA, 1991.
12. Getty, P.R. Excavated and in situ dinosaur footprints from the Murray Quarry (Early Jurassic, East Berlin Formation), Holyoke, Massachusetts, USA. *Ichnos* **2005**, *12*, 163–178. [CrossRef]
13. Getty, P.R.; Hardy, L.; Bush, A.M. Was the *Eubrontes* track maker gregarious? Testing the herding hypothesis at Powder Hill Dinosaur Park, Middlefield, Connecticut. *Bull. Peabody Mus. Nat. Hist.* **2015**, *56*, 95–106. [CrossRef]
14. Olsen, P.E. On the use of the term Newark for Triassic and Early Jurassic rocks of eastern North America. *Newsl. Stratigr.* **1978**, *7*, 90–95.
15. Olsen, P.E.; McDonald, N.G.; Huber, P.; Cornet, B. Stratigraphy and paleoecology of the Deerfield Rift Basin (Triassic–Jurassic, Newark Supergroup), Massachusetts. In *Guidebook for Field Trips in the Connecticut Valley Region of Massachusetts and Adjacent States, Volume 2, Proceedings of the New England Intercollegiate Geology Conference, Amherst, MA, USA, 9–11 October 1992*; Robinson, P., Brady, J.B., Eds.; University of Massachusetts: Amherst, MA, USA, 1992; pp. 488–535.
16. Withjack, M.O.; Schlische, R.W.; Olsen, P.E. Diachronous rifting, drifting, and inversion on the passive margin of central eastern North America: An analog for other passive margins. *Am. Assoc. Pet. Geol.* **1998**, *82*, 817–835.
17. Olsen, P.E.; Kent, D.V. Long-period Milankovitch cycles from the Late Triassic and Early Jurassic of eastern North America and their implications for the calibration of the Early Mesozoic time-scale and the long-term behaviour of the planets. *Philos. Trans. R. Soc. Lond. A* **1999**, *357*, 1761–1786. [CrossRef]
18. Olsen, P.E.; Kent, D.V. High-resolution early Mesozoic Pangean climactic transect in lacustrine environments. *Zentralbl. Geol. Paläont. A* **2000**, *1998*, 1475–1495.
19. McDonald, N.G. Paleontology of the early Mesozoic (Newark Supergroup) rocks of the Connecticut Valley. *Northeast. Geol.* **1992**, *14*, 185–200.
20. Olsen, P.E. A 40-million-year lake record of early Mesozoic orbital climactic forcing. *Science* **1986**, *234*, 842–848. [CrossRef] [PubMed]
21. Getty, P.R.; Fox, N. An isolated *Eubrontes giganteus* trackway from the Gary Gaulin dinosaur track site (Early Jurassic, East Berlin Formation), Holyoke, Massachusetts. *Northeast. Geosci.* **2015**, *33*, 16–21.

22. Olsen, P.E. Field Guide for non-marine boundary events in the Newark Basin (New Jersey, Pennsylvania, and Connecticut), eastern United States and their litho-, chrono-, and biostratigraphic context. In Proceedings of the International Geological Correlation Project 458, Orangeburg, NY, USA, 7–12 June 2002; State University of New York: Cortland, NY, USA; Columbia University: New York, NY, USA, 2002; pp. 1–181.
23. Olsen, P.E.; Smith, J.B.; McDonald, N.G. Type material of the type species of the classic theropod footprint genera *Eubrontes*, *Anchisauripus*, and *Grallator* (Early Jurassic, Hartford and Deerfield Basins, Connecticut and Massachusetts, USA). *J. Vert. Paleol.* **1998**, *18*, 586–601. [CrossRef]
24. Nichols, G. *Sedimentology & Stratigraphy*; Blackwell Science: Malden, MA, USA, 1999.
25. MacDonald, D.I.M.; Jefferson, T.H. Orientation studies of waterlogged wood: A paleocurrent indicator? *J. Sediment. Petrol.* **1985**, *55*, 235–239.
26. Hammer, Ø.; Harper, D.A.T.; Ryan, P.D. PAST: Paleontological statistics software package for education and data analysis. *Palaeontol. Electron.* **2001**, *4*, 9. Available online: http://palaeo-electronica.org/2001_1/past/past.pdf (accessed on 30 January 2017).
27. Getty, P.R.; Judge, A.; Csonka, J.; Bush, A.M. Were Early Jurassic dinosaurs gregarious? Reexamining the evidence from Dinosaur Footprint Reservation in Holyoke, Massachusetts. In *Guidebook for Field Trips in Connecticut and Massachusetts, Proceeding of the 47th Annual Meeting of the Northeastern Section of the Geological Society of America, Hartford, CT, USA, 18–20 March 2012*; Thomas, M.A., Ed.; State Geological and Natural History Survey of Connecticut: Hartford, CT, USA, 2012; pp. A1–A18.
28. Smith, J.B.; Smith, J.R.; Sweeny, J.P. Mt Tom revisited: A new fossil horizon at Dinosaur Footprint Reservation, Northampton, Massachusetts. *J. Vert. Paleol.* **1996**, *16*, 67A.
29. Thibodeau, K.N. *Images of America: Holyoke: The Skinner Family and Wistariahurst*; Arcadia Publishing: Charleston, SC, USA, 2005.
30. Demicco, R.V.; Gierlowski-Kordesch, E. Facies sequences of a semi-arid closed basin: The Lower Jurassic East Berlin Formation of the Hartford Basin, New England, USA. *Sedimentology* **1986**, *33*, 107–118. [CrossRef]
31. Dalman, S.G.; Weems, R.E. A new look at morphological variation in the ichnogenus *Anomoepus*, with special reference to material from the Lower Jurassic Newark Supergroup: Implications for ichnotaxonomy and ichnodiversity. *Bull. Peabody Mus. Nat. Hist.* **2015**, *54*, 67–124. [CrossRef]
32. Getty, P.R.; McCarthy, T.D.; Hsieh, S.; Bush, A.M. A new reconstruction of continental *Treptichnus* based on exceptionally preserved material from the Jurassic of Massachusetts. *J. Paleol.* **2016**, *90*, 269–278. [CrossRef]
33. Dalman, S.; Lucas, S. Lower Jurassic arthropod resting trace from the Hartford Basin of Massachusetts, USA. *Ichnos* **2015**, *22*, 177–182. [CrossRef]
34. Atkins, T. *History of Middlefield and Longhill*; Case, Lockwood, and Brainerd: Hartford, SC, USA, 1883.
35. Ostrom, J.H. Peabody paleontologists assist new dinosaur-track park. *Discovery* **1967**, *2*, 21–24.
36. Lockley, M.G.; Gillette, D.D. Dinosaur tracks and traces: An overview. In *Dinosaur Tracks and Traces*; Gillette, D.D., Lockley, M.G., Eds.; Cambridge University Press: New York, NY, USA, 1989; pp. 3–10.
37. Farlow, J.O.; Galton, P.M. Dinosaur trackways of Dinosaur State Park, Rocky Hill, Connecticut. In *The Great Rift Valleys of Pangea in Eastern North America*; Letourneau, P.M., Olsen, P.E., Eds.; Columbia University Press: New York, NY, USA, 2003; pp. 248–263.
38. Galton, P.M.; Farlow, J.O. Dinosaur State Park, Connecticut, USA: History, footprints, trackways, exhibits. *Zubía* **2003**, *21*, 129–173.
39. McDonald, N.G. *Window into the Jurassic World*; Friends of Dinosaur State Park and Arboretum: Rocky Hill, VT, USA, 2010.
40. Coombs, W.P. Swimming ability of carnivorous dinosaurs. *Science* **1980**, *207*, 1198–1200. [CrossRef] [PubMed]
41. Colbert, E.H. The Triassic dinosaur *Coelophysis*. *Mus. North. Ariz. Bull.* **1989**, *57*, 1–160.
42. Lockley, M.G.; Matsukawa, M. Some observations on trackway evidence for gregarious behavior among small bipedal dinosaurs. *Palaeogeogr. Palaeoclimatol. Palaeoecol.* **1999**, *150*, 25–31. [CrossRef]
43. Lingham-Soliar, T.; Broderick, T.; Ahmed, A.A.K. Closely associated theropod trackways from the Jurassic of Zimbabwe. *Naturwissenschafen* **2003**, *90*, 572–576. [CrossRef] [PubMed]
44. Smith, J.B.; Farlow, J.O. Osteometric approaches to trackmaker assignment for the Newark Supergroup ichnogenera *Grallator*, *Anchisauripus*, and *Eubrontes*. In *The Great Rift Valleys of Pangea in Eastern North America*; Letourneau, P.M., Olsen, P.E., Eds.; Columbia University Press: New York, NY, USA, 2003; pp. 273–292.

45. Lucas, S.G.; Klein, H.; Lockley, M.G.; Spielmann, J.A.; Gierlinski, G.D.; Hunt, A.D.; Tanner, L.H. Triassic-Jurassic stratigraphic distribution of the theropod footprint ichnogenus *Eubrontes*. *New Mex. Mus. Nat. Hist. Sci. Bull.* **2006**, *37*, 86–93.
46. Weems, R.E. A Late-Triassic footprint fauna from the Culpeper Basin Northern Virginia (USA). *Trans. Am. Philos. Soc.* **1987**, *77*, 1–79. [CrossRef]
47. Weems, R.E. A re-evaluation of the taxonomy of Newark Supergroup saurischians dinosaur tracks, using extensive statistical data from a recently exposed tracksite near Culpeper, Virginia. In Proceedings of the 26th Forum on the Geology of Industrial Minerals, Charlottesville, VA, USA, 15–18 May 1990; Sweet, P.C., Ed.; Division of Mineral Resources Publications: Charlottesville, VA, USA, 1992; Volume 119, pp. 113–126.
48. Weems, R.E. *Plateosaurus* foot structure suggests a single trackmaker for *Eubrontes* and *Gigandipus* footprints. In *The Great Rift Valleys of Pangea in Eastern North America*; Letourneau, P.M., Olsen, P.E., Eds.; Columbia University Press: New York, NY, USA, 2003; pp. 293–313.
49. Miller, W.E.; Britt, B.B.; Stadtman, K.L. Tridactyl trackways from the Moenave Formation of southwestern Utah. In *Dinosaur Tracks and Traces*; Gillette, D.D., Lockley, M.G., Eds.; Cambridge University Press: New York, NY, USA, 1989; pp. 209–215.

Permissions

All chapters in this book were first published in GEOSCIENCES, by MDPI; hereby published with permission under the Creative Commons Attribution License or equivalent. Every chapter published in this book has been scrutinized by our experts. Their significance has been extensively debated. The topics covered herein carry significant findings which will fuel the growth of the discipline. They may even be implemented as practical applications or may be referred to as a beginning point for another development.

The contributors of this book come from diverse backgrounds, making this book a truly international effort. This book will bring forth new frontiers with its revolutionizing research information and detailed analysis of the nascent developments around the world.

We would like to thank all the contributing authors for lending their expertise to make the book truly unique. They have played a crucial role in the development of this book. Without their invaluable contributions this book wouldn't have been possible. They have made vital efforts to compile up to date information on the varied aspects of this subject to make this book a valuable addition to the collection of many professionals and students.

This book was conceptualized with the vision of imparting up-to-date information and advanced data in this field. To ensure the same, a matchless editorial board was set up. Every individual on the board went through rigorous rounds of assessment to prove their worth. After which they invested a large part of their time researching and compiling the most relevant data for our readers.

The editorial board has been involved in producing this book since its inception. They have spent rigorous hours researching and exploring the diverse topics which have resulted in the successful publishing of this book. They have passed on their knowledge of decades through this book. To expedite this challenging task, the publisher supported the team at every step. A small team of assistant editors was also appointed to further simplify the editing procedure and attain best results for the readers.

Apart from the editorial board, the designing team has also invested a significant amount of their time in understanding the subject and creating the most relevant covers. They scrutinized every image to scout for the most suitable representation of the subject and create an appropriate cover for the book.

The publishing team has been an ardent support to the editorial, designing and production team. Their endless efforts to recruit the best for this project, has resulted in the accomplishment of this book. They are a veteran in the field of academics and their pool of knowledge is as vast as their experience in printing. Their expertise and guidance has proved useful at every step. Their uncompromising quality standards have made this book an exceptional effort. Their encouragement from time to time has been an inspiration for everyone.

The publisher and the editorial board hope that this book will prove to be a valuable piece of knowledge for researchers, students, practitioners and scholars across the globe.

List of Contributors

Simone Santoro
Dipartimento di Scienze della Terra e del Mare, Università di Palermo, 90123 Palermo, Italy
Nuclear Fusion and Safety Technologies Department, ENEA (Italian National Agency for New Technologies, Energy and Sustainable Economic Development), 00044 Frascati, Italy

Stefano Parracino
Nuclear Fusion and Safety Technologies Department, ENEA (Italian National Agency for New Technologies, Energy and Sustainable Economic Development), 00044 Frascati, Italy
Department of Industrial Engineering, University of Rome "Tor Vergata", 00173 Rome, Italy

Luca Fiorani, Enzo Di Ferdinando and Marcello Nuvoli
Nuclear Fusion and Safety Technologies Department, ENEA (Italian National Agency for New Technologies, Energy and Sustainable Economic Development), 00044 Frascati, Italy

Roberto D'Aleo
Dipartimento di Scienze della Terra e del Mare, Università di Palermo, 90123 Palermo, Italy

Gaetano Giudice
Istituto Nazionale di Geofisica e Vulcanologia, 90146 Palermo, Italy

Giovanni Maio
ARES Consortium, 00100 Rome, Italy

Alessandro Aiuppa
Dipartimento di Scienze della Terra e del Mare, Università di Palermo, 90123 Palermo, Italy
Istituto Nazionale di Geofisica e Vulcanologia, 90146 Palermo, Italy

Andrea Donnellan
Jet Propulsion Laboratory, California Institute of Technology, Pasadena, CA 91109-8099, USA

Ramón Arrowsmith
School of Earth and Space Exploration, Arizona State University, Tempe, AZ 85287-6004, USA

Stephen DeLong
United States Geological Survey, Menlo Park, CA 94025, USA

Dolores Pereira
Department of Geology, University of Salamanca, Plaza de la Merced s/n, Salamanca 37008, Spain

Brian Marker
Independent consultant, 40 Kingsdown Avenue, London W13 9PT, UK

Tomás Fernández, José Luis Pérez and Javier Cardenal
Department of Cartographic, Geodetic and Photogrammetric Engineering, University of Jaén, Campus de las Lagunillas s/n, 23071 Jaén, Spain
Centre for Advanced Studies in Earth Sciences (CEACTierra), University of Jaén, Campus de las Lagunillas s/n, 23071 Jaén, Spain

Carlos Colomo and Jorge Delgado
Department of Cartographic, Geodetic and Photogrammetric Engineering, University of Jaén, Campus de las Lagunillas s/n, 23071 Jaén, Spain

José Antonio Palenzuela, Clemente Irigaray and José Chacón
Department of Civil Engineering, University of Granada, Campus de Fuentenueva s/n, 18071 Granada, Spain

Daniel Hölbling, Florian Albrecht, Barbara Friedl and Elisabeth Weinke
Department of Geoinformatics–Z_GIS, University of Salzburg, Schillerstrasse 30, 5020 Salzburg, Austria

Clemens Eisank
GRID-IT—Gesellschaft für angewandte Geoinformatik mbH, Technikerstraße 21a, 6020 Innsbruck, Austria

Filippo Vecchiotti and Arben Kociu
Geological Survey of Austria (GBA), Neulinggasse 38, 1030 Vienna, Austria

Christos Chalkias, Christos Polykretis and Efthimios Karymbalis
Department of Geography, Harokopio University, El. Venizelou 70, Athens 17671, Greece

Maria Ferentinou
Department of Geological Sciences, University of KwaZulu-Natal, Westville, Private Bag X54001, South Africa

Yu Nakayama
Institute of Industrial Science, The University of Tokyo, 4-6-1 Komaba, Meguro-ku, Tokyo, Japan

Kazuhiko Nakamura
Center for Spatial Information Science, The University of Tokyo, 5-1-5 Kashiwanoha, Kashiwa-shi, Chiba, Japan

Hitoshi Saito
College of Economics, Kanto Gakuin University, 1-50-1 Mutsuura-higashi, Kanazawa-ku, Yokohama, Japan

Rui Fukumoto
Open Concierge, #307 3-21-5 Kandasakumacho, Chiyoda-ku, Tokyo, Japan

Christina Stouraiti, Iakovos Pantziris, Charalampos Vasilatos, Panagiotis Mitropoulos and Panagiotis Pomonis
Department of Geology and Geoenvironment, National and Kapodistrian University of Athens, Panepistimiopolis, Ano Ilissia, 15784 Athens, Greece

Robert Moritz and Massimo Chiaradia
Department of Earth Sciences, University of Geneva, Rue des Maraîchers 13, 1205 Genève, GE, Switzerland

Christos Kanellopoulos
Department of Geology and Geoenvironment, National and Kapodistrian University of Athens, Panepistimiopolis, Ano Ilissia, 15784 Athens, Greece
Department of Earth Sciences, University of Geneva, Rue des Maraîchers 13, 1205 Genève, GE, Switzerland

Dennis R. Ruez Jr.
Department of Geological Sciences, The University of Texas at Austin, Austin, TX 78712, USA
Department of Environmental Studies, University of Illinois at Springfield, Springfield, IL 62703, USA
Research and Collections Center, Illinois State Museum, Springfield, IL 62703, USA

Roberta Bonì, Claudia Meisina and Davide Notti
Department of Earth and Environmental Sciences, University of Pavia, Via Ferrata 1, 27100 Pavia, Italy

Francesca Cigna and Stephanie Bricker
British Geological Survey, Natural Environment Research Council, Nicker Hill, Keyworth, Nottinghamshire NG12 5GG, UK

Pablo Ezquerro
Geohazards InSAR Laboratory and Modeling Group, Instituto Geológico y Minero de España (IGME), C/. Alenza 1, 28003 Madrid, Spain

Gerardo Herrera
Geohazards InSAR Laboratory and Modeling Group, Instituto Geológico y Minero de España (IGME), C/. Alenza 1, 28003 Madrid, Spain
Unidad Asociada de Investigación IGME-UA de Movimientos del Terreno Mediante Interferometría Radar (UNIRAD), Universidad de Alicante, Alicante, Spain
Earth Observation and Geohazards Expert Group (EOEG), Euro Geo Surveys, the Geological Surveys of Europe, 36–38, Rue Joseph II, 1000 Brussels, Belgium

Harry McCormack
CGG, NPA Satellite Mapping, Crockham Park, Edenbridge Kent TN8 6SR, UK

Roberto Tomás
Geohazards InSAR Laboratory and Modeling Group, Instituto Geológico y Minero de España (IGME), C/. Alenza 1, 28003 Madrid, Spain
Unidad Asociada de Investigación IGME-UA de Movimientos del Terreno Mediante Interferometría Radar (UNIRAD), Universidad de Alicante, Alicante, Spain
Departamento de Ingeniería Civil, Escuela Politécnica Superior, Universidad de Alicante. 03080 Alicante, Spain

Marta Béjar-Pizarro and Joaquín Mulas
Geohazards InSAR Laboratory and Modeling Group, Instituto Geológico y Minero de España (IGME), C/. Alenza 1, 28003 Madrid, Spain
Unidad Asociada de Investigación IGME-UA de Movimientos del Terreno Mediante Interferometría Radar (UNIRAD), Universidad de Alicante, 03080 Alicante, Spain

Dickson Cunningham
Department of Environmental Earth Science, Eastern Connecticut State University, Willimantic, CT 06226, USA

Patrick R. Getty
Center for Integrative Geosciences, University of Connecticut, 354 Mansfield Road, U-1045, Storrs, CT 06269, USA

Christopher Aucoin
Department of Geology, University of Cincinnati, 500 Geology Physics Building, Cincinnati, OH 45221, USA

Nathaniel Fox
Environmental Systems Graduate Group, University of California, 5200 North Lake Road, Merced, CA 95340, USA

Aaron Judge
14 Carleton Street, South Hadley, MA 01075, USA

Laurel Hardy
1476 Poquonock Avenue, Windsor, CT 06095, USA

Andrew M. Bush
Center for Integrative Geosciences, University of Connecticut, 354 Mansfield Road, U-1045, Storrs, CT 06269, USA
Department of Evolutionary Biology, University of Connecticut, 75 North Eagleville Road,U-3403, Storrs, CT 06269, USA

Index

R

S

T

U

V

W